Security and Data Reliability in Cooperative Wireless Networks

Security and Data Reliability in Cooperative Wireless Networks

Emad S. Hassan

CRC Press
Taylor & Francis Group
Boca Raton London New York

CRC Press is an imprint of the
Taylor & Francis Group, an **informa** business

CRC Press
Taylor & Francis Group
6000 Broken Sound Parkway NW, Suite 300
Boca Raton, FL 33487-2742

© 2018 by Taylor & Francis Group, LLC
CRC Press is an imprint of Taylor & Francis Group, an Informa business

No claim to original U.S. Government works

Printed on acid-free paper

International Standard Book Number-13: 978-1-138-09279-2 (Hardback)

Visit the Taylor & Francis Web site at
http://www.taylorandfrancis.com

and the CRC Press Web site at
http://www.crcpress.com

Contents

SECTION II SECURITY AND DATA RELIABILITY IN WIRELESS SENSOR NETWORKS

Preface

Broadcast nature is one of the main characteristics of the wireless medium with a double-edged arm; the first is beneficial while the other is harmful. With respect to its beneficial side, it allows applying what is called cooperative communications in wireless systems. Cooperative communication is a mechanism that aims to achieve transmission diversity performance enhancements in terms of increased capacity and improved transmission reliability in a new and interesting way. It enables many wireless devices in a multi-user environment, which are limited by size or hardware complexity to one antenna, to share their antennas for forwarding their messages to the destination together.

On the other hand, the harmful side of the wireless medium broadcast nature lies in its negative effect on the system security. Due to the broadcast nature of wireless communication networks, the adversarial "eavesdropper" nodes in their coverage area can intercept transmissions and try to recover parts of the transmitted message. Therefore, a resurgence of interest has been given recently for studying the security of data transmission in wireless systems from a physical layer point of view. The main objective behind physical layer security is to enable the exchange of confidential messages over a wireless medium in the presence of unauthorized eavesdroppers, without relying on higher-layer encryption.

This book provides new solutions for these problems, and its main objective is to enhance the security and data reliability in cooperative wireless networks.

A major attraction of the book is the presentation of MATLAB® simulations and the inclusion of MATLAB codes to help readers understand the topic under discussion and to be able to carry out extensive simulations. The book is structured into ten chapters and broadly covers two important parts as follows:

Part one: Security in cooperative wireless networks
Part two: Security and data reliability in wireless sensor networks (WSNs)

- In the first part, we first give a detailed overview about both cooperative communications and the physical layer security, the two main topics on which our book relies. We firstly introduce cooperative communications, the innovative approach which exploits wireless medium broadcast nature to achieve multiple-input–multiple-output (MIMO) gains in a distributed manner in order to be suitable for application in small wireless devices. Different cooperative protocols concerned with the processing of the signal received from the source node at the relay node are discussed. Furthermore, different relay selection metrics concerned with selecting the best relay among the available N relays with an indication to the entity which evaluates these metrics and selects the relay are also given. Finally, we end this point by presenting cooperative communication applications and the pros and cons.

■ In addition to the foregoing, an overview for physical layer security is also given in this first part of the thesis. The main objective behind physical layer security is to improve the secrecy rate of a given transmitter–receiver pair in the presence of unauthorized eavesdroppers. This can be accomplished by using some relay nodes as jamming nodes to transmit artificial interference and confuse the eavesdroppers. Firstly, we give an overview about both key-based and keyless security. This is followed by discussing such cooperating approaches that help in achieving secrecy at the physical layer of a multi-user system through introducing the cooperative jamming concept. Then, the idea of employing cooperative jammers in a multiple relay network in order to improve security is adopted. Finally, the interactions arising between cooperation and secrecy in the channel models with untrusted relays are studied.

■ Then, different relay and jammer selection schemes are proposed in order to achieve security in one-way cooperative networks. It should be obvious that selecting the best relay is necessary for applying the cooperative communication idea through its assistance to the source in forwarding its message to the corresponding destination besides its own message. Moreover, selecting jammers is necessary for achieving physical layer security through their intentional interference at the eavesdroppers' nodes. The selection schemes without jamming, the selection schemes with conventional jamming, the selection schemes with controlled jamming, and the hybrid switching schemes are the four different proposed selection schemes presented through this part. The numerical results shown at the end of this part illustrate the effectiveness of the different proposed schemes in improving both ergodic secrecy capacity and secrecy outage probability performance metrics. The proposed selection schemes with jamming outperform the corresponding nonjamming selection schemes, and the hybrid schemes which switch between jamming and nonjamming selection schemes further improve both performance metrics.

■ Because of two-way relay channel bandwidth efficiency and potential application to cellular networks and peer-to-peer networks, different relay and jammer selection schemes are proposed to improve physical layer security in two-way cooperative networks. The obtained results show that the selection schemes with jamming cannot outperform the schemes without jamming in all cases. Therefore, a hybrid scheme which switches between both jamming and nonjamming selection schemes is introduced as an efficient solution in such cases. In addition to the foregoing, the obtained results show the ongoing effectiveness of our proposed selection schemes in improving both the secrecy rate and the secrecy outage probability of the two-way cooperative networks despite the presence of multiple cooperating eavesdroppers. At the end of this part, a comparison between relay and jammer selection schemes in both one-way and two-way cooperative networks is given in terms of both secrecy metrics.

■ The second part of this book focuses on data security and reliability in unattended wireless sensor networks (UWSNs) in the presence of a mobile adversary. In this part, we explore the different challenges of UWSNs, such as compromising probability, probability of BSe to be compromised, and data reliability. During this part, several self-healing algorithms are developed to provide data security and reliability in UWSNs. In the second part of this book, we cover the following points:

– Overview of WSNs followed by an overview of UWSNs.

– A proposal called the cooperative hybrid self-healing randomized distributed (CHSFRD) scheme is introduced to provide self-healing in UWSNs. Self-healing algorithms are developed to increase the likelihood for data reliability and data security in homogeneous UWSNs, without implementing cryptography. In addition, the UWSN model is defined in a way that encompasses all common WSN assumptions and characterizes the unattended operation mode that involves periodic visits by an itinerant sink. Also, we define a new adversarial model geared for UWSNs, delineating its capabilities and identifying many adversary subtypes based on its specific goals. The proposed scheme is based mainly on the hybrid cooperation principal between healthy and compromised (sick) sensors and that sensor collaboration is necessary to mislead an adversary. The proposed scheme proves its ability to enhance the UWSN security by improving the data reliability and compromising probability and probability of backward secrecy.

– A proposal called self-healing controlled mobility within a cluster (SH-CMC) scheme is developed for self-healing enhancement in UWSNs, in which the clustering and mobility of some sensors were used beside the hybrid cooperation. Both of them can enhance the self-healing capability of UWSNs. In addition, different mobility models available for wireless networks are discussed in detail. The proposed scheme proves that using the mobility within a cluster of sick sensors is the best and complementary solution for the problem of the leakage of healthy sponsors. The proposed scheme proves that the use of mobility beside the hybrid cooperation can enhance the self-healing capability over the scheme that does not consider mobility.

– A proposal called self-healing single flow cluster controlled mobility (SH-SFCCM) scheme is introduced for self-healing enhancement considering energy consumption due to mobility. The trade-off between energy consumption in both mobility and communication is estimated. The energy consumption cost functions for both communication and mobility are estimated. In addition, the influence of sensor mobility on self-healing capability and other network aspects is studied.

■ Finally, MATLAB codes for all simulation experiments are included in Appendices A-E at the end of the book.

MATLAB® is a registered trademark of The MathWorks, Inc. For product information, please contact:

The MathWorks, Inc.
3 Apple Hill Drive
Natick, MA 01760-2098 USA
Tel: (508) 647 7000
Fax: (508) 647-7001
E-mail: info@mathworks.com
Web: http://www.mathworks.com

List of Abbreviations

AF	Amplify-and-forward
AWGN	Additive white Gaussian noise
BPCU	Bits per channel use
CF	Compress-and-forward
CRC	Cyclic redundancy check
CS	Conventional selection
CJ	Cooperative jamming
CRBC	Cooperative relay broadcast channel
CSI	Channel state information
DF	Decode-and-forward
ECC	Error control coding
EGC	Equal gain combiner
GSM	Global system for mobile communications
GWT	Gaussian wiretap channel
ISI	Intersymbol interference
i.i.d	Independent and identically distributed
MIMO	Multiple-input–multiple-output
MISO	Multiple-input–single-output
MRC	Maximal ratio combiner
MAC-GF	Multiple access channel with generalized feedback
NE	Nash equilibrium
NF	Noise forwarding
OR	Opportunistic relaying
OS	Optimal selection
OSJ	Optimal selection with jamming
OW	Optimal switching
OSCJ	Optimal selection with controlled jamming
OS-MMISR	Optimal selection with max–min instantaneous secrecy rate
OSJ-MMISR	Optimal selection with jamming with max–min instantaneous secrecy rate
PL	Path loss
PU	Primary user

PHY	Physical layer
QoS	Quality of service
SC	Selection combiner
SNR	Signal-to-noise ratio
SU	Secondary user
SIMO	Single-input–multiple-output
SS	Suboptimal selection
SSJ	Suboptimal selection with jamming
SW	Suboptimal switching
s.d.o.f.	Secure degrees of freedom
SINR	Signal to interference-plus-noise ratios
SS-MMISR	Suboptimal selection with max–min instantaneous secrecy rate
SSJ-MMISR	Suboptimal selection with jamming with max–min instantaneous secrecy rate
VAA	Virtual antenna array
WSN	Wireless sensor network
WANET	Wireless ad hoc network
ZF	Zero-forcing

List of Symbols

A	Best passive helpers
B_d	Doppler-frequency spread
B_c	Channel coherence bandwidth
B_s	Transmitted signal bandwidth
$b(\cdot)$	Function depends on the processing strategy implemented at the relay node
c	Speed of propagation of the electromagnetic field in the medium
C	Cryptogram
C_S	Secrecy capacity
C_B	Channel capacity of the legitimate link
C_E	Channel capacity of the eavesdropping link
C_d	Decoding set
d_0	Reference distance
$d_{a,b}$	Euclidean distance between node a and node b
D	Destination
d	Distance between the transmitter and the receiver
E	Eavesdropper
$\mathrm{E}[.]$	Expectation operator
$\mathrm{E}\left[R_S^{\lvert C_d \rvert}(R, J_1, J_2)\right]$	Ergodic secrecy rate
F	Number of subcarriers
f_o	Transmitted frequency
f_d	Doppler shift
g	Channel gains vector to the eavesdropper
G_d	Diversity gain
$h_{s,d}$	Channel fades between the source and destination
$h_{s,r}$	Channel fades between the source and the relay
$h_l(t)$	Attenuation of the l-th path at time t
$h_{R_i,D}$	Channel gain between the i-th relay node and the destination node
h_{S,R_i}	Channel gain between the source node and the i-th relay node

h	Channel gains vector to the intended receiver
$h_{a,b}$	Channel coefficient for each channel $a \rightarrow b$
$H(.)$	Entropy
$I(.\,;.)$	Mutual information
$I_{S,Ri}$	Source–relay channel mutual information
J_1	First phase jammer
J_2	Second phase jammer
J_1^*	First selected jammer
J_2^*	Second selected jammer
K	Number of helpers
$\mathbb{K}$	Secret key
L	Number of resolvable paths at the receiver
L'	Ratio between the relay power to the jammer power
m	Eavesdropper number
M	Number of eavesdroppers
m_g	Multiplexing gain
N_t	Number of transmitting antennas
N_r	Number of receiving antennas
$n_{s,d}$	Additive noise at the destination
$n_{s,r}$	Additive noise at the relay
$\check{N}$	Noise variance
N_p	Noise power spectral density
N	Number of intermediate nodes (relays)
n	Relay number
$PL_{(dB)}$	Path loss measured in dB
P_{SER}	Probability of symbol error
P	Source and relay nodes equal transmitted power
P_l	Transmitted power of the legitimate transmitter
P_{out}	Secrecy outage probability
$P^{(S)}$	Transmitted power of the source node
$P^{(R)}$	Transmitted power of the relay node
$P^{(J)}$	Transmitted power of the jamming nodes
Q	Covariance matrix
R^*	Best relay
R_T	Target secrecy rate
R_0	Transmission spectral efficiency/required transmission rate
R	Conventional relay
$R_{S_i}^{\|C_d\|}(R, J_1, J_2)$	The instantaneous secrecy rate with the decoding set C_d for source S_i
S_{eves}	Eavesdroppers set
S_1	First source
S_2	Second source
S	Source

S_{relay}	Intermediate node set
T_m	Channel delay spread
T_s	Duration of the symbols
T_c	Channel coherence time
$\mathcal{F}$	Number of transmission hops
v	Speed of the vehicle
V	Message carrying signal
W	Confidential message
x	Jamming signals vector emitted by the helpers
X	Legitimate transmitter information signal
$x(t)$	Transmitted signal
$y(t)$	Received signal
$y_{s,d}$	Received signal at the destination in the first phase
$y_{s,r}$	Received signal at the relay
$y_{r,d}$	Received signal at the destination in the first phase
Y	Intended receiver observed signal
Z	Unauthorized eavesdropper received signal
β	Path-loss exponent
ω	Shadow loss
Γ_i	Signal to interference-plus-noise ratio of link $S_j \rightarrow S_i$
Γ_{E_j}	Signal to interference-plus-noise ratio of link $S_j \rightarrow E$
δ	Constant related to the antenna gain and the average channel attenuation
$\sigma_{a,b}^2$	Channel variance
$(.)^T$	Transpose
$\tau_l(t)$	Corresponding path delay
θ	Angle between the direction of propagation of the electromagnetic wave and the direction of motion
α	Coding length
$\gamma_{a,b}$	Instantaneous signal-to-noise ratio for the link $a \rightarrow b$
γ	Signal-to-noise ratio
σ_B^2	Power of ambient Gaussian noise at the intended receiver
σ_E^2	Power of ambient Gaussian noise at the unauthorized eavesdropper
ψ_0	Global instantaneous knowledge for all the links
ψ_1	Average channel knowledge for the eavesdropper links

About the Author

Dr. Emad S. Hassan received his BSc (Honors), MSc, and PhD from the Electronics and Electrical Communications Engineering Department, Faculty of Electronic Engineering, Menoufia University, Egypt, in 2003, 2006, and 2010, respectively. In 2008, he joined the Communications Research Group at Liverpool University, Liverpool, UK, as a visitor researcher to finish his PhD research. He is Associate Professor at the Electronics and Electrical Communications Engineering Department, Faculty of Electronic Engineering, Menoufia University, Egypt.

Dr. Hassan was a full-time demonstrator (2003–2006) and Assistant Lecturer (2007–2010) at the Faculty of Electronic Engineering, Menoufia University. He has been a visitor researcher at Liverpool University, Liverpool, UK, (2008–2009); a teaching assistant at the University of Liverpool, UK (2008–2009); Assistant Professor (2010–2015) and Associate Professor (2015–present) at the Faculty of Electronic Engineering, Menoufia University; and a part-time lecturer at many respectable private engineering universities in Egypt (2010–2011). He co-supervises many MSc and PhD students, (2010–present). Currently, he is Associate Professor at the Faculty of Engineering, Jazan University, Saudi Arabia.

Dr. Hassan is a reviewer for many international journals and conferences. He was a Technical Program Committee (TPC) member for many international conferences. He has published more than 66 scientific papers in national and international conference proceedings and journals and three books under CRC Press, Taylor & Francis. His current research areas of interest include image processing, digital communications, cooperative communication, cognitive radio networks, OFDM, SC-FDE, MIMO, WSNs, and CPM-based systems.

Acknowledgments

First and foremost, I am thankful to *God*, the most gracious most merciful, for helping me finish this work.

I wish to express my sincere thanks to *Prof. Fathi E. Abd El-Samie, Prof. Moawad Dessouky,* and *Prof. Sami El-Dolil.* I am deeply indebted to them for their valuable comments, continuous encouragement, useful suggestions, and active help during the course of this work.

Many thanks are extended to the authors of all journals and conference papers, articles, and books that have been consulted in writing this book. I would also like to extend my gratitude to all my past and current MSc and PhD students for their immense contributions to knowledge in the area of security and data reliability in cooperative wireless networks. Their contributions have undoubtedly enriched the content of this book.

Also, I greatly appreciate the support received through the collaborative work undertaken with the *Engineering College, Jazan University, KSA.*

Finally, I remain extremely grateful to my family who has continued to be supportive and provide the needed encouragement. In particular, my very special thanks go to my wife, *Samah A. Ghorab,* for her continuous patience and unconditional support that have enabled me to finally complete this challenging task. Her support has been fantastic. Also, my three wonderful children, Mahmoud, Omar and Joury, who provide unending inspiration.

Chapter 1

Introduction

1.1 Cooperative Wireless Networks

Future generations of cellular communications require higher data rates and a more reliable transmission link with the growth of multimedia services, while keeping satisfactory quality of service. Multiple-input–multiple-output (MIMO) antenna systems have been considered as an efficient approach to address these demands by offering significant multiplexing and diversity gains over single-antenna systems without increasing bandwidth and power. Although MIMO systems can unfold their huge benefit in cellular base stations, they may face limitations when it comes to their deployment in mobile handsets. To overcome this drawback, an innovative approach known as cooperative communication has been suggested to exploit MIMO's benefit in a distributed manner. Such a technique is also called a virtual MIMO, since it allows single-antenna mobile terminals in a multi-user environment to share their antennas and therefore reap some of the benefits of MIMO systems.

Because of the inherent openness of the wireless transmission medium, wireless communication systems are particularly vulnerable to security attacks. It is therefore necessary to focus on guaranteeing confidentiality against eavesdropping attacks where an unauthorized entity aims to intercept an ongoing wireless communication. From a physical layer point of view, the transmission secrecy can be achieved by exploiting the inherent randomness of noise and communication channels to limit the amount of information that can be extracted at the 'bit' level by an unauthorized receiver. Adopting this point of view, in a multi-user network, focusing on a specific transmitter–receiver pair, other (independent) transmitters can act as helpers that can improve the individual secrecy rate of this specific pair by cooperatively jamming the eavesdropper.

1

1.1.1 Cooperative Communications Idea

The increasing number of users demanding service has encouraged intensive research in wireless communications. However, the problem with wireless communications is the unreliable medium through which the signal has to travel. To mitigate the effects of wireless channel on the transmitted signal, the idea of diversity has been deployed in many wireless systems [1–3]. Space diversity, for example, is a communication technique where the transmitted signal travels through various independent paths, and thus the probability that all the wireless paths are in fade is made negligible. Time diversity, frequency diversity, and space diversity are the three basic techniques for providing diversity to wireless communication systems.

Multiple-input–multiple-output (MIMO) systems, where the transmitters as well as the receivers are equipped with multiple antennas, proved to be a breakthrough in wireless communication systems, which offered a new degree of freedom, in spatial domain, to wireless communications. However, due to size, cost, or hardware limitations of many wireless agents, it became a challenge to support them with multiple transmitting antennas. To address this challenge, the idea of cooperative communications came into existence to implement the idea of MIMO in a distributed manner. This concept says that transmitting users share each others' antennas to forward their messages to the destination and form a virtual MIMO.

1.1.2 Physical Layer Security Idea

Away from the traditional cryptographic techniques for ensuring security in a wireless system, nowadays, most researchers started studying secrecy from a physical layer point of view. The idea of achieving security in a physical layer depends mainly on maximizing the capacity of the main channel, i.e., channel between the legitimate source–destination pair, about that of the wiretap channel, i.e., channel between the source and the eavesdropper. The degradation of the wiretap channel can be achieved by using some jammers to transmit intentional interference on the eavesdroppers and confuse them.

1.2 Wireless Sensor Networks

The goal of a wireless sensor network (WSN) is to provide the end user with a more intelligent understanding of its life and environment. WSN is a class of special wireless ad-hoc networks. An ad hoc network is a group of wireless nodes that interconnect directly over a common wireless channel. There is no extra structure needed for ad hoc networks. A strength of this type of networks is their capability of self-organizing the infrastructure after they are installed. There are many differences between common ad hoc networks and WSNs; they are outlined below [4–6]:

- Different areas of application.
- The number of sensor nodes in a WSN is several orders higher than that in an ad hoc network.
- Sensor nodes are closely deployed.
- The topology of this network changes frequently.
- WSNs use broadcast while most ad hoc networks use point-to-point communications.
- Sensors are limited in computational capacities, power, and memory.
- Sensors may not take global identification (ID) because this will cause a large overhead communication.

A WSN is formed from distributed autonomous sensor nodes to monitor environmental conditions, such as temperature, pressure, and sound, and cooperatively send their data through the network to a main site. For example, sensors can be deployed underwater to monitor ocean currents or on top of a mountain to monitor pollution at high altitudes, within the foundation of a building to acquire information on vibration, or be attached to animals to supervise migration patterns. WSNs are used in many industrial and civilian applications, such as environment and habitat monitoring, industrial process control, home automation, health care, and traffic regulation.

The typical size of a WSN ranges between tens to thousands of sensors. WSNs are managed through a powerful device, usually referred to as the sink that represents the gateway between the WSN and the external world (e.g., the Internet). The sink is considered to be a trusted, tamper-resistant, online device. It is responsible for providing commands to sensors and collecting data.

In some application, it is required that the sink is not existent all the time. This is due to the fact that the area is too large to be covered by the sink, or it is not required for the sink to exist all the time due to the environment or the measurement. We therefore have what is called an unattended wireless sensor network (UWSN).

1.2.1 Unattended Wireless Sensor Networks

A UWSN [7–10] (Figure 1.1) is a specific type of WSN where the sink is absent most of the time. Data sensed by the sensor nodes is not collected continuously by the data sink. Data has to be stored and secured by every node until the subsequent visit of the mobile data sink. This inability to communicate with the sink might be for reasons such as power constraints, limited transmission ranges, or signal propagation difficulties [11]. The concept of UWSNs with a mobile sink looks realistic if we consider the environments where the sensing under consideration is too far

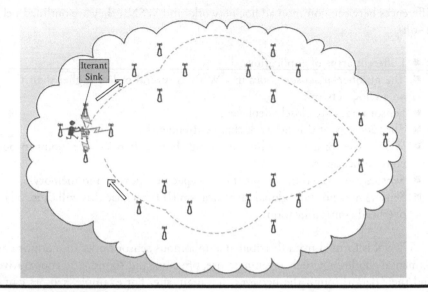

Figure 1.1 Unattended wireless sensor network with mobile sink collecting the data.

from the base station. Sending data through midway nodes may therefore result in a weakening of the network security (e.g., for example the midway nodes may alter the data) or an increase in the energy consumption of the nodes close to the base station. In normal multi-hop WSN, the power of the sensors placed nearby the sink will be exhausted earlier than that of the other sensor nodes.

The reason behind it is that all the sensors have to transmit their data to the data sink through the sensor nodes placed nearby the data sink. Therefore, the UWSN can save the battery of these sensor nodes, and, as a result, an increase in the life-time of the network can be achieved.

Unattended environments as mentioned in [12,13] include sensor networks for monitoring airborne networks for tracking adversary aircrafts, vibration and sound produced by army troop motion, and LANdroids [14], which retain information until soldiers move close to the network. In addition, it can be used for monitoring nuclear emissions, national parks for firearm discharge and illicit cultivation, and along an international border to record illegal crossings. The scale, in terms of both the number of sensors and the size of the coverage area, might make it too costly to install a multitude of fixed sinks—one per border segment.

The common feature in the above-mentioned examples is that constant physical access to the entire sensor network is impossible, and a periodic visit by an itinerant sink may be more realistic. Consequently, sensors cannot off-load their data in real-time fashion: they have to collect data and wait for an upload signal. Sensors' lack of ability to off-load their data in real time exposes them to a great risk. Moreover, this rules out all security protocols that rely on the constant presence of the sink.

Another common assumption in prior WSN's security is that an adversary can compromise a set of sensors during the entire operation of the network. In contrast, we envision a powerful UWSN adversary that can compromise up to a certain number of sensors within a specific time interval. This interval can be smaller than the time between two consecutive visits of the data sink. Given more intervals, the adversary can undermine the whole network as it moves from a set of compromised sensors to another set, thus gradually undermining network security.

1.3 Motivation

In many real-world applications, critical data is collected and stored in the unattended nodes in hostile environments. The data must be stored until the next visit of the data sink. The unattended nature of the network and the lack of tamper-resistant hardware increase the susceptibility of attacks over the data collected by the sensors [15].

Since the UWSN scenario is different from that of a traditional WSN, defense solutions from WSN security literature are not suitable for coping with a mobile adversary in UWSNs. Security needs should be taken into account to ensure data protection (also called data reliability) in these sensors at the time of design. Data security and data authentication are a major concern in UWSNs.

Most cryptographic techniques provide data authenticity and integrity but do not ensure data reliability. This implies that if an adversary compromises a sensor and destroys the data contained therein, the data is lost forever. Another drawback of cryptographic schemes is that they are computationally costly, and this is not suitable for resource-constrained sensors. For these reasons, non-cryptographic techniques can be considered over cryptographic ones. In the past few years, techniques for data authentication have been proposed [15,16] as well as cryptographic techniques for data protection [9,13,17–19].

The above-mentioned schemes assume that the sensors are stationary between consecutive visits from the data sink. However, this assumption is not practical in some real applications, so it must be relaxed and allow nodes to move between two visits from the data sink. Another important concern in UWSNs is that a mechanism is needed to ensure that the data received at the sink is authentic.

The main goal of some of the adversaries is to inject fraudulent data into the information collected by the nodes and remain undetected. The mobile adversary can compromise K out of N nodes during each round; also, it can switch to other sets of K nodes during the next rounds. Authentication schemes for UWSN against a mobile adversary presented in [15] and [16] guarantee good security but suffer from high communication cost relative to the level of security achieved. The problems identified above motivates us to find the best possible way for securing the data in UWSNs.

1.4 Problem Statement

Along with the growing popularity of WSN, it continues to be plagued by issues of data security, a situation which has prompted considerable research during the past period. WSNs are vulnerable to many kinds of attacks due to their inherent features, such as being self-organized and lacking tamper resistance [20,21]. Most of the previous research has focused on data security in the presence of a static online sink, which is considered as follows:

- A trusted party (i.e., tamper resistant)
- Can obtain collected data from sensors in real time (or near real time)
- Can take appropriate action instantly to cease the further effects of an adversary if one is detected [22]

However, our focus is on security in UWSNs, which is more challenging than for WSNs because most of the time sensors' activities are left unattended. Sensors are not able to communicate with the sink in a real-time manner and do not off-load data immediately after collecting them. An adversary can easily take advantage of the time between sink visits to roam the network with impunity and thereby learn the network topology and security strategy, compromise sensors, alter or delete the collected data in the sensors' storage, and leave the network without leaving a trace to the collector. In light of such potential overwhelming and pervasive threats, the issue here is how to protect data against an adversary or how to maintain data survival until a collector arrives [23].

1.5 Book Objectives and Contributions

This book addresses the following:

1. **Overview of the wireless channel impairments**. At first, the path loss, shadowing, and fading effects are introduced as the main wireless channel impairments that impede the achievement of future generations' requirements.
2. **Study of MIMO antenna systems**. The basic idea of the spatial-diversity-based MIMO systems is briefly explained. Then, the efficiency of MIMO systems in dealing wireless channel impairments is shown.
3. **Overview of cooperative communication in wireless systems**. The great role of the emerged cooperative communication paradigm in overcoming the difficulties of applying MIMO in mobile handsets is illustrated. In cooperative communication, a virtual MIMO system is formed by sharing antennas of the single-antenna mobiles in the multi-user environment. A hint to the relay channel concept which considers the basis for a cooperative communication working principle is also given.

4. **Study of different cooperation protocols**. Both fixed relaying schemes and adaptive relaying schemes focused on studying the processing strategy implemented on the source information at the relay node are introduced.

5. **Study of different relay selection metrics**. In this work, different relay selection metrics concerned with selecting the best relay among the available N relays are presented. The selected relay helps in applying the cooperative communications concept by assisting the source in forwarding its message to the destination besides its own message.

6. **Overview of the physical layer security in wireless systems**. The side effect of the wireless transmission broadcast nature which led to the application of data security in the wireless systems is firstly illustrated. Then, the main vulnerabilities of many implemented traditional cryptographic schemes which have recently motivated many researchers to study the data security from a physical layer point of view are shown. Physical layer security depends mainly on the inherent randomness of noise and communication channels to limit the amount of information that can be extracted by the eavesdropper.

7. **Study of cooperative secrecy techniques for physical layer**. Various cooperating approaches helping in achieving secrecy at the physical layer of a multi-user system through introducing the cooperative jamming concept are given. All the nodes in the coverage area of the transmission except the legitimate source–destination pair can act as jammers to confuse the eavesdroppers and prevent them from extracting the source information.

8. **Proposal of joint relay and jammer selection schemes for secure one-way cooperative networks**. Different proposed relay and jammer selection schemes focused on selecting one relay and two jamming nodes are introduced for ensuring physical layer security in one-way cooperative networks. In one-way cooperative networks, the signal is transmitted in one direction from the source to the destination. The obtained results showed the effectiveness of the different proposed schemes in improving both ergodic secrecy capacity and secrecy outage probability metrics.

9. **Proposal of joint relay and jammer selection schemes for secure two-way cooperative networks**. Due to the great benefits of the two-way relay channel into which the legitimate transmission pair has the ability to both transmit and receive messages, various relay and jammer selection schemes have been proposed in order to improve physical layer security in this type of networks. Then, through the numerical results, the integration between different categories of the proposed relay and jammer selection schemes is shown, and the ability of each category to improve performance metrics under certain network conditions is illustrated.

10. **Performance comparison of different proposed selection schemes in different network models**. A comparison between the proposed relay and jammer selection schemes in both one-way and two-way cooperative networks is presented in terms of ergodic secrecy capacity and secrecy outage probability.

The obtained results showed that when the relays are distributed dispersedly between the sources and the eavesdropper, all the proposed two-way schemes outperform the proposed one-way schemes, especially when the transmitted power is increased.

11. **Study of the effect of the multiple eavesdroppers' presence on the secrecy of both one-way and two-way network models**. The obtained results showed that even if there are multiple eavesdroppers, the different proposed selection schemes still have the ability to improve the performance metrics.

12. **Overview of WSNs** followed by an overview of UWSNs.

13. **A proposal called the cooperative hybrid self-healing randomized distributed (CHSFRD) scheme** is introduced to provide self-healing in UWSN. Self-healing algorithms is developed to increase the likelihood for data reliability and data security in homogeneous UWSNs, without implementing cryptography. In addition, the UWSN model is defined in a way that encompasses all common WSN assumptions and characterizes the unattended operation mode that involves periodic visits by an itinerant sink. Also, we define a new adversarial model geared for UWSNs, delineating its capabilities and identifying many adversary subtypes based on its specific goals. The proposed scheme is based mainly on the hybrid cooperation principal between healthy and compromised (sick) sensors; sensor collaboration is necessary to mislead an adversary. The proposed scheme proves its ability to enhance the UWSN security by improving the data reliability and compromising probability and probability of backward secrecy.

14. **A proposal called the self-healing controlled mobility within a cluster (SH-CMC) scheme** is developed for self-healing enhancement in UWSNs, in which the clustering and mobility of some sensors were used beside the hybrid cooperation. Both of them can enhance the self-healing capability of UWSNs. In addition, different mobility models available for wireless networks were discussed in detail. The proposed scheme proves that using the mobility within a cluster of sick sensor is the best and complementary solution for the problem of the leakage of health sponsors. The proposed scheme proves that the use of mobility beside the hybrid cooperation can enhance the self-healing capability more than the scheme that does not consider mobility.

15. **A proposal called the self-healing single flow cluster controlled mobility (SH-SFCCM) scheme** is introduced for self-healing enhancement considering energy consumption due to mobility. The trade-off between energy consumption in both mobility and communication is estimated. The energy consumption cost functions for both communication and mobility are estimated. In addition, the influence of sensor mobility on self-healing capability and other network aspects is studied.

1.6 Book Outline

Chapter 2 gives a general overview of the cooperative communications through handling the first five objectives of the book in detail.

Chapter 3 highlights many issues concerned with achieving secrecy in the physical layer through discussing the sixth and seventh objectives of the book in detail.

Chapter 4 focuses on achieving the book's eighth objective by presenting the different proposed relay and jammer selection schemes for ensuring secrecy in one-way cooperative networks. Also, the book's eleventh objective concerned with studying the effect of the presence of multiple eavesdroppers on the network performance metrics is discussed.

Chapter 5 illustrates the efficiency of the proposed relay and jammer selection schemes in improving the physical layer security of two-way cooperative networks, the book's ninth objective. Moreover, a comparison between the different proposed relay and jammer selection schemes in both one-way and two-way cooperative networks (the book's tenth objective) is provided in the presence of one or multiple eavesdroppers.

Chapter 6 presents an overview on the WSN, composition of WSN, types, modes, application, and factors influencing WSN design. This is followed by an overview on the UWSN, as well as security research applied to the field, expounding on the unattended feature of this network, together with the benefits and impacts. An explanation of the network composition, the strong and weak points and application, the mobile adversary, security goals, and challenge, and the possible attacks on nodes is given.

Chapters 7 proposes a novel cooperative hybrid self-healing randomized distributed (CHSFRD) scheme for self-healing in UWSNs. The proposed scheme is based mainly on the hybrid cooperation principal; it proves its ability to enhance the UWSN security by improving data reliability and security in UWSNs.

Chapters 8 presents a novel proposal of a self-healing controlled mobility within a cluster (SH-CMC) scheme. This scheme uses the clustering and mobility beside the hybrid cooperation to enhance the self-healing capability. Different mobility models are discussed. Also, we define a new powerful adversarial model to attack the UWSN.

Chapters 9 proposes a self-healing single flow cluster controlled mobility (SH-SFCCM) scheme; it is a novel scheme for self-healing. The trade-off between energy consumption in both mobility and communication is estimated. The energy cost functions for communication and motion are assessed. The impact of sensor mobility on network aspects is studied.

Chapter 10 presents the concluding remarks and the future work.

SECURITY IN COOPERATIVE WIRELESS NETWORKS

I

Dr. Emad S. Hassan
Eng. Doaa H. Ibrahim
Prof. Sami A. El-Dolil

Chapter 2

Overview of Cooperative Communications in Wireless Systems

2.1 Introduction

The next-generation wireless systems are supposed to handle high data rate as well as large coverage area. It should consume less power and utilize bandwidth efficiently. At the same time, the mobile terminals must be simple, cheap, and smaller in size. In wireless environment, the quality of the received signal level degraded due to path loss, shadowing, and fading impairments. The effect of fading can be suppressed by any diversity technique. For example, spatial diversity can be achieved with the help of a multiple-input–multiple-output system (MIMO). However, implementing multiple antennas at a wireless terminal is not practical due to size, power, cost, and weight constraints. Thus, a virtual MIMO known as cooperative diversity is introduced. In cooperative wireless networks, the single-antenna mobiles share their antennas to transmit their data to the destination and therefore reap some of the MIMO benefits without physically deploying antenna elements.

This chapter highlights the main key contributions presented in the literature relevant to the problems studied in this book. Section 2.2 briefly describes the main characteristics of the wireless channels. Section 2.3 presents the different common diversity techniques in addition to an introduction to cooperative diversity. The concept of relay channel which considers the basis for cooperative communications is introduced in Section 2.4. A general overview on the cooperative communications working principle is given in Section 2.5. Section 2.6 describes the different cooperative protocols concerned with the processing of received source information at the relay nodes. Section 2.7 discusses the cooperative diversity based on relay selection. Section 2.8 highlights some of the areas where the cooperative relaying strategies can be applied. Finally, the pros and cons of cooperation are given in Section 2.9.

2.2 Characteristics of Wireless Channels

Over the past few decades, there has been great development in wireless communications due to advances in wireless hardware technology and the large demand for mobile access. However, communication through a wireless channel is a challenging task because the medium introduces much impairment to the signal. Path loss, shadowing, and fading are the main impairments which affect the signal transmitted over wireless channels. In this section, these three impairments will be briefly discussed, and their effect on the performance of wireless communication systems will be illustrated.

2.2.1 Path Loss

Path loss is the attenuation in the transmitted signal power when it traverses the medium to the receiver. This attenuation is increased as the propagation distance increases. The value of the path loss is highly dependent on numerous factors related to the entire transmission setup. In general, the path loss is usually represented in the decibel scale [24].

$$PL_{(dB)} = 10\,\beta\log(d/d_0) + \delta \tag{2.1}$$

where $PL_{(dB)}$ is the path loss (PL) measured in dB, d is the distance between transmitter and receiver, β is the path loss exponent, δ is a constant related to the antenna gain and the average channel attenuation, and d_0 is the reference distance. The constant δ can be obtained from the empirical average of the receive power at the reference distance d_0. The reference distance is usually 1–10 m indoors and 10–100 m outdoors [25]. The value of the path loss exponent β depends on the propagation environment and usually ranges between 2 and 6. Table 2.1 provides path loss exponents for different propagation environments [24–30].

Table 2.1   Typical Values of the Path Loss Exponent, β, for Several Environments [24–30]

Environment	β
Free space	2
Urban macro cells	3.7 to 6.5
Urban microcells	2.7 to 3.5
Office building (same floor)	1.6 to 3.5
Office building (multiple floors)	2 to 6
Store	1.8 to 2.2
Factory	1.6 to 3.3
Home	≈3

2.2.2 Shadowing

In addition to the power loss attenuation, the radio waves may also be distorted by the obstacles that appear along the transmission paths. These obstacles may absorb part of the signal energy, resulting in signal strength degradation or random scattering. This type of impairment has been named shadow loss or shadow fading. Since the nature and location of the obstacles causing shadow loss cannot be known in advance, the path loss introduced by this effect is a random variable that follows a log-normal distribution. Denoting the value of the shadow loss as ω, we may simply combine log-normal distributed shadowing effect with the average path loss as [31]

$$PL_{(dB)} = 10\,\beta \log(d/d_0) + \omega + \delta \qquad (2.2)$$

Since the effects of both path loss and shadow fading are noticeable over relatively long distances, they are classified as large-scale propagation effects.

2.2.3 Fading

Fading loss is classified as a small-scale propagation effect because its effect is noticeable at distances in the order of the signal wavelength. This type of impairment occurs as a result of both multipath propagation and Doppler frequency shift phenomena [32], whose combination generates random fluctuations in the received power. In the following subsections, an explanation for these two phenomena is given in some detail.

2.2.3.1 Multipath Propagation

In wireless communication systems, a single transmitted signal encounters random reflectors, scatters, and attenuators during propagation, resulting in multiple copies of the signal arriving at the receiver after each has travelled through a different path [33]. The multiple copies of the transmitted signal, each having different amplitude, phase, and delay, are added at the receiver, creating either constructive or destructive interference with each other. This results in a received signal whose shape changes over time. This is the so-called multipath fading effect, which results in fast and small-scale amplitude and phase distortion, as shown in Figure 2.1. If we denote the transmitted signal by $x(t)$ and the received signal by $y(t)$, then we can write their relation as

$$y(t) = \sum_{l=1}^{L} h_l(t)x(t - \tau_l(t)) \tag{2.3}$$

where $h_l(t)$ is the attenuation of the *l*-th path at time t, $\tau_l(t)$ is the corresponding path delay, and L is the number of resolvable paths at the receiver. In many situations, it is convenient to consider the discrete-time baseband-equivalent model of the channel, for which the input–output relation derived from Equation 2.3 for sample ε can be written as

$$y[\varepsilon] = \sum_{f} h_f[\varepsilon]x[\varepsilon - f] \tag{2.4}$$

The conversion to a discrete-time model combines all the paths with arrival time within one sampling period into a single channel response coefficient $h_f[\varepsilon]$.

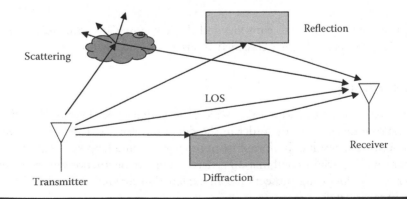

Figure 2.1   Wireless multipath fading channel.

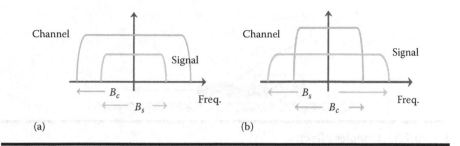

Figure 2.2 Flat fading (a) and frequency selective (b) channels.

Since the multipath propagation results in time spreading in the channel, it is then important to discuss two main time dispersion parameters which are used to both characterize and classify different multipath channels.

- Channel delay spread (T_m) is the time difference between the arrival of the earliest significant multipath component and the latest, i.e., $T_m = \max_l \tau_l - \min_l \tau_l$. If the duration of the symbols (T_s) used for signaling over the channel exceeds the channel delay spread, then the symbols will suffer from inter-symbol interference (ISI).
- Channel coherence bandwidth (B_c) is related to the inverse of the delay spread, and it is used to provide a measurement of the range of frequencies over which the channel shows a flat frequency response, in the sense that all the spectral components have approximately the same amplitude and a linear change of phase. If the transmitted signal bandwidth (B_s) is less than the channel coherence bandwidth, then all the spectral components of the signal will be affected by the same attenuation and by a linear change of phase, and the channel is said to be a flat fading channel, as shown in Figure 2.2a. However, if the transmitted signal bandwidth is more than the channel coherence bandwidth, then the spectral components of the signal will be affected by different attenuations, and the channel is said to be a frequency selective channel, as shown in Figure 2.2b.

2.2.3.2 Doppler Frequency Shift

Due to the relative motion between the transmitter and the receiver, the received signal frequency is displaced from the transmitted one by an amount called the Doppler shift, as shown in Figure 2.3. Assuming that f_o is the transmitted frequency, v is the speed of the vehicle, and θ is the angle between the direction of propagation of the electromagnetic wave and the direction of motion, the Doppler shift can therefore be expressed as [34]

$$f_d = f_o \frac{v}{c} \cos\theta \tag{2.5}$$

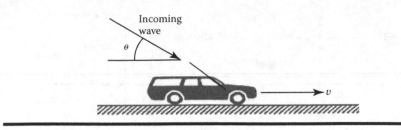

Figure 2.3   Doppler effect.

where c is the speed of propagation of the electromagnetic field in the medium. Depending on the direction of the transmitter movement with respect to the receiver, the Doppler frequency shift is either greater or lower than 0.

Since the Doppler shift results in frequency spreading, it is then important to discuss two main frequency dispersion parameters, namely, the Doppler-frequency spread and channel coherence time.

- The Doppler-frequency spread (B_d) characterizes the spreading of transmitted frequency due to different Doppler shifts.
- Channel coherence time (T_c) is the inverse of the Doppler spread and used to characterize the time over which the channel is time invariant.

According to the value of the Doppler-frequency spread as compared to the signal bandwidth (B_s), the channel is said to have fast fading (see Figure 2.4a) or slow fading (see Figure 2.4b). If the Doppler spread is smaller than the signal bandwidth or, equivalently, if the channel coherence time is larger than the transmitted symbol period, the channel will be changing over a period of time longer than the input symbol duration. In this case, the channel is said to have slow fading. If the converse applies, the channel is said to have fast fading.

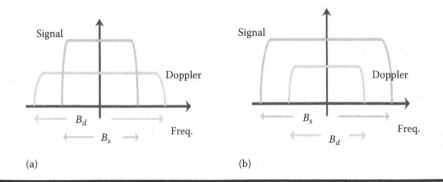

Figure 2.4   (a) Fast and (b) slow fading channels.

2.3 Common and Cooperative Diversity

Acquiring a high data rate together with reliable transmission over error-prone wireless channels is a major challenge for wireless system designers due to the wireless channel impairments. These impairments can be compensated by various ways such as increasing transmit power and bandwidth and/or applying powerful error control coding (ECC). However, power and bandwidth are very scarce and expensive radio resources whereas ECC yields reduced transmission rate. The following subsections highlight some relevant work on the popular diversity techniques, multiple antenna systems and cooperative diversity, the most effective solutions used for handling channel fading impairment.

2.3.1 Common Diversity Techniques

With a view to mitigate the fading effect of a wireless channel and thus improve the overall channel reliability, more than one signal path can be provided between the source and the destination. Since each path exhibits a fading process as independent from the others as possible, the chance that there is at least one sufficiently strong path is improved, and the probability that all the wireless paths are in fade is made negligible. The communication techniques that aim at providing multiple, ideally independent signal paths are collectively known as diversity techniques [35].

At the receiver, the signals arriving through the multiple paths are constructively combined in order to obtain a resulting signal of better quality or with better probability of successful reception than each of the received ones. There are multiple combination techniques that differ in the nature of the processing applied to each signal during combining.

The popular combination techniques are as follows [36–38]:

- Selection combiner (SC), where the output is the input with the best signal-to-noise ratio (SNR)
- Threshold combiner, where the combiner sequentially scans the received signals and outputs the first one with SNR exceeding a threshold
- Maximal ratio combiner (MRC), where the combiner firstly co-phases the multiple received signals, followed by weighting each sample proportionally to the corresponding path SNR and finally adding them. The SNR of the resulting signal at the output of the MRC is equal to the sum of the SNRs corresponding to each path.
- Equal gain combiner (EGC), where the signals are co-phased and added (maximal ratio with equal weights).

For any diversity technique, the diversity gain (G_d) is used to measure the performance improvement of the system. It is defined as the rate of decrease in the communication error probability at high channel SNR. When using log–log scales, the diversity gain is defined as [39]

$$G_d = -\lim_{\gamma \to \infty} \frac{\log P_{SER}}{\log \gamma} \tag{2.6}$$

where γ is the SNR and P_{SER} is the probability of symbol error.

There are many different forms of diversity, including time diversity, frequency diversity, and spatial diversity. Next, a succinct introduction to each diversity technique is provided.

- **Time Diversity.** Using time diversity the same symbol is transmitted at different time instants provided that the time separation exceeds the coherence time (T_c) of the channel. This implies that the different transmitted symbols will experience channel realizations that are highly uncorrelated and can be used to obtain diversity. The simplest way to achieve this type of diversity is using a repetition coding scheme. Also, in order to guarantee that the repeated symbols will be transmitted over uncorrelated channel realization, an appropriate interleaver is applied to the stream of symbols to be transmitted. From the diversity gain point of view, the time-diversity system with repetition coding achieves full diversity gain. Nevertheless, the use of repetition coding sacrifices the total bit rate.
- **Frequency Diversity.** In this form of diversity, the same symbol is transmitted over multiple carrier frequencies provided that the frequency separation exceeds the channel coherence bandwidth (B_c). This approach is applicable in multicarrier systems, where transmission is implemented by dividing the wideband channel into nonoverlapping narrowband subchannels. The symbol used for transmission in each subchannel has a transmission period long enough for the subchannel to appear as a flat fading channel. Different subchannels, each separated in the frequency domain from the rest by more than the coherence bandwidth, are used together to achieve frequency diversity since the fading processes among the different subchannels will show a small cross-correlation.
- **Spatial Diversity.** This type of diversity uses multiple antennas at the transmitter or receiver or at both ends. The configuration of deploying multiple antennas is often referred to as multiple-input–single-output (MISO) systems if only a single antenna is deployed at the receiver side, single-input–multiple output (SIMO) systems if a single transmit antenna is used, or, in general, multiple-input–multiple-output (MIMO) systems if multiple antennas are used at the transceiver. The greater the number of antenna pairs, the greater

the redundancy (diversity) of the received signals, i.e., the higher the reliability of the transceiver detection. The antennas should be spaced far apart enough so that different received copies of the signal undergo independent fading. The necessary antenna separation at each side depends on the scattering in the neighborhood of the antenna and on the signal carrier frequency. The typical antenna separation is between half to one carrier wavelength for a mobile and is in the order of tens of wavelengths for base stations.

2.3.2 MIMO Systems

MIMO communication systems proved to be a breakthrough in wireless communication system because of their abilities in providing a high data rate together with a reliable transmission over error-prone wireless channels [40–42]. Aiming at improving the system reliability, the multiple antenna elements at each side are placed with appropriate separation between them so that different independent and low-correlated channels between each pair of transmit and receive antennas establish "spatial diversity." The presence of multiple transmit/receive antenna pairs improves the chance that there is at least one sufficiently strong path, and thus the reliability of the transceiver detection is increased. This improvement in reliability translates into performance improvement—measured as diversity gain. Aiming at improving the system data rate, the different portions of the data are placed on different propagation paths ("spatial multiplexing"), and this results in capacity gain measured by the number of degrees of freedom in the channel, or the multiplexing gain with no additional power or bandwidth.

Figure 2.5 shows a MIMO system with N_t transmit antennas and N_r receive antennas; assuming the path gains between individual antenna pairs are independent

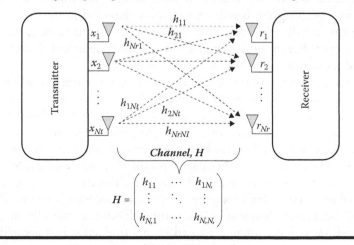

Figure 2.5   Block diagram of a MIMO system.

and identically distributed (i.i.d.), the maximal diversity gain and the multiplexing gain can be expressed as

$$G_d = N_t \times N_r \tag{2.7}$$

$$Multiplexing\ gain = \min (N_t, N_r) \tag{2.8}$$

For a given MIMO channel, both diversity and multiplexing gains can be achieved simultaneously but with a fundamental trade-off between the two gains. For example, as shown by Zheng and Tse [43], the optimal diversity gain that can be achieved by any coding scheme having a multiplexing gain m_g is $(N_t - m_g) \times (N_r - m_g)$. This implies that out of the total resources, m_g transmit and m_g receive antennas (m_g integer) are used for multiplexing and the remaining $N_t - m_g$ transmit and $N_r - m_g$ receive antennas provide the diversity. In summary, the higher spatial diversity gain comes at the price of a lower spatial multiplexing gain, and vice versa.

2.3.3 Cooperative Diversity

Despite the gains of MIMO systems in improving both data rate and reliability of the wireless link, they may result in degrading the system quality of service (QoS) due to the correlation between multiple co-located antennas. Moreover, it is difficult for small handheld wireless devices to support multiple antennas due to size, cost, or hardware limitations.

To overcome the above MIMO limitations, new techniques known as cooperative communications are introduced. The basic idea of these innovative approaches depends on exploiting the broadcast nature of the wireless channel to achieve MIMO gains in a distributed manner. Adopting this point of view, the network nodes have been thought of as a set of antennas that cooperate with each other for distributed transmission and processing of information.

As shown in Figure 2.6, the cooperating node acts as a relay node for forwarding the source node information to the intended destination besides its own information. Since the relay node is usually several wavelengths away from the source, the relay channel is guaranteed to fade independently from the direct channel, which introduces a MIMO channel between the source and the destination. The direct channel information and the relayed information are subsequently combined at a destination node so as to create spatial diversity. This creates a network that can be regarded as a system implementing a distributed multiple antenna where collaborating nodes create diverse signal paths for each other. In the following section, the relay channel concept will be discussed as an introduction to study in detail the design and analysis of cooperative communications.

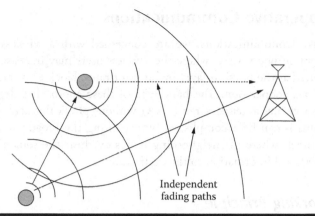

Figure 2.6   Cooperative communication.

2.4 Classical Relay Channel

Indeed, the relay channel concept was firstly presented by van der Meulen [44]. Figure 2.7 shows the basic three-terminal relay channel model consisting of a source, a destination, and a relay. In this model, when source A sends a signal X, a noisy, attenuated version is received by both the destination C and a relay B. Based on what the relay received is, another signal X_1 is forwarded to the destination. Since it was assumed that all nodes operate in the same band, the system can be decomposed into a broadcast channel (A transmitting, B and C receiving) from the viewpoint of the source and a multiple access channel (A and B transmitting, C receiving) from the viewpoint of the destination.

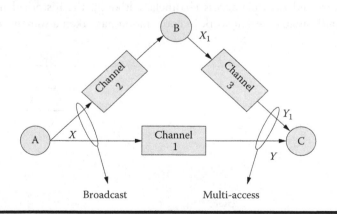

Figure 2.7   Relay channel.

2.5 Cooperative Communications

In cooperative communications, we are concerned with a wireless network, of the cellular or ad hoc variety, where the wireless users may increase the effective QoS (measured at the physical layer by bit error rates, block error rates, or outage probability) via cooperation. The basic idea of these approaches depends mainly on creating a virtual antenna array (VAA) without physically deploying antenna elements. This is can be accomplished by exploiting the broadcast nature of the wireless channels where the neighboring nodes overhear the source's signals and relay the overheard information to the destination.

2.5.1 Working Principle

Figure 2.8 illustrates a basic cooperation system that consists of two mobile agents communicating with the same destination. Each mobile has its own antenna and cannot individually generate spatial diversity. However, it may be possible for each mobile to act as a relay to the other and forward some version of the "overheard" information along with its own data. Therefore, there is more than one signal path at the destination, each exhibiting a fading process as independent from the other as possible; this generates transmit diversity.

2.5.2 Historical Background

The basic ideas behind cooperative communication are traced back to the groundbreaking work of Cover and El Gamal on the information theoretic capacity of the previously introduced relay channel [45]. They found that the capacity of this channel is bounded by the minimum of the rates of transmission of the constituent broadcast and multiple access channels. Although the historical importance of [45] is indisputable, recent work in cooperation has taken a somewhat different

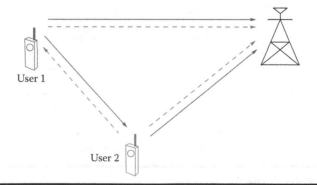

Figure 2.8 In cooperative communication, each user is both a source and a relay.

emphasis for the following two reasons. Firstly, Cover and El Gamal mostly analyze capacity in an additive white Gaussian noise (AWGN) channel, while recent developments are motivated by the concept of diversity in a fading channel. Secondly, in a relay channel, the relay's sole purpose is to help the main channel, whereas in cooperation, each wireless user is assumed to both transmit its own data and act as a cooperative relay for another user. Since the strategy of relaying the source information to the destination is a key aspect for the cooperative communication process, the following section discusses in detail different cooperative relaying techniques which result in different cooperation protocols.

2.6 Cooperation Protocols

Cooperative communication protocols can be generally categorized into two main schemes: fixed relaying schemes and adaptive relaying schemes. In this section, both schemes will be described along with a single relay under a half-duplex constraint, i.e., the relay cannot transmit and receive simultaneously at the same time. Therefore, the typical cooperative strategy is performed in two different phases:

- In phase 1, a source sends its information to the intended destination, and due to the broadcast nature of the transmission, the information is also received by the relay node.
- In phase 2, the relay forwards the source information received in the first phase to the destination.

Figure 2.9 shows a simplified cooperation model, where the relay node helps the source to deliver its information to the destination. For simplicity, we assume that both the source and the relay nodes transmit with equal power P. In the first phase, the source broadcasts its information to both the destination and the relay. The received signals at the destination and the relay are denoted respectively by $y_{s,d}$ and $y_{s,r}$ and can be written as

$$y_{s,d} = \sqrt{P}h_{s,d}x + n_{s,d} \qquad (2.9)$$

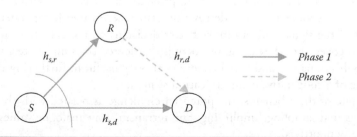

Figure 2.9   Simplified cooperation model.

$$y_{s,r} = \sqrt{P} h_{s,r} x + n_{s,r} \qquad (2.10)$$

where x is the transmitted information symbol, and $n_{s,d}$ and $n_{s,r}$ are additive noise modeled as zero-mean complex Gaussian random variables with variance $\check{N}$. In Equations 2.9 and 2.10, $h_{s,d}$ and $h_{s,r}$ are the channel fades between the source and destination and the relay, respectively, and are modelled as Rayleigh flat fading channels.

In the second phase, the relay forwards a processed version of the source's signal to the destination, and therefore, the received signal at the destination denoted by $y_{r,d}$ and can be written as

$$y_{r,d} = \sqrt{P} b(y_{s,r}) + n_{r,d} \qquad (2.11)$$

where the function $b(\cdot)$ depends on the processing strategy implemented at the relay node [46].

2.6.1 Fixed Cooperation Strategies

In fixed relaying, the channel resources are divided between the source and the relay nodes in a fixed (deterministic) manner. The processing at the relay differs according to the employed protocol. The most common fixed cooperation techniques are the fixed amplify-and-forward (AF) relaying protocol and the fixed decode-and-forward (DF) relaying protocol.

2.6.1.1 Fixed AF Relaying Protocol

Firstly, the AF relaying protocol was proposed and analyzed by Laneman *et al.* [47]. In this relaying technique, as the name implies, the relay amplifies the signal received from the source and forwards it to the destination. The destination then combines the information sent by the source and the relay and makes a final decision on the transmitted bit, as shown in Figure 2.10.

From the discussion of this relaying protocol, it is clear that although noise is amplified by cooperation, the destination still receives two independently faded versions of the signal and can make better decisions on the detection of information. Moreover, the AF relaying protocol is considered the simplest technique that lends itself to analysis and has therefore been very useful in furthering the understanding of a cooperative communication system.

In spite of these benefits, the potential challenge associated with this relaying scheme is that sampling, amplifying, and retransmitting analogue values are technologically nontrivial.

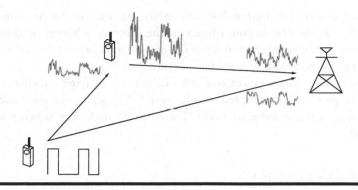

Figure 2.10   Amplify-and-forward protocol.

2.6.1.2 Fixed DF Relaying Protocol

The first work proposing a DF protocol for user cooperation was by Sendonaris *et al.* [48]. In this relaying technique, the relay decodes the received signal from the source, re-encodes it, and then retransmits it to the destination, as shown in Figure 2.11. Comparing with the previously discussed fixed AF relaying technique, fixed DF relaying is most often the preferred method used to process the data in the relay for the following reasons:

- Nowadays, a wireless transmission is very seldom analogue.
- The relay has enough computing power.
- The effects of additive noise are eliminated at the relay.

In spite of the benefits of the fixed DF relaying technique, degradation in the performance of the system may result if the decoded signal at the relay is incorrect.

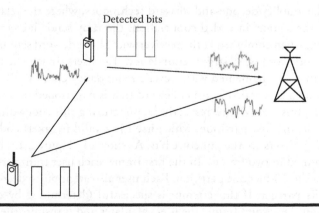

Figure 2.11   Decode-and-forward protocol.

Forwarding erroneously detected signals to the destination results in meaningless decoding at the destination. Moreover, the diversity achieved by this scheme becomes one, because the performance of the system is limited by the worst link from the source–relay and source–destination.

In addition to the two most common techniques for fixed relaying, there are other techniques, such as compress-and-forward (CF) cooperation and coded cooperation, which deserve some attention. Therefore, a quick view is taken for these techniques.

2.6.1.3 CF Cooperation

The CF protocol was first described by Cover and El Gamal [45]. Unlike in the amplify/decode-and-forward techniques where the relay transmits a copy of the received message to the destination, in the CF protocol, the relay transmits a quantized and compressed version of the received message to the destination. Therefore, the destination combines the received message from the source node and its quantized and compressed version from the relay node. It is important to note that the source information received at the destination can be used, as side information, while decoding the message from the relay. This will allow for encoding at a lower source encoding rate [46]. As a final note to CF cooperation, we note that the quantization and compression process at the relay node is a process of source encoding implemented with a set of coding techniques known as distributed source coding, Sleppian–Wolf coding, or Wyner–Ziv coding [49].

2.6.1.4 Coded Cooperation

Coded cooperation differs from the previously discussed schemes in that the cooperation is implemented at the level of the channel coding subsystem [50–52]. Unlike in the amplify/decode-and-forward techniques, where the relay repeats the bits sent by the source, in coded cooperation, the relay sends incremental redundancy, which, when combined at the receiver with the codeword sent by the source, results in a codeword with larger redundancy. The users divide their source data into blocks that are augmented with a cyclic redundancy check (CRC) code. Each of the users' data is encoded into a codeword that is partitioned into two segments containing N_1 bits and N_2 bits, respectively; puncturing this codeword down to N_1 bits, we obtain the first partition, which itself is a valid (weaker) codeword while the remaining N_2 bits are the puncture bits. As shown in Figure 2.12, the transmission is performed in two frames. In the first frame, each user transmits a codeword consisting of the N_1-bit code partition. Each user also attempts to decode the transmission of its partner. If this attempt is successful (determined by checking the CRC code), in the second frame, the user calculates and transmits the second code partition of its partner, containing N_2 code bits. Otherwise, the user transmits its

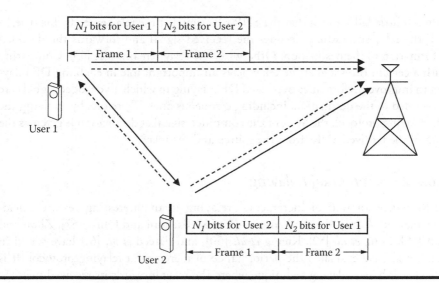

Figure 2.12   Coded cooperation framework.

own second partition, again containing N_2 bits. Thus, each user always transmits a total of $N_T = N_1 + N_2$ bits per source block over the two frames. We define the level of cooperation as N_2/N_T, the percentage of the total bits for each source block the user transmits for its partner.

2.6.2 *Adaptive Cooperation Strategies*

Although fixed relaying techniques have the advantage of easy implementation, they suffer from low bandwidth efficiency. This is because half of the channel resources are allocated for relay transmission, which reduces the overall rate, especially in the scenario where the source–destination channel is not very bad and therefore the packets transmitted by the source to the destination could be received correctly by the destination and the relay's transmissions would be wasted. Adaptive relaying techniques, comprising both DF selective and incremental relaying, try to overcome this limitation.

2.6.2.1 *Selective DF Relaying*

In addition to the major drawback of the fixed relaying techniques in reducing the transmission rate, fixed DF relaying may also reduce the diversity gains to one due to the fact that the performance is limited by the weakest source–relay and relay–destination channels. In order to overcome these problems, selective decode-and-forward relaying protocol or simply (DF) relaying can be developed to improve the inefficiency. This relaying technique depends on comparing the SNR of the signal at the relay with a certain threshold. If the channel between the source and the relay suffers a little fading such that the SNR of a signal at the relay exceeds this threshold, the relay successfully decodes the received signal and forwards the decoded information to the destination. Otherwise, the relay remains idle [53]. Comparing with a certain threshold at the relay plays an important role in enabling DF relaying to improve performance over fixed DF relaying in which all decoded signals are forwarded to the destination, including erroneous ones. Therefore, by applying the DF relaying protocol, the SNR of the combined signal at destination is given as the sum of the received SNRs from the source and the relay.

2.6.2.2 Incremental Relaying

Performance analysis of incremental relaying is an interesting research field. Leneman *et al.* [54], Ikki and Ahmed [55,56], Bastami and Olifat [57], Zhou and Lau [58], Long *et al.* [59], Kuang *et al.* [60], and Fareed *et al.* [61] have stated in detail and have evaluated the performance of incremental relaying protocol. It is a feedback-based relaying technique where the relay forwards its received message only upon request from the destination. If the destination receives the source's message correctly in the first transmission phase, it acknowledges the relay so that the relay does not need to transmit. This protocol has the best spectral efficiency among the previously described protocols because the relay does not always need to transmit, and hence, the second transmission phase becomes opportunistic depending on the channel state condition of the direct channel between the source and the destination. If the source transmission in the first phase was successful, then there is no second phase, and the source transmits new information in the next time slot. This results in a transmission rate equals R_t. On the other hand, if the source transmission was not successful in the first phase, the relay can use any of the fixed relaying protocols to transmit the source signal from the first phase. This results in a lower transmission rate that is equals $R_t/2$ as in fixed relaying.

2.7 Cooperative Diversity Based on Relay Selection

After discussing different cooperative protocols concerned with the processing of the signal received from the source node at the relay, it is now important to show the different relay selection metrics concerned with selecting the best relay among the available *N* relays, which plays an important role in implementing cooperative diversity in distributed systems [62–66]. Moreover, it is also important to know which entity evaluates these metrics and selects the relay.

2.7.1 Relay Selection Metrics

Opportunistic relaying techniques have been developed to choose the best relay which aids the source in its communication with the destination. The selection of

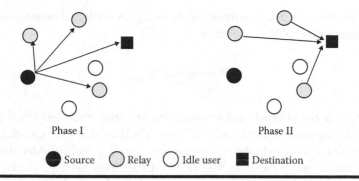

Phase I   Phase II

● Source ◐ Relay ○ Idle user ■ Destination

Figure 2.13   Illustration of a two-phase cooperative diversity system with multiple relays. Phase I is the broadcasting phase, and phase II is the cooperation phase.

the best relay could be done proactively before the source transmission or reactively after the reception of the source signal by the relays. In this section, both proactive and reactive relaying modes are discussed. In order to show the main difference between these relaying modes, a half-duplex dual-hop communication scenario in a cluttered environment is considered, as shown in Figure 2.13, where the direct path between the source and the destination is blocked by an intermediate wall, while relays are located at the periphery of the obstacle (around-the-corner). Therefore, the communication between source and destination endpoints is implemented through the relays in two successive phases.

2.7.1.1 Reactive Opportunistic Relaying

This type of relaying is shown in Figure 2.14 and is implemented among DF relays. During the first communication phase, the source transmits its message to the relays. The relays that successfully decode the source message form a decoding set C_d. During the second phase, the best relay (R^*) among the decoding set relays

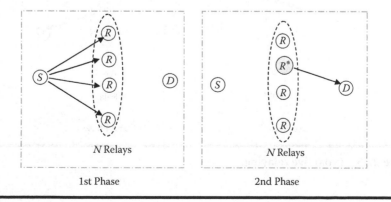

Figure 2.14   Reactive relaying.

which maximizes the relay–destination channel gain is selected in order to forward the source message to the destination.

$$R^* = \arg \max_{R_i \in C_d} \left| h_{R_i,D} \right|^2 \tag{2.12}$$

where $h_{R_i,D}$ is the channel gain between the i-th relay node and the destination node. From the reactive mode point of view, it is clear that the best selected relay is the one that can decode the source message correctly and provides the greatest information to the destination. The required condition for a particular relay to belong to the decoding set C_d and thus successfully decode the source message is that its source–relay channel mutual information $I_{S,Ri}$ is higher than the required transmission rate R_0, and this is given by

$$I_{S,R_i} = \frac{1}{2} \log_2 \left(1 + \frac{P^{(S)}}{N_p} \left| h_{S,R_i} \right|^2 \right) > R_0 \tag{2.13}$$

where $P^{(S)}$ is the power transmitted by the source, N_p is the noise power spectral density, and h_{S,R_i} is the channel gain between the source node and the i-th relay node. The factor $1/2$ is referred to as the half-duplex constraint where the relaying is performed through two phases.

2.7.1.2 Proactive Opportunistic Relaying

This type of relaying is shown in Figure 2.15 and is implemented among DF or AF relays. In the first phase, the best relay is selected prior to the source transmission.

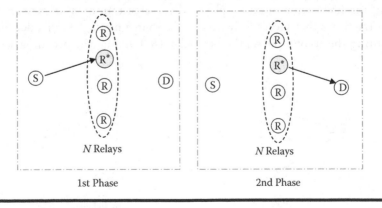

1st Phase   2nd Phase

Figure 2.15　Proactive relaying.

Then, in the second phase, the best relay forwards the source message to the destination. The best participating relay is selected according to the following criteria:

$$R^* = \arg \max_{R_i \in \{1,2,\ldots,N\}} \min\left(\left|h_{S,R_i}\right|^2, \left|h_{R_i,D}\right|^2\right) \qquad (2.14)$$

Unlike the previously discussed DF reactive opportunistic relaying, in DF proactive opportunistic relaying, the relay is picked from the entire set of available relays, not just from the decoding set. Moreover, the constraint of successful decoding is reintroduced after the relay selection, i.e., the relay transmits only if the source–relay and relay–destination mutual information is above the required transmission rate

$$\frac{1}{2}\log_2\left(1 + \frac{P}{N_P}\min\left(\left|h_{S,R_i}\right|^2, \left|h_{R_i,D}\right|^2\right)\right) > R_0 \qquad (2.15)$$

where $P^{(S)} = P^{(R)} = P$. Comparing Equations 2.13 and 2.15, which represent the constraints for successful decoding for both DF reactive and DF proactive opportunistic relaying, respectively, it is concluded that the overly constrained relay selection in Equation 2.15 leads to some performance degradation in DF proactive relaying as compared with the DF reactive one.

2.7.2 Relay Selection Implementation

After discussing the metrics used to select the relay, it is important in this subsection to know which entity evaluates these metrics and selects the relay. There are two different protocols known as destination driven and relay driven, each of which is used to implement the relay selection.

2.7.2.1 Destination-Driven Protocol

It is a simple, practical, and more suitable implementation for DF reactive opportunistic relaying. When the source distributes its data in the first phase, both the destination and the N available relays decode the source information. The relay which correctly decodes the source information sends a feedback bit to the destination to indicate its participation in the decoding set C_d. Using this feedback bit for channel estimation, the destination selects the best relay which maximizes the relay–destination channel [62].

2.7.2.2 Relay-Driven Protocol

This protocol is suitable for implementing the relay selection in both reactive and proactive opportunistic relaying. In this scenario, each relay evaluates the selected metric and starts a timer with a certain initial value, inversely proportional to the metric. The timer of the relay with the largest metric is reduced to zero first. This relay, thus having selected itself, sets off a flag and transmits the source information to the destination. The remaining relays overhear the flag and do not initiate their own transmissions. This implementation necessitates channel feedback to the relays. Furthermore, there is a nonzero probability that the trimmers of multiple relays will expire essentially simultaneously, causing a correlation at the destination [63].

2.8 Application Scenarios

This section highlights some of the areas where the different cooperative relaying strategies can be applied.

2.8.1 Virtual Antenna Array

Acquiring a high data rate together with reliable transmission over error-prone wireless channels is currently receiving much attention. It is a well-known fact that the use of a MIMO antenna system improves the diversity gain of wireless systems, which therefore helps in eliminating the effect of multipath fading phenomena. However, a multi-antenna technique is not attractive for small wireless nodes due to limited hardware and signal processing capability. Diversity can be achieved through user cooperation, whereby mobile users share their physical resources in order to create a virtual array, which thus removes the burden of applying actually multiple antennas on wireless terminals.

2.8.2 Wireless Sensor Network

A wireless sensor network (WSN) is an energy-constrained network whose nodes are typically powered by batteries for which replacement or recharging is difficult if not impossible [67]. For such networks, minimizing the energy consumption per unit information transmission becomes a very important design consideration in order to increase the lifetime of these networks. Cooperative communications work on increasing the WSN lifetime by eliminating the energy consumption where incorporation of cooperating relay nodes into routing process selects only the better communication links, which results in saving more power than when communicating through weaker channels.

2.8.3 Wireless Ad Hoc Network

A wireless ad hoc network (WANET) is a decentralized type of wireless network [68,69]. The network is ad hoc because it does not rely on a pre-existing infrastructure. Instead, distributed nodes participate in routing by forwarding data for other nodes, and the determination of which nodes forward data is made dynamically on the basis of network connectivity.

2.8.4 Vehicle-to-Vehicle Communication

Future vehicles will provide platooning (automated steering within a group of cars), in-vehicle Internet access, and inter-vehicle communication [70–72]. The increasing density of vehicles enables the deployment of cooperative vehicle systems. The main advantage behind applying the cooperative techniques is that the redundant links established in cooperation offer high link stability in such volatile and dynamic propagation conditions.

2.8.5 Cooperative Sensing for Cognitive Radio

Cognitive radio is emerging as a means to improve the wireless spectrum utilization [73,74], by allowing a secondary user (SU) to access a licensed spectrum simultaneously with a primary user (PU). In the time when the PUs want to use their licensed resources, SUs have to vacant these resources. Therefore, SUs have to constantly sense the presence of the PU. By using spatially distributed nodes, the probability of false alarming is reduced because the channel sensing reliability is improved by sharing the information [75].

2.9 Pros and Cons of Cooperation

2.9.1 Cooperation Advantages

- **Balanced QoS**. Unlike the traditional systems where the users at the cell edge or in shadowed areas suffered from capacity and/or coverage problems, relaying overcomes this discrepancy and gives almost an equal QoS to all users.
- **Performance Gains**. Large systems-wide performance gains can be achieved due to path loss, diversity, and multiplexing gains. These gains translate into decreased transmission powers, higher capacity, or better cell coverage.
- **Cost Reduction**. Taking into account a purely cellular approach, relaying is a more cost-effective solution in order to provide a given level of QoS to all users.
- **Infrastructure-Less Deployment**. By using the relays, it is possible to roll out a system that has minimal or no infrastructure available prior to deployment. For example, in disaster-struck areas, where the cellular system is non-functioning, relaying can be used to facilitate communications.

2.9.2 Cooperation Disadvantages

Despite numerous advantages of cooperation, helping out users in a cooperative fashion has its price. In this subsection, the challenges that incur in systems with cooperative communication incorporated [76,77] will be described.

■ **Complex Schedulers.** In cooperative communications, since not only traffic of different users and applications needs to be scheduled but also the relayed data flows, relaying requires more sophisticated schedulers.

■ **Increased Overhead.** A full system functioning requires handovers, synchronization, extra security, etc. This clearly induces an increased overhead with respect to a system that does not use relaying.

■ **Partner Choice.** The selection process of the optimum relaying and cooperative partner(s) is a fairly intricate task with respect to noncooperative relaying.

■ **Extra Relay Traffic.** From a system throughput point of view, the relayed traffic is a redundant traffic which decreases the effective system throughput since in most cases resources in the form of extra frequency channels or time slots need to be provided.

■ **More Channel Estimates.** The use of relays effectively increases the number of wireless channels. This requires the estimation of more channel coefficients, and hence, more pilot symbols need to be provided if coherent modulation was to be used.

■ **Increased Interference.** In addition to boosting capacity or coverage, if the offered power savings are not used to decrease the transmission power of the relay nodes, then relaying will certainly generate extra intra- and inter-cell interference, which potentially degrades the system performance.

■ **Increased End-to-End Latency.** Relaying typically involves the reception and decoding of the entire data packet before retransmitting it. If delay-sensitive services are being supported, such as voice or the increasingly popular multimedia web services, then the latency induced by the decoding may be detrimental. Latency increases with the number of relays and also with the use of interleavers, such as those utilized in GSM voice traffic. To overcome this latency, either simple transparent relaying or novel decoding methods need to be used.

Chapter 3

Physical Layer Security in Wireless Networks

3.1 Introduction

Due to the broadcast nature of wireless communications, it is difficult to shield transmitted signals from adversarial users (eavesdroppers) in the coverage area of the transmission. Therefore, it becomes too urgent to achieve the security of data transmission. Traditionally, the security of data transmission has been entrusted to key-based enciphering (cryptographic) techniques at higher layers of the protocol stack. Recently and only in the past decade or so, a resurgence of interest in aspects of secrecy at the physical layer has been experienced. The main objective behind physical layer security is to enable the exchange of confidential messages over a wireless medium in the presence of unauthorized eavesdroppers, without relying on higher-layer encryption.

This chapter highlights many issues concerned with achieving secrecy in the physical layer. In the next section, the main reasons which led us to really study secrecy from a physical layer point of view are illustrated. Section 3.3 gives an overview of the foundations dating back to the pioneering key-based Shannon and keyless Wyner works on information-theoretic security. Multiformity of cooperative jamming techniques have been introduced in Section 3.4 to show the effectiveness of jamming signals transmitted by some legitimate transmitters in a network of multiple transmitters and receivers in achieving physical layer security for a certain legitimate pair. In Section 3.5, the idea of employing cooperative jammers in a multiple trusted/untrusted relay networks in order to improve security is given, and the interactions arising between cooperation and secrecy in the presence of untrusted relays are adopted.

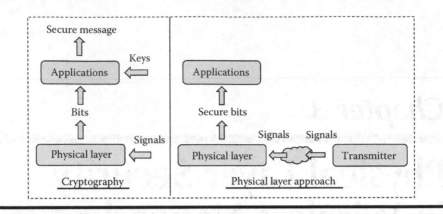

Figure 3.1 Cryptographic and information-theoretic approaches for wireless communication security.

3.2 Why Physical Layer Security

Although the current standard cryptographic algorithms proved to withstand exhaustive attacks [78], there are some vulnerabilities shown by many implemented cryptographic schemes [79,80]. Key distribution for symmetric cryptosystems and high computational complexity of asymmetric cryptosystems are the main disadvantages of the cryptographic schemes. There is also the complete dependence of all cryptographic measures on the premise that it is computationally infeasible for them to be deciphered without knowledge of the secret key, which remains mathematically unproven. In addition to the previously mentioned vulnerabilities, the ciphers that were considered virtually unbreakable in the past can be surmounted due to the relentless growth of computational power. Therefore, a trend for studying the properties of the physical layer in order to ensure some level of information-theoretic security came into existence [81–84].

Figure 3.1 illustrates the difference between cryptographic and information-theoretic approaches in order to ensure communication security in wireless networks. The basic idea behind physical layer security is to exploit the inherent randomness of noise and communication channels to limit the amount of information that can be extracted at the 'bit' level by an unauthorized receiver. In these schemes, no limitations are assumed for the eavesdropper in terms of computational resources or network parameter knowledge.

3.3 Secrecy Fundamentals

The simplest network where the problems of secrecy arise consists of a transmitter that wishes to communicate a private message to an intended receiver in the presence of an unauthorized receiver often called the eavesdropper or the wiretapper.

Usually, the nomenclature Alice, Bob, and Eve are used to refer to the legitimate transmitter, intended receiver, and unauthorized eavesdropper, respectively. The eavesdropper is assumed to be a passive entity, i.e., does not transmit in order to conceal its presence. Recently, research in information-theoretic secrecy can be classified into two main branches. The first branch depends on using a secret key in order to ensure secrecy while the other branch ensures secrecy without using a secret key. Therefore, in the following subsections, an overview about both key-based and keyless security is given.

3.3.1 Key-Based Security for Wireless Channels

In 1949, Shannon presented the first approach to information-theoretic security when he describes a special case of what is now known as the wiretap channel in his ground-breaking work [85]. Through this model, Shannon considered noiseless bit-pipes between Alice, Bob, and Eve entities and showed that in the presence of Eve, information-theoretically secure communications can be achieved between Alice and Bob provided they share a nonreusable secret key between them, as shown in Figure 3.2.

Denoting the confidential message as W and the secret key as K, Shannon's one-time pad approach requires the legitimate transmitter to send $C = W \oplus K$, which both the legitimate receiver and the eavesdropper receive. The legitimate receiver further XORs the received signal with the preshared key K to retrieve the message W. Based on Shannon one-time pad; each key is uniform and used only once, and then the signal C leaks no information about the message W to the eavesdropper because it is statistically independent of the message. Unfortunately, it should be clear that the perfect secrecy can be guaranteed only if the key length is as large as the size of the message (the secret key has at least as much entropy as the message to be encrypted), i.e., $H(K) \geq H(W)$, which is often too costly to implement efficiently.

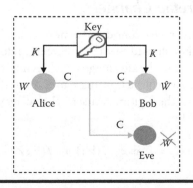

Figure 3.2 Shannon's model with noiseless bit-pipes between the legitimate transmitter, legitimate receiver, and the eavesdropper.

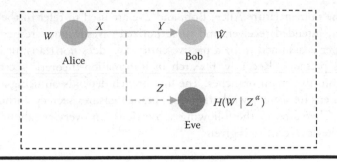

Figure 3.3 **Wyner's model with noisy channels from the legitimate transmitter to the legitimate receiver and the eavesdropper.**

3.3.2 Keyless Security for Wireless Channels

In 1975, Wyner's model shown in Figure 3.3 has been introduced where both links from Alice to Bob and Eve are noisy and the eavesdropper gets a degraded version of the legitimate receiver's observation [86]. For this model, Wyner has determined the secrecy capacity, defined as the supreme of communication rates to the legitimate receiver at which one can guarantee reliability and information-theoretic security against the eavesdropper.

Wyner's notion of information-theoretic security relaxes Shannon's definition by requiring that $\lim_{\alpha \to \infty} \left[\left(\frac{1}{\alpha} \right) I(W; Z^{\alpha}) \right] = 0$, i.e., the eavesdropper's observation Z^{α} leaks a vanishing rate of information about the message W in the limit of large coding length α. Therefore, the valuable result from Wyner's work is that if the eavesdropper's observation is a degraded version of the legitimate user's observation, information-theoretically secure communication between the legitimate users is possible while keeping the eavesdropper completely ignorant of the secure message, without using any secret keys.

3.3.3 General Wiretap Channel

In 1987, Csiszár and Körner have considered a more general nondegraded version of Wyner's channel [87]. They showed that the inherent randomness in the communication channel can be used to the advantage of achieving information-theoretically secure communication between the legitimate users even if the eavesdropper is not degraded with respect to the legitimate user. The secrecy capacity for a general wiretap channel is given by

$$C_s = \max_{V \to X \to Y, Z} I(V; Y) - I(V; Z) \tag{3.1}$$

where the mapping from V, the message carrying signal, to X, the channel input, is called channel prefixing. When the wiretap channel is degraded, i.e., the channel input X, Bob's channel output Y, and Eve's channel output Z satisfy the Markov chain $X \to Y \to Z$, there is no need for channel prefixing, and $V = X$ selection is optimal [87], and the secrecy capacity reduces to

$$C_s = \max_X I(X;Y) - I(X;Z) \qquad (3.2)$$

Equations 3.1 and 3.2 show that secrecy is a relative concept, involving the difference of rates going to Bob and Eve. The secrecy capacity in Equation 3.2 is achieved with what is known as stochastic encoding, where every message is associated with multiple codewords to confuse the eavesdropper. With the message rate C_s and confusion rate $I(X;Z)$, Bob is able to decode both the secure message and the confusion message, since his channel can resolve combined messages at rates up to $I(X;Y)$; on the other hand, all messages look equally likely to Eve because her channel's resolvability is limited to $I(X;Z)$. Channel prefixing in Equation 3.1 shows another aspect of relativeness of the concept of secrecy. From the data processing inequality, using a prefixed channel reduces both the useful rate from $I(X;Y)$ to $I(V;Y)$ and the leakage rate from $I(X;Z)$ to $I(V;Z)$. However, a careful selection of V may increase the difference in Equation 3.1, by decreasing $I(V;Z)$ relatively more than $I(V;Y)$, hence the need to use channel prefixing, in general.

3.3.4 Gaussian Wiretap Channel

In 1978, Leung-Yang-Cheong and Hellman considered the Gaussian wiretap channel [88]. For this wiretap channel, which models a nonfading wireless communication channel, a Gaussian channel input maximizes both mutual information terms as well as the difference of the mutual information terms in Equation 3.2, and hence, the secrecy capacity equals the difference of the channel capacities of the legitimate link C_B and the eavesdropping link C_E. Assuming $C_B \geq C_E$,

$$C_S = C_B - C_E = \frac{1}{2} \log_2 \left(1 + \frac{P_l}{\sigma_B^2} \right) - \frac{1}{2} \log_2 \left(1 + \frac{P_l}{\sigma_E^2} \right) \qquad (3.3)$$

It is noted that the secrecy capacity in Equation 3.3 does not scale with the transmit power P_l of the legitimate transmitter. That is, as P_l goes to infinity, C_s converges to a constant. The cost of providing information-theoretic secrecy is often measured in terms of secure degrees of freedom (*s.d.o.f.*), defined as the ratio of the secure communication rate R_s to $(1/2) \log P_l$ in the limit of infinitely large P_l, i.e., $s.d.o.f. = \lim_{P_l \to \infty} \left[\left(R_S / (1/2) \log P_l \right) \right]$, relative to its counterpart, the degrees of freedom, i.e., the same asymptotic behavior of rate without the secrecy constraint. Thus, we observe that the Gaussian wiretap channel incurs a severe penalty for secrecy, having reduced its degrees of freedom from one to zero for secrecy.

3.4 Cooperative Secrecy Techniques for the Physical Layer

In a network of multiple transmitters and receivers, if security is our only concern, the independent transmitters can be used to improve the secrecy rate of a given transmitter–receiver pair by transmitting jamming signals. This idea was primarily proposed by Tekin and Yener in [89] and then developed in [90] and [91] in order to describe such cooperating approaches helping in achieving secrecy at the physical layer of a multiuser system through introducing the cooperative jamming concept. When a sender transmits signals that are independent of the intended message, these signals create interference for both the legitimate receiver and the eavesdropper, limiting both of their decoding capabilities and reducing both of their reliable decoding rates. However, the net effect of this jamming may be an increase in the difference between legitimate channel and the eavesdropper's channel rates shown in Equation 3.3 and hence an increase in the achievable secrecy rate between the legitimate pair. There are several effective cooperative techniques illustrating the manner in which jamming can be implemented. Cooperative jamming with Gaussian noise, cooperative jamming with noise forwarding, cooperative jamming with structured codes, and cooperative jamming with interference alignment are the most popular applied techniques. Therefore, they will be discussed in some detail in the following subsections.

3.4.1 Cooperative Jamming with Gaussian Noise

Cooperative jamming [89–91] was originally implemented by using independent and identically distributed (i.i.d.) Gaussian signals over a multiple access channel [92], where multiple legitimate users want to establish simultaneous secure communications with an intended receiver in the presence of an eavesdropper.

Figure 3.4 shows a simplified multiple access scenario, where two users (Alice and Charlie) wish to have simultaneous secure communication with the intended

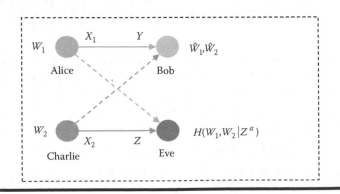

Figure 3.4 Two-user multiple access wiretap channel.

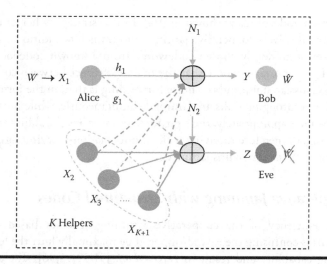

Figure 3.5 Gaussian wiretap channel with *K* helpers.

receiver (Bob) over a multiple access channel in the presence of Eve. In this situation, the user (Charlie) who has a stronger channel to the eavesdropper (Eve) than to the intended receiver (Bob) should stop sending his information message and operate as a cooperative jammer for Alice and send (i.i.d.) Gaussian noise signals at the eavesdropper. Since Charlie has a strong link with Eve than Bob, his jamming is more harmful to Eve than Bob, thus increasing Alice's achievable secrecy rate.

While the cooperative jamming [89–91] was originally designed for a multiple access channel, Figure 3.5 illustrates the application of the cooperative jamming in a multiuser system. Denoting $h = (h_2,\ldots,h_{K+1})$ as the channel gains vector to the intended receiver "Bob," $g = (g_2,\ldots,g_{K+1})$ as the channel gains vector to "Eve," and $x = (x_2,\ldots,x_{K+1})$ as the jamming signals vector emitted by the helpers, the secrecy capacity with i.i.d. Gaussian helper signals can be written as

$$C_S = \frac{1}{2}\log_2\left(1+\frac{h_1^2 P_I}{\sigma_B^2 + h^T Q h}\right) - \frac{1}{2}\log_2\left(1+\frac{g_1^2 P_I}{\sigma_E^2 + g^T Q g}\right) \tag{3.4}$$

where Q is the covariance matrix of x and $(.)^T$ refers to the transpose. In the presence of multiple independent helpers, all of the transmitted helper signals need to be independent, and this means that the covariance matrix Q must be diagonal. In this case, the denominators will reduce to sum of powers of the helpers multiplied by squared channel gains plus the power of the ambient Gaussian noise.

3.4.2 Cooperative Jamming with Noise Forwarding

Instead of transmitting i.i.d. Gaussian noise signals as in the previously mentioned technique, Lai and El Gamal [93] introduced a related strategy called noise forwarding for networks of relays shown in [94]. In this strategy, a helper relay terminal acts as an effectively cooperative jammer and transmits additional randomness in the form of randomly chosen codewords from a known codebook to the eavesdropper. It is important to note that the main difference between cooperative jamming with Gaussian noise and the noise forwarding is that, in the former, both legitimate and eavesdropping links are jammed simultaneously, while in the latter, by choosing the rates appropriately, the legitimate receiver has the ability to decode the confusion signal, therefore receiving a clean information-carrying signal while the eavesdropper's channel remains jammed.

3.4.3 Cooperative Jamming with Structured Codes

Although the efficiency of the cooperative jamming schemes based on i.i.d. Gaussian signals is confusing the eavesdropper, they maximally hurt the legitimate user's decoding capability and result in zero *s.d.o.f.* [91,92,95,96]. Therefore, we expect that strictly positive *s.d.o.f.* may be achieved with some *weak* jamming signals. He and Yener [97–99] confirmed this intuition and achieved positive *s.d.o.f.* by using nested lattice codes in a Gaussian wiretap channel with a helper, as shown in Figure 3.6.

Considering that the binary codeword and the jamming signals $a_L, a_{L-1}, \ldots, a_1$ and $b_L, b_{L-1}, \ldots, b_1$ are modified by inserting a "0" in between each digit so that we have $X_1 = a_L, 0, a_{L-1}, 0, \ldots, 0, a_1$ and $X_2 = b_L, 0, b_{L-1}, 0, \ldots, 0, b_1$. The received signal at Bob is given by $Y = X_1 + 2X_2 + N_1$, and Bob can reliably receive $b_L, a_L, b_{L-1}, a_{L-1}, \ldots, b_1, a_1$ and hence $a_L, a_{L-1}, \ldots, a_1$ whereas Eve having received $Z = X_1 + X_2 + N_2$ can only see $(b_L + a_L), 0, (b_{L-1} + a_{L-1}), 0, \ldots, (b_1 + a_1)$. This simple structure in the transmitted and jamming codewords allows covering of the legitimate communication by

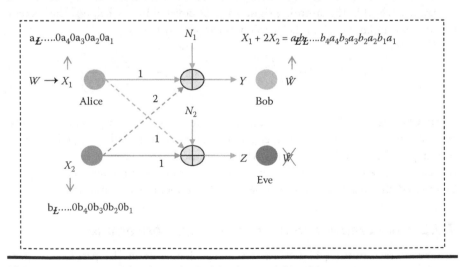

Figure 3.6 Cooperative jamming with structured codes.

the cooperative jammer, enabling a positive scaling of the secrecy rate. Specifically, seeing the sum of digits, Eve can recover a_i only when $b_i = a_i$, which happens with probability 0.5. Therefore, each a_i could support a secrecy rate of 0.5 bit per channel use.

3.4.4 Cooperative Jamming by Alignment

Figure 3.7 illustrates the interference alignment idea for the Gaussian wiretap channel with two helpers ($K = 2$). The legitimate transmitter divides its message into K parts. Each helper sends a cooperative jamming signal. All of the K cooperative jamming signals are aligned in the same dimension at the legitimate receiver to occupy the smallest signal space to allow for maximum signal space that can be used by the messages. All of the K submessages are separable at the legitimate receiver because they are in different irrational dimensions. On the other hand, each cooperative jamming signal is aligned with a message signal at the eavesdropper to protect it. This alignment makes sure that the information leakage to the eavesdropper is upper-bounded by a constant. Therefore, each message signal is protected by one of the cooperative jamming signals at the eavesdropper. In this achievable scheme, both the legitimate receiver's and the eavesdropper's channel state information are used to align message signals and cooperative jamming signals simultaneously at the legitimate receiver and the eavesdropper in the desired manner.

Koyluoglu *et al.* [100] studied the frequency/time selective K-user Gaussian interference channel with secrecy constraints and showed the positive impact of interference on the secrecy capacity region of wireless networks. Moreover, Motahari *et al.* attained the total degree of freedom of K-user Gaussian interference channel by incorporating real interference alignment in the signaling [101].

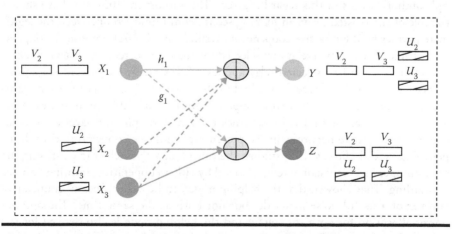

Figure 3.7 Illustration of interference alignment for the Gaussian wiretap channel with K helpers. Here, $K = 2$.

3.5 Cooperative Jamming for Secure Relay Networks

In this section, the idea of employing cooperative jammers in multiple relay networks in order to improve security is adopted. Consider a network model that consists of a legitimate transmitter who wants to have secure communication with a legitimate receiver in the presence of a malicious eavesdropper who wants to intercept transmission and a set of cooperating partner "relays" ready to help the legitimate pair. Generally, the cooperating relays can be classified into two main classes. The first class is concerned with trusted relays, where the transmitted messages need not be kept secret from them, while the other class deals with untrusted relays, where the transmitted messages must be kept confidential from them even if they are used for relaying source messages. In the following subsections, the security issues will be discussed in terms of both trusted and untrusted relays.

3.5.1 Secrecy in View of Trusted Relays

There are two possible cooperating techniques in which the trusted relays help the legitimate pair in achieving a secure communication. The first technique is called the passive cooperation, where the cooperating relay forwards the legitimate transmitter signal to the intended receiver without hearing it or, even if it hears it, ignores it. The other technique is called the active cooperation, where the cooperating relay listens to the source messages and utilizes its overheard information to improve the achievable secrecy rate. In the following discussion, the attention is restricted to cooperation schemes with decode-and-forward (DF) relays, although other relaying schemes previously mentioned in Chapter 2 are also possible.

In the secrecy context, relay selection is one of the main problems that have a great importance in multiple relay networks. Therefore, some attention to the multiple authors' works in this regard is given. The authors in [102–106] have shown the role of the trusted relays in making secure communication between the legitimate parities by their passive cooperation. Krikidis *et al.* [102] proposed a cooperative scheme in which two relays are selected for ensuring secrecy against a malicious eavesdropper. The first selected relay applies a DF strategy to forward the source message to the intended receiver while the other relay operates as a cooperative jammer in order to confuse the eavesdropper. Ibrahim *et al.* [103] proposed multiple joint relay and jammer selection schemes for achieving physical layer security in one-way cooperative networks in the presence of multiple cooperating eavesdroppers. Bassily and Ulukus [104] studied the role of a passive helper in confusing the eavesdropper in a Gaussian wiretap channel by either cooperative jamming or noise forwarding. They showed that the helping node can be either a useful cooperative jammer or a useful noise forwarder but not both at the same time. The optimal power allocation for both the source and the helping node is also derived for the two modes of passive helping. Moreover, a suboptimal strategy for selecting the best A passive helpers among N system relays ($A \leq N$) is considered for achieving the maximum possible achievable secrecy rate. In [105] and [106], as a result of the two-way relay channel bandwidth efficiency, Ibrahim *et al.* proposed different categories of relay and jammer selection schemes to assist the legitimate sources to securely deliver their data to the corresponding destinations in the presence of multiple cooperating and noncooperating eavesdroppers.

On the other side, Bassily and Ulukus [107] considered the role of active cooperation for secrecy in multiple relay networks through DF strategies. They proposed three different strategies based on DF with the zero-forcing (ZF) technique and obtained the achievable secrecy rate by each strategy. In the first strategy, all the relays decode the source message at the same time and then perform beamforming by transmitting scaled versions of the same signal to the destination (see Figure 3.8a). One of the main advantages of this strategy is that all the relays' signal components can be eliminated from the eavesdropper's observation, i.e., full ZF can be achieved. However, on the other side, one clear drawback of the above strategy is the requirement that all relays must decode the source message in a single hop at the same time, and thus, the furthest relay from the source creates a bottleneck in the achievable secrecy rate.

To overcome this drawback, a second strategy based on multihop DF has been introduced. In this strategy, the relays are ordered with respect to their distances from the source, and they perform DF in a multihop fashion (see Figure 3.8b), i.e., the closest relay decodes the source message first, forwards it (with the help of the source) to the second closest relay, and so forth until it reaches the destination. Thus, for $\mathcal{F}$ relays, the transmission of each message block is done in $\mathcal{F}$ hops. This strategy overcomes the bottleneck drawback of the first strategy. However, given that all the relays transmit fresh information in every transmission block, only half of the relays' signal components can be forced to zero at the eavesdropper "partial zero-forcing."

In order to achieve full zero-forcing in this second strategy, half of the relays' signal components are set to zero. Based on this observation, the ($\mathcal{F}/2$)-hop third strategy is proposed to efficiently combine the advantages of the two aforementioned strategies in an efficient way. That is, the achievable rate is not limited by the worst source–relay channel as in the first strategy; also, all the relays' signals can be eliminated from the eavesdropper's observation. In this strategy, the relays are ordered with respect to their distances from the source and then grouped into clusters of two relays per each cluster. The source transmits the message to the relays in the first cluster (closest to the source), which decode the message and forward it (with the help of the source) to the relays in the second cluster and so on and so forth until the message is forwarded to the destination (see Figure 3.8c). The relays in each cluster do not have any direct communication between them. By properly adjusting the signal coefficients at the relays, we can zero-force all the relays' signals at the eavesdropper.

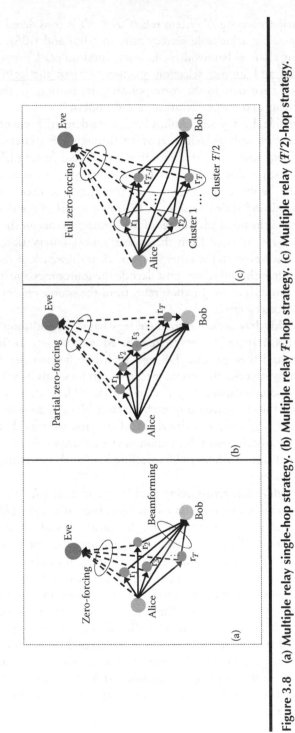

Figure 3.8 (a) Multiple relay single-hop strategy. (b) Multiple relay T-hop strategy. (c) Multiple relay (T/2)-hop strategy.

3.5.2 Secrecy in View of Untrusted Relays

The untrusted relays work on helping the source in forwarding its messages to the intended destination besides eavesdropping them. Therefore, it becomes too important in this subsection to study the interactions arising between cooperation and secrecy in these channel models. Most research studies concentrate on the knowledge of whether there is a tradeoff or a combined effort between untrusted relay cooperative role and secrecy, i.e., does the untrusted relay cause additional leakage of information as a result of overheard information from the source, or can it improve secrecy by limiting or reversing the leakage of information?

Oohama [108] deals with this question by considering a three-node relay network, as shown in Figure 3.9a. In this model, the transmitter sends two messages; a common message to both the legitimate receiver and the relay in addition to a confidential message directed to the legitimate receiver, which needs to be kept hidden from the relay node. The relay uses a DF strategy where the common message and a part of the confidential message are decoded and forwarded to the receiver. The extracted conclusion from the achievable scheme in [108] is that as long as the relay node uses a DF-type cooperation strategy, it cannot increase the secrecy rate of the transmitter, even though it can increase its achievable rate.

He and Yener [109] extend studying the interaction between cooperation and secrecy for a special class of relay channels. In this special class, there is an

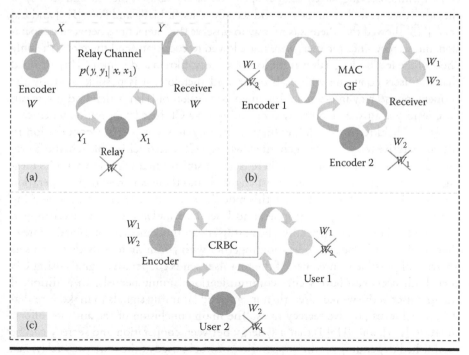

Figure 3.9 (a) Relay channel. (b) MAC-GF. (c) CRBC.

orthogonal link between the relay and the receiver, and the transmitter has a broadcast channel to the relay and the receiver. Based on the assumption that the relay uses a compress-and-forward (CF) strategy for analyzing the performance of this channel, the CF-based relay proves its efficiency in improving both the transmitter rate and the transmitter secrecy rate.

In [110], He and Yener considered the case where there is no direct link between the legitimate nodes, i.e., the untrusted relay is the only path for passing Alice's messages to Bob. It has been shown that a combination of cooperative jamming by the destination (Bob) and CF relaying scheme results in a secure communication.

He and Yener [111] considered the case of a malicious relay, which impacts the signals it receives, and show that it is still possible to achieve secure communication and detection via a combination of structured codes and structured cooperative jamming.

Following this, it is important to refer to the so called "two-sided version of the relay channel," where the cooperating partners have their own information to send as well. Figure 3.9b shows a multiple access channel with generalized feedback (MAC-GF) first studied in [112] and [113], where there are two users that each has its own messages to send and that each receives feedback signals that are correlated with the message of the other user. Although these signals can be used to cooperate and increase the rates, they still form the basis for secrecy loss. In this setup, each user considers the other user as a cooperating partner and also as an eavesdropper. This model was concerned with observing the effect of cooperation of one user on the rate and secrecy of the other user, as well as on the rate and secrecy of itself. Liang and Poor [112] and Liu *et al.* [113] showed that there was no way to observe the interactions between cooperation and secrecy since the users were not allowed to cooperate, and therefore, the only effect of the feedback signals was the loss of secrecy. Ekrem and Ulukus [114] assumed that the users cooperate with each other, and therefore, it is possible to study the cooperation–secrecy interaction. The main conclusion of [114] is that both users can have secrecy against each other by cooperating via a CF-based cooperation scheme.

In [115], Ekrem and Ulukus further investigate secrecy versus cooperation in the cooperative relay broadcast channel, where a CF-based achievable scheme is presented. As shown in Figure 3.9c, there is a transmitter that wishes to send its messages to two receivers over a broadcast channel, and there is a one-sided cooperation link from User 1 to User 2. During this model, User 2 treats User 1 as a cooperating partner and also as an eavesdropper, and User 1 treats User 2 as an eavesdropper. This model shows that, generally, the first user can implement a combined strategy that combines CF-based cooperation together with jamming to provide both users an array of possible secure rates. If the first user sends cooperative signals using CF, then both users can have secure communications simultaneously. In addition, if the first user is the weaker user, then it can send jamming signals to make sure that it itself achieves positive secrecy rates. The main conclusion of the authors' efforts in [109], [114], and [115] is that a synergy between cooperation and secrecy can be established depending on the cooperation protocol used at the untrusted relay, i.e., the efficient role of CF-based cooperation strategy in improving secrecy.

Chapter 4

Relay and Jammer Selection Schemes for Secure One-Way Cooperative Networks

4.1 Introduction

Information privacy in wireless networks has been given considerable attention for several years due to the broadcast nature of the wireless medium which allows all users in the coverage area of a transmission to overhear the source message. Traditionally, security in wireless networks has been mainly focused on higher layers using cryptographic methods [116]. However, as the implementation of secrecy at higher layers becomes the subject of increasing potential attacks, physical layer (PHY)–based security has drawn increasing attention recently [86,117,118,119]. The main objective of PHY-based approaches is to enable source and destination exchanging secure messages at a nonzero rate when the eavesdropper channel is a degraded version of the main channel. In Liang *et al.* [120], joint optimal power control and optimal scheduling schemes were proposed to enhance the secrecy rate of the intended receiver against cooperative and noncooperative eavesdropping models. In Dong *et al.* [121], decode-and-forward (DF) cooperative protocol was considered to improve the performance of secure wireless communications in the presence of one or more eavesdroppers. The interaction of the cooperative diversity concept with secret communications has also recently been reported as an interesting solution [54,64,122]. In [123], Dong *et al.* proposed a variety of cooperative schemes to achieve secure data transmission with the help of multiple relays in the presence of one or multiple eavesdroppers. In Lai and El Gamal [124], a four-node model was proposed, and it was shown that if the relay is closer to the destination than the eavesdropper, a positive secrecy rate can be achieved even if the source–destination rate is zero. In Krikidis [125], some relay selection metrics have been proposed with different levels of feedback overhead. Krikidis *et al.* in [102] extended the work presented in [125] for cooperative networks with jamming protection without taking direct links into account. In Ibrahim *et al.* [66] and Beres and Adve [126], different relay selection strategies were introduced for improving the secrecy rate in [125]. In Tekin and Yener [91], Simeone and Popovski [127], and Popovski and Simeone [128], jamming schemes which produce an artificial interference at the eavesdropper node in order to reduce the capacity of the related link seem to be an interesting approach for practical applications. In Wang and Giannakis [129], the interaction between relay and jammer is introduced as a noncooperative game where both nodes have conflicting objectives, and the Nash equilibrium (NE) of the system was derived.

The main contribution of this chapter is to investigate relay and jammer selection schemes to increase one-way cooperative networks' security in the presence of one or more eavesdroppers. In contrast to [102], the selection in the proposed schemes is made with the presence of direct links, the assumption that broadcast phase is unsecured, and when one or more eavesdroppers are present in the system. A single source–destination cooperative network with one or more eavesdroppers and multiple intermediate nodes is considered to increase security against malicious eavesdroppers. In the proposed schemes, an intermediate node is selected to operate in the conventional DF relay mode and assists the source to deliver data to the corresponding destination. Meanwhile, another two intermediate nodes that perform as jamming nodes are selected and transmit artificial interference in order to degrade the eavesdroppers' links in the first and second phases of data transmission, respectively. The proposed schemes are analyzed for different complexity requirements based on global instantaneous knowledge of all links and average knowledge of the eavesdroppers' links.

The obtained results reveal that the proposed schemes with cooperative jamming can improve the secrecy capacity and the secrecy outage probability of the cooperative network. In addition to the investigation of these jamming-based selection schemes, it is shown that jamming is not always beneficial for security. According to this observation, a switching scheme between jamming and nonjamming relay selection is proposed. This hybrid scheme overcomes jamming limitations and seems to be an efficient solution for practical application with critical secrecy constraints. Moreover, the impact of changing both the eavesdroppers and the relays' location on the system performance is also discussed in this chapter. Finally, the impact of the presence of multiple eavesdroppers is studied.

The rest of this chapter is organized as follows. Section 4.2 introduces the system model and formulates the problem. Section 4.3 presents the proposed selection schemes. Numerical results are shown and discussed in Section 4.4, followed by concluding remarks in Section 4.5.

4.2 System Model and Problem Formulation

In this chapter, two different scenarios of eavesdropper are considered. The first scenario discusses the effect of presence of one eavesdropper while the second scenario studies the effect of presence of more eavesdroppers on cooperative network performance as follows.

4.2.1 Presence of One Eavesdropper

4.2.1.1 System Model

A single source–destination cooperative network is considered with one eavesdropper E and an intermediate node set $S_{relay} = \{1, 2,..., N\}$ with N nodes, as shown in Figure 4.1. The intermediate nodes operate in half-duplex mode; therefore, they cannot transmit and receive simultaneously, and the communication process is performed in two phases. During the broadcasting phase, the source S transmits its data to the destination D, and due to the broadcasting nature of the transmission, the intermediate nodes and eavesdropper overhear the transmitted information. In addition, according to security schemes, one node J_1 is selected from the S_{relay} set to operate as a jammer and transmits intentional interference to degrade the eavesdropper links in this phase. In the cooperative phase, according to security schemes, an intermediate node R is selected to operate as

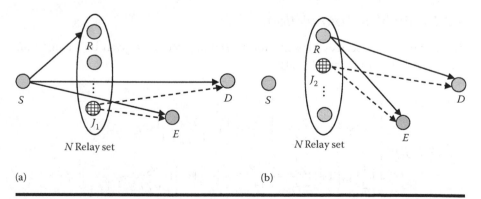

(a)   (b)

Figure 4.1 System model with one eavesdropper. (a) Broadcasting phase. (b) Cooperative phase.

a conventional relay and forwards the source message to the corresponding destination (R must belong to the decoding set $C_d \subseteq S_{relay}$). A second jammer J_2 is selected from S_{relay}, for the same reason as in J_1. Note that the destination D is not able to mitigate the artificial interference from the jamming nodes (interference is unknown at the destination) and refers to applications with critical secrecy constraints.

The main objective of this chapter is the investigation of relay and jammer selection criteria for cooperative systems with secrecy constraints. In this work, the following assumptions have been made:

■ As in most existing cooperative network topologies [130], the direct links ($S \to D$ and $S \to E$) are available.
■ The broadcasting phase is unsecured. Therefore, the eavesdropper can overhear the transmitted information in this phase.
■ In both phases, a slow, flat, and block Rayleigh fading environment is assumed, i.e., the channel remains static for one coherence interval and changes independently in different coherence intervals with a variance $\sigma_{a,b}^2 = d_{a,b}^{-\beta}$, where $d_{a,b}$ is the Euclidean distance between node a and node b, and β is the pathloss exponent.
■ Furthermore, additive white Gaussian noise (AWGN) is assumed with zero mean and unit variance.

Let $P^{(S)}$, $P^{(R)}$, and $P^{(J)}$ denote the transmitted power for the source node, the relay node, and the jamming nodes, respectively. In order to protect the destination from the artificial interference and maximize the benefits of the proposed schemes, the jamming nodes transmit with a lower power than the relay nodes, and thus, their transmitted power is defined as $P^{(J)} = P^{(R)}/L'$ (with $P^{(S)} = P^{(R)}$), where $L' > 1$ denotes the ratio of the relay power to the jammer power.

4.2.1.2 Problem Formulation

The instantaneous secrecy capacity for a decoding set C_d is given by Liang *et al.* [120] and Lai and El Gamal [124]:

$$C_S^{|C_d|}(R, J_1, J_2) = \begin{cases} \left[\frac{1}{2}\log_2\left(1 + \frac{\gamma_{S,D}}{1+\gamma_{J_1,D}}\right) - \frac{1}{2}\log_2\left(1 + \frac{\gamma_{S,E}}{1+\gamma_{J_1,E}}\right) \right]^+ & \text{for } |C_d| = 0 \\ \left[\frac{1}{2}\log_2\left(1 + \frac{\gamma_{S,D}}{1+\gamma_{J_1,D}} + \frac{\gamma_{R,D}}{1+\gamma_{J_2,D}}\right) - \frac{1}{2}\log_2\left(1 + \frac{\gamma_{S,E}}{1+\gamma_{J_1,E}} + \frac{\gamma_{R,E}}{1+\gamma_{J_2,E}}\right) \right]^+ & \text{for } |C_d| > 0 \end{cases}$$

$$(4.1)$$

where $R \in C_d$, $J_1 \in S_{relay}$ and $J_2 \in \{S_{relay} -R^*\}$, $\gamma_{a,b} \triangleq P^{(a)} |f_{a,b}|^2$ is the instantaneous signal-to-noise ratio (SNR) for the link $a \to b$ modeled as a zero-mean, independent, circularly symmetric complex Gaussian random variable with variance $\sigma_{a,b}^2$, $[\chi]^+ \triangleq \max\{0, \chi\}$, and $|\mathcal{F}|$ denotes the cardinality of a set $\mathcal{F}$. Sometimes, secrecy performance of the system is characterized by the secrecy outage probability, which is defined as the probability that the instantaneous secrecy capacity is less than a target secrecy rate $R_T > 0$ [119]. The secrecy outage probability is written as

$$P_{out} = \sum_{n=1}^{N} P_r\left\{C_S^n(R, J_1, J_2) < R_T\right\} P_r\left\{|C_d| = n\right\} \qquad (4.2)$$

The main objective is to select appropriate nodes R, J_1, and J_2 in order to maximize the instantaneous secrecy capacity for different types of channel feedback. The optimization problem can be formulated as

$$\left(R^*, J_1^*, J_2^*\right) = \arg \max_{\substack{J_1 \in S_{relay} \\ R \in |C_d| \\ J_2 \in \{S_{relay} - R^*\}}} \left\{C_s^{|C_d|}(R, J_1, J_2)\right\}, \text{s.t. } \psi_u\left(for\ u = 0,1\right) \qquad (4.3)$$

where R^* and J_1^* and J_2^* denote the selected relay and jamming nodes, respectively (note that selected jammers J_1^* and J_2^* in the two phases may be the same node, which is determined by the instantaneous secrecy capacity), $\psi_0 = \{\gamma_{n,D}, \gamma_{n,E}\}$ and $\psi_1 = \{\gamma_{n,D}, E[\gamma_{n,E}]\}$ with $n \in S_{relay}$ are the feedback knowledge sets, E[.] stands for the expectation operator, ψ_0 denotes a global instantaneous knowledge for all the links, and ψ_1 denotes an average channel knowledge for the eavesdropper links.

4.2.2 Presence of Multiple Eavesdroppers

4.2.2.1 System Model

In this scenario, M eavesdroppers set was considered, $S_{eves} = \{1, 2,..., M\}$, as shown in Figure 4.2.

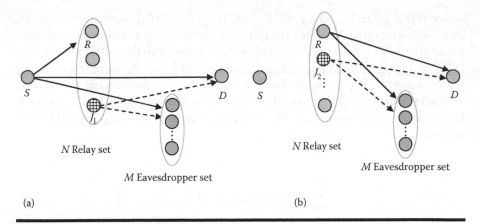

Figure 4.2 System model with multiple eavesdroppers. (a) Broadcasting phase. (b) Cooperative phase.

4.2.2.2 Problem Formulation

The instantaneous secrecy capacity for a decoding set C_d is given by Liang *et al.* [120] and Lai and El Gamal [124]:

$$C_S^{|C_d|}(R, J_1, J_2) = \begin{cases} \left[\dfrac{1}{2}\log_2\left(1+\dfrac{\gamma_{S,D}}{1+\gamma_{J_1,D}}\right) - \dfrac{1}{2}\log_2\left(1+\displaystyle\sum_{m=1}^{M}\left(\dfrac{\gamma_{S,E_m}}{1+\gamma_{J_1,E_m}}\right)\right)\right]^{+} & \text{for } |C_d| = 0 \\[4mm] \left[\dfrac{1}{2}\log_2\left(1+\dfrac{\gamma_{S,D}}{1+\gamma_{J_1,D}}+\dfrac{\gamma_{R,D}}{1+\gamma_{J_2,D}}\right)\right. \\[2mm] \left. -\dfrac{1}{2}\log_2\left(1+\displaystyle\sum_{m=1}^{M}\left(\dfrac{\gamma_{S,E_m}}{1+\gamma_{J_1,E_m}}+\dfrac{\gamma_{R,E_m}}{1+\gamma_{J_2,E_m}}\right)\right)\right]^{+} & \text{for } |C_d| > 0 \end{cases}$$

(4.4)

Note that in Equation 4.4, the malicious eavesdroppers cooperate together in order to overhear the source message, and this complicates the process of exchanging secure information between the source and the destination.

4.3 Relay and Jammer Selection Schemes

In order to investigate the effect of jamming, it is necessary to distinguish between the following three cases: no jammer selection, conventional jamming (where the jamming signal is unknown at the destination), and controlled jamming (where the jamming signal is known at the destination), as will be explained in the next subsections.

4.3.1 Presence of One Eavesdropper

4.3.1.1 Selection Schemes without Jamming

This category of solutions does not involve a jamming process, and therefore, only a conventional relay accesses the channel during the second phase of the protocol. The existing selections are summarized as follows:

■ **Conventional Selection (CS).** This solution does not take the eavesdropper channels into account, and the relay node is selected according to the instantaneous quality of the $S \rightarrow D$ link [125]. Although it is an effective solution for noneavesdropper environments, it cannot support systems with secrecy constraints. The conventional selection is written as

$$R^* = \arg \max_{R \in C_d} \left\{ 1 + \gamma_{S,D} + \gamma_{R,D} \right\} \tag{4.5}$$

■ **Optimal Selection (OS).** This solution takes into account the relay–eavesdropper links and selects the relay node according to the knowledge set ψ_0. The optimal selection is given by Popovski and Simeone [125]:

$$R^* = \arg \max_{R \in C_d} \left\{ \frac{1 + \gamma_{S,D} + \gamma_{R,D}}{1 + \gamma_{S,E} + \gamma_{R,E}} \right\} \tag{4.6}$$

■ **Suboptimal Selection (SS).** It avoids the instantaneous estimate of the relay eavesdropper links by selecting the appropriate relay based on the knowledge set ψ_1. It is a solution which efficiently fills the gap between optimal and conventional selection with a low implementation/complexity overhead. The suboptimal selection is expressed by Popovski and Simeone [125]:

$$R^* = \arg \max_{R \in C_d} \left\{ \frac{1 + \gamma_{S,D} + \gamma_{R,D}}{1 + \mathrm{E}\left[\gamma_{S,E}\right] + \mathrm{E}\left[\gamma_{R,E}\right]} \right\} \tag{4.7}$$

4.3.1.2 Selection Schemes with Conventional Jamming

In this subsection, an extension to the above eavesdropper-based relay selection schemes is presented for systems with jamming via several relay and jammer selection schemes based on the optimization problem given by Equation 4.3.

■ Optimal Selection with Jamming (OSJ)

The optimal selection with jamming assumes the knowledge of the set ψ_0 and ensures a maximization of the instantaneous secrecy capacity given in Equation 4.1. The selection policy which maximizes Equation 4.1 assuming that $|C_d| > 0$ is given as

$$R^*, J_1^*, J_2^* = \arg \max_{\substack{J_1 \in S_{relay} \\ R \in C_d \\ J_2 \in \{S_{relay} - R^*\}}} \left\{ \frac{1 + \dfrac{\gamma_{S,D}}{1 + \gamma_{J_1,D}} + \dfrac{\gamma_{R,D}}{1 + \gamma_{J_2,D}}}{1 + \dfrac{\gamma_{S,E}}{1 + \gamma_{J_1,E}} + \dfrac{\gamma_{R,E}}{1 + \gamma_{J_2,E}}} \right\} \qquad (4.8)$$

The selection process in Equation 4.8 requires a large number of comparisons and algebraic computations. Therefore, a simpler selection policy can be proposed considering a high SNR, where all the relay nodes can decode the source message $(P_r \{|C_d| = |S_{relay}| = N\} = 1)$ and the assumption that the jammer power is much less than the source and relay powers [102]; it can be shown that

$$\left(R^*, J_1^*, J_2^* \right) = \arg \max_{\substack{J_1 \in S_{relay} \\ R \in C_d \\ J_2 \in \left\{ S_{relay} - R^* \right\}}} \left\{ \frac{\dfrac{\gamma_{S,D}}{\gamma_{J_1,D}} + \dfrac{\gamma_{R,D}}{\gamma_{J_2,D}}}{\dfrac{\gamma_{S,E}}{\gamma_{J_1,E}} + \dfrac{\gamma_{R,E}}{\gamma_{J_2,E}}} \right\} \qquad (4.9)$$

According to Equation 4.9, we can obtain

$$J_1^* = \arg \max_{J_1 \in S_{relay}} \left\{ \frac{\gamma_{J_1,E}}{\gamma_{J_1,D}} \right\} \quad \text{and} \quad R^* = \arg \max_{R \in C_d} \left\{ \frac{\gamma_{R,D}}{\gamma_{R,E}} \right\}$$

$$J_2^* = \arg \max_{J_2 \in \left\{ S_{relay} - R^* \right\}} \left\{ \frac{\gamma_{J_2,E}}{\gamma_{J_2,D}} \right\} \qquad (4.10)$$

With regard to the relay and the jamming nodes, the relay selection tries to maximize the ratio $\gamma_{n,D}, \gamma_{n,E}$, while the jamming nodes try to minimize the same function; consequently, the selection policy is independent of the selection order and will always select different relay terminals.

■ Suboptimal Selection with Jamming (SSJ)

Although the assumption ψ_0 provides some optimal selection metrics, its practical interest is limited to some special applications (e.g., military applications), where the instantaneous quality of the eavesdropper links can be measured by some specific protocols. In practice, only an average knowledge of eavesdropper links would be available from long-term supervision of the eavesdropper transmission. The selection metric is modified as

$$J_1^* = \arg \max_{J_1 \in S_{relay}} \left\{ \frac{E[\gamma_{J_1,E}]}{\gamma_{J_1,D}} \right\} \quad \text{and} \quad R^* = \arg \max_{R \in C_d} \left\{ \frac{\gamma_{R,D}}{E[\gamma_{R,E}]} \right\}$$

$$J_2^* = \arg \max_{J_2 \in \{S_{relay} - R^*\}} \left\{ \frac{E[\gamma_{J_2,E}]}{\gamma_{J_2,D}} \right\} \tag{4.11}$$

4.3.1.3 Selection Schemes with Controlled Jamming

The previous selection schemes are proposed based on the assumption that jamming signal is unknown at the destination. This assumption avoids the initialization period in which the jamming sequence is defined, and thus, it reduces the risk of giving out the artificial interference to the eavesdropper. For comparison reasons, a control scheme is proposed, in which the jamming signal can be decoded at the destination D, but not at the eavesdropper E. This scheme is called optimal selection with controlled jamming (OSCJ). In this scheme, the secrecy capacity expression given in Equation 4.1 is modified as follows:

$$C_S^{|C_d|}(R, J_1, J_2) = \left[\frac{1}{2} \log_2 \left(\frac{1 + \gamma_{S,D} + \gamma_{R,D}}{1 + \dfrac{\gamma_{S,E}}{1 + \gamma_{J_1,E}} + \dfrac{\gamma_{R,E}}{1 + \gamma_{J_2,E}}} \right) \right]^+ \quad for \ |C_d| > 0 \tag{4.12}$$

Following the same assumptions in obtaining Equation 4.9, the selection policy that maximizes the secrecy capacity given in Equation 4.12 is given by

$$J_1^* = \arg \max_{J_1 \in S_{relay}} \{\gamma_{J_1,E}\} \quad \text{and} \quad R^* = \arg \max_{R \in C_d} \left\{ \frac{\gamma_{R,D}}{\gamma_{R,E}} \right\}$$

$$J_2^* = \arg \max_{J_2 \in \{S_{relay} - R^*\}} \left\{ \gamma_{J_2,E} \right\} \tag{4.13}$$

4.3.1.4 Hybrid Selection Schemes

■ Optimal Switching (OW)

The original idea of using jamming nodes is to introduce interference on the eavesdropper links. Based on the assumption that the destination cannot mitigate artificial interference from the jamming nodes, continuous jamming in both phases is not always beneficial for the system. More specifically, for some operational scenarios (i.e., jammers are close to the destination), the continuous use of jamming decreases secrecy and acts as a bottleneck for the system. In order to overcome this limitation, an intelligent switching technique between optimal selection with jamming and optimal selection without jamming is proposed. This hybrid selection scheme overcomes problems of "negative jamming," which leads to excessive interference at the destination and is introduced as the optimal general solution for the problem under consideration. The required condition for the participation of the jamming nodes is

$$\frac{1}{2}\log_2 \left[\frac{1 + \dfrac{\gamma_{S,D}}{1+\gamma_{J_1,D}} + \dfrac{\gamma_{R,D}}{1+\gamma_{J_2,D}}}{1 + \dfrac{\gamma_{S,E}}{1+\gamma_{J_1,E}} + \dfrac{\gamma_{R,E}}{1+\gamma_{J_2,E}}} \right]^+ > \frac{1}{2}\log_2 \left[\frac{1+\gamma_{S,D}+\gamma_{R,D}}{1+\gamma_{S,E}+\gamma_{R,E}} \right]^+ \tag{4.14}$$

For high SNRs and based on an appropriate power allocation [102], the secrecy capacity for the OSJ and OS schemes can be simplified, and therefore, Equation 4.14 can be rewritten as

$$\frac{1}{2}\log_2 \left[\frac{\dfrac{\gamma_{S,D}}{\gamma_{J_1,D}} + \dfrac{\gamma_{R,D}}{\gamma_{J_2,D}}}{\dfrac{\gamma_{S,E}}{\gamma_{J_1,E}} + \dfrac{\gamma_{R,E}}{\gamma_{J_2,E}}} \right]^+ > \frac{1}{2}\log_2 \left[\frac{\gamma_{S,D}+\gamma_{R,D}}{\gamma_{S,E}+\gamma_{R,E}} \right]^+ \tag{4.15}$$

If the above condition is achieved, the OSJ scheme provides a higher instantaneous secrecy capacity than OS does and is preferred. Otherwise, the OS scheme is more efficient in promoting the system's secrecy performance and should be employed. Because of the uncertainty of the channel coefficient $h_{a,b}$ for each channel $a \rightarrow b$, OW should outperform either the continuous jamming scheme or the nonjamming one.

■ Suboptimal Switching (SW)

Given the fact that jamming is not always a positive process for the performance of the system, the suboptimal switching refers to the practical application of the intelligent switching between the SSJ and SS schemes. The basic idea is the same as OW, but the switching criterion uses the available knowledge set ψ_1. More specifically, the required condition for switching from SS to SSJ is

$$\frac{1}{2}\log_2\left[\frac{1+\dfrac{\gamma_{S,D}}{1+\gamma_{J_1,D}}+\dfrac{\gamma_{R,D}}{1+\gamma_{J_2,D}}}{1+\dfrac{E[\gamma_{S,E}]}{1+E[\gamma_{J_1,E}]}+\dfrac{E[\gamma_{R,E}]}{1+E[\gamma_{J_2,E}]}}\right]^+ > \frac{1}{2}\log_2\left[\frac{1+\gamma_{S,D}+\gamma_{R,D}}{1+E[\gamma_{S,E}]+E[\gamma_{R,E}]}\right]^+ \tag{4.16}$$

Following the same assumptions in obtaining Equation 4.15,

$$\frac{1}{2}\log_2\left[\frac{\dfrac{\gamma_{S,D}}{\gamma_{J_1,D}}+\dfrac{\gamma_{R,D}}{\gamma_{J_2,D}}}{\dfrac{E[\gamma_{S,E}]}{E[\gamma_{J_1,E}]}+\dfrac{E[\gamma_{R,E}]}{E[\gamma_{J_2,E}]}}\right]^+ > \frac{1}{2}\log_2\left[\frac{\gamma_{S,D}+\gamma_{R,D}}{E[\gamma_{S,E}]+E[\gamma_{R,E}]}\right]^+ \tag{4.17}$$

4.3.2 Presence of Multiple Eavesdroppers

4.3.2.1 Selection Schemes without Jamming

As described by Al-nahari *et al.* [131], the optimal relay selection is given by

$$R^* = \arg\max_{R \in C_d}\left\{\frac{1+\gamma_{S,D}+\gamma_{R,D}}{1+\displaystyle\sum_{m=1}^{M}\left(\gamma_{S,E_m}+\gamma_{R,E_m}\right)}\right\} \tag{4.18}$$

4.3.2.2 Selection Schemes with Conventional Jamming

As described by Al-nahari *et al.* [131], the selection policy which maximizes the instantaneous secrecy capacity given in Equation 4.4 assuming that $|C_d| > 0$ is given as

$$\left(R^*, J_1^*, J_2^*\right) = \arg \max_{\substack{J_1 \in S_{relay} \\ R \in C_d \\ J_2 \in \{S_{relay} - R^*\}}} \left\{ \frac{1 + \dfrac{\gamma_{S,D}}{1 + \gamma_{J_1,D}} + \dfrac{\gamma_{R,D}}{1 + \gamma_{J_2,D}}}{1 + \displaystyle\sum_{m=1}^{M} \left(\dfrac{\gamma_{S,E_m}}{1 + \gamma_{J_1,E_m}} + \dfrac{\gamma_{R,E_m}}{1 + \gamma_{J_2,E_m}} \right)} \right\} \tag{4.19}$$

Following the same assumptions in obtaining Equation 4.15, it can be shown that a simpler selection policy can be proposed as follows:

$$\left(R^*, J_1^*, J_2^*\right) = \arg \max_{\substack{J_1 \in S_{relay} \\ R \in C_d \\ J_2 \in \{S_{relay} - R^*\}}} \left\{ \frac{\dfrac{\gamma_{S,D}}{\gamma_{J_1,D}} + \dfrac{\gamma_{R,D}}{\gamma_{J_2,D}}}{\displaystyle\sum_{M} \left(\dfrac{\gamma_{S,E_m}}{\gamma_{J_1,E_m}} + \dfrac{\gamma_{R,E_m}}{\gamma_{J_2,E_m}} \right)} \right\} \tag{4.20}$$

According to Equation 4.20, we can obtain

$$J_1^* = \arg \min_{J_1 \in S_{relay}} \left\{ \frac{\gamma_{J_1,D}}{\sum \gamma_{J_1,E_m}} \\ \forall m \in S_{eves} \right\} \quad \text{and} \quad R^* = \arg \max_{R \in C_d} \left\{ \frac{\gamma_{R,D}}{\sum \gamma_{R,E_m}} \\ \forall m \in S_{eves} \right\}$$

$$J_2^* = \arg \min_{J_2 \in \{S_{relay} - R^*\}} \left\{ \frac{\gamma_{J_2,D}}{\sum \gamma_{J_2,E_m}} \\ \forall m \in S_{eves} \right\} \tag{4.21}$$

4.3.2.3 Selection Schemes with Controlled Jamming

In this scheme, the secrecy capacity expression given in Equation 4.4 is modified by Al-nahari *et al.* [131]:

$$C_S^{|C_d|} (R, J_1, J_2) = \left[\frac{1}{2} \log_2 \frac{1 + \gamma_{S,D} + \gamma_{R,D}}{1 + \displaystyle\sum_{m=1}^{M} \left(\dfrac{\gamma_{S,E_m}}{1 + \gamma_{J_1,E_m}} + \dfrac{\gamma_{R,E_m}}{1 + \gamma_{J_2,E_m}} \right)} \right]^+ \quad for\ |C_d| > 0 \tag{4.22}$$

Following the same assumptions in obtaining Equation 4.15, the selection policy that maximizes the secrecy capacity given in Equation 4.22 is given by

$$J_1^* = \arg \max_{J_1 \in S_{relay}} \left(\sum_{m=1}^{M} \gamma_{J_1, E_m} \right) \quad \text{and} \quad R^* = \arg \max_{R \in C_d} \left(\frac{\gamma_{R,D}}{\displaystyle\sum_{m=1}^{M} \gamma_{R,E_m}} \right)$$

$$J_2^* = \arg \max_{J_2 \in \{S_{relay} - R^*\}} \left(\sum_{m=1}^{M} \gamma_{J_2, E_m} \right) \tag{4.23}$$

4.4 Numerical Results and Discussion

In this section, the effectiveness of the proposed selection schemes is investigated via computer simulations. The simulation environment follows the model explained in Figure 4.1 and consists of a 2-D square topology where the nodes S, D, and E are located as $\{X_S, Y_S\} = \{0, 0\}$, $\{X_D, Y_D\} = \{1, 0\}$, $\{X_E, Y_E\} = \{0, 1\}$, respectively, and the direct paths $S \rightarrow D$, $S \rightarrow E$ are available. For simplicity, the source and relay nodes transmit with the same power, i.e., $P^{(S)} = P^{(R)}$. The relay and jammer nodes transmit with a relay–jammer power ratio of $L' = 100$. The number of the relays $N = 4$ and the relays are located randomly in the 2-D space considered; their exact location is given for each example considered. The path-loss exponent is set to $\beta = 3$, the area of the network is a 1×1 unit square, the transmission spectral efficiency is equal to $R_0 = 2$ bits per channel use (BPCU), and the target secrecy rate is equal to $R_T = 0.1$ BPCU. In this chapter, the adopted performance metrics are the ergodic secrecy capacity and the secrecy outage probability.

4.4.1 Impact of Changing the N-Relays Set Location with Respect to the Destination and the Eavesdropper

To study the effect of changing the relays location on system performance, three different scenarios are considered as follows:

First scenario: When the *N*-relays are located in the middle of the space between *D* and *E*

Figure 4.3 shows the topology of this scenario where the four relays have comparable links with *D* and *E*.

Figure 4.4 shows the ergodic secrecy capacity of the different proposed selection schemes versus the transmitted power *P*. From this figure, it is clear that selection with jamming outperforms the corresponding nonjamming

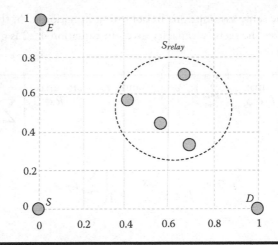

Figure 4.3 1×1 simulation environment with $N = 4$, $\beta = 3$.

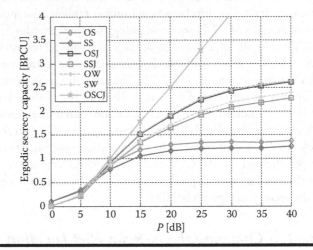

Figure 4.4 Ergodic secrecy capacity versus transmitted power (P) for the different selection schemes.

schemes. The integration of jamming in the optimal selection increases the ergodic secrecy capacity to 2.6087 BPCU compared with 1.3702 BPCU for the nonjamming case. This significant gain introduces jamming selection as an efficient technique to support secrecy constraints. The integration of jamming also improves the suboptimal selection protocols based on average channel knowledge; the ergodic secrecy capacity for the SSJ scheme converges to 2.2717 BPCU compared with 1.2551 BPCU for the SS scheme, with a gain higher for the higher SNRs.

Regarding the hybrid schemes, it can be seen that the OW scheme outperforms all the other selection schemes and provides the best performance where its secrecy capacity converges to 2.6413 BPCU. This result validates that an appropriate mechanism for switching between OS and OSJ overcomes the cases where the interference decreases the secrecy. For the suboptimal case, SW outperforms SS and SSJ selection schemes (its secrecy capacity converges to 2.3876 BPCU). An observation of the OSCJ scheme performance shows that it outperforms all the other selection schemes, providing the highest ergodic secrecy capacity when the transmitted power increases due to the ability of the destination to decode the artificial interference in this scheme.

Figure 4.5 presents the secrecy outage probability metric for the considered selection schemes using the above configuration. The presented results are in line with the above ergodic secrecy capacity results and show that jamming provides lower secrecy outage probability for both the OS and SS schemes. Regarding the hybrid schemes, the OW outperforms all schemes, and SW has a higher gain than SSJ in terms of outage probability. We also note that the OSCJ scheme gives the best performance because the effect of jamming signals is removed at the main destination node.

Second scenario: When the *N*-relays are located close to the eavesdropper

Figure 4.6 shows the considered topology, as well as the ergodic secrecy capacity of the different selection schemes. It is clear that the nonjamming approaches are inefficient, as the relays have a strong link with the eavesdropper. On the other hand, jamming schemes confuse the eavesdropper and significantly increase the ergodic secrecy capacity. For this configuration, the proposed hybrid schemes (OW, SW) have a similar performance to the jamming schemes (OSJ, SSJ), as jamming is always beneficial in this case.

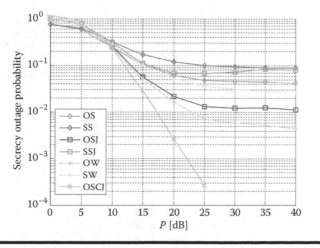

Figure 4.5 Secrecy outage probability versus *P* for the different selection schemes with $R_T = 0.1$ BPCU and $R_0 = 2$ BPCU.

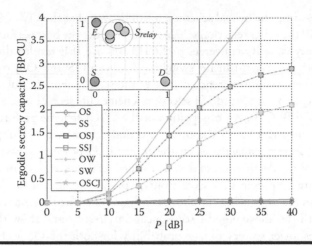

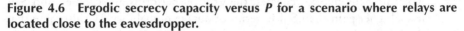

Figure 4.6 Ergodic secrecy capacity versus *P* for a scenario where relays are located close to the eavesdropper.

Figure 4.7 shows the secrecy outage probability metric for the considered selection schemes using the above configuration. The obtained results show that jamming provides lower secrecy outage probability for both the OS and SS schemes. Regarding the hybrid schemes, the OW outperforms all schemes. We also note that the OSCJ scheme gives the best performance because the effect of jamming signals is removed at the main destination node.

Figure 4.8 presents a comparison between the proposed schemes and the schemes presented by Krikidis *et al.* [102] in terms of secrecy outage

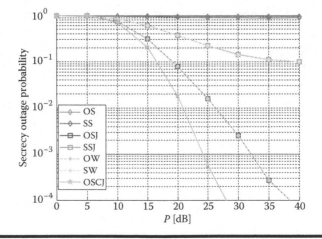

Figure 4.7 Secrecy outage probability versus *P* for the different selection schemes with $R_T = 0.1$ BPCU and $R_0 = 2$ BPCU.

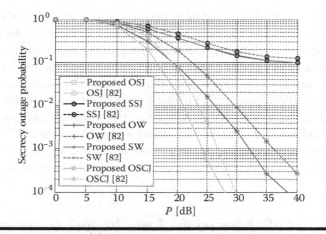

Figure 4.8 Secrecy outage probability versus *P* for the proposed selection schemes and the schemes presented by Krikidis *et al.* [102].

probability. The obtained results show that the proposed schemes outperform the schemes presented in [102] when the relays are located close to the eavesdropper *E*.

Third scenario: When the *N*-relays are located close to the destination

Figure 4.9 shows the considered topology, as well as the ergodic secrecy capacity of the different selection schemes. As can be seen, for this scenario, continuous jamming schemes introduce high interference at the destination and become less efficient. The main reason for this result is that for strong

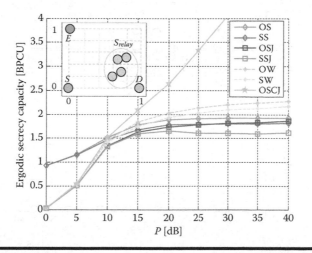

Figure 4.9 Ergodic secrecy capacity versus *P* for a scenario where relays are located close to the destination.

relay destination links, jamming becomes stronger at high SNRs and can decrease the secrecy performance achieved. On the other hand, nonjamming schemes significantly increase the ergodic secrecy capacity. As far as the hybrid schemes is concerned, appropriate switching significantly improves the performance compared with the continuous jamming schemes. Both OW and SW schemes yield a higher gain than previous configurations.

From Figures 4.6 and 4.9, it is shown that although jamming is an interesting solution for scenarios with strong relay–eavesdropper links, the hybrid schemes avoid strong interference at the destination in strong relay–destination scenarios and are thus promising solutions to maximize secrecy capacity.

4.4.2 Impact of Changing the Eavesdropper Location with Respect to the Source and the Destination

Assuming the locations of the source, relays and the destination are fixed, as in Figure 4.3, and the eavesdropper location has two different scenarios.

First scenario: $\{X_E, Y_E\} = \{0.2, 0.2\}$, i.e., the eavesdropper is close to the source

Figure 4.10 shows the ergodic secrecy capacity versus P for the different selection schemes when the eavesdropper is close to the source. It is clear that the nonjamming approaches are inefficient due to the strong link between the source and

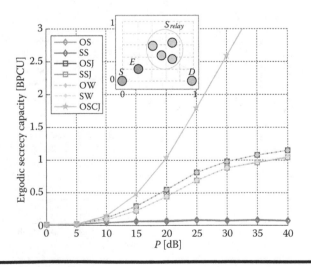

Figure 4.10 **Ergodic secrecy capacity versus P for the different selection schemes when the eavesdropper is close to the source.**

the eavesdropper. On the other hand, jamming schemes confuse the eavesdropper and significantly increase the ergodic secrecy capacity. The ergodic secrecy capacity for the OSJ scheme converges to 1.1433 BPCU compared with 0.0765 BPCU for the OS scheme and converges to 1.0417 BPCU for SSJ compared with 0.0679 BPCU for the SS scheme. For hybrid schemes (OW, SW), they have a similar performance to the jamming schemes (OSJ, SSJ), as jamming is always beneficial in this case. For the OSCJ scheme, it provides the highest ergodic secrecy capacity than all the other selection schemes.

Second scenario: $\{X_E, Y_E\} = \{0.8, 0.2\}$, i.e., the eavesdropper is close to the destination

Figure 4.11 shows the ergodic secrecy capacity versus P for the different selection schemes when the eavesdropper is close to the destination. From this figure, it is clear that the nonjamming approaches are inefficient, and almost no improvement in the ergodic secrecy capacity is achieved. On the other hand, jamming schemes significantly increase the ergodic secrecy capacity; ergodic secrecy capacity for the OSJ scheme converges to 2.3998 BPCU compared with 0.2375 BPCU for the OS scheme and converges to 1.1275 BPCU for SSJ compared with 0.0867 BPCU for the SS scheme. For hybrid schemes (OW, SW), they have a similar performance to the jamming schemes (OSJ, SSJ), as jamming is always beneficial in this case. For the OSCJ scheme, it provides the highest ergodic secrecy capacity than all the other selection schemes.

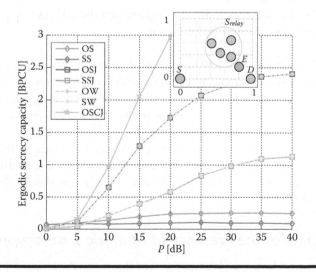

Figure 4.11 Ergodic secrecy capacity versus P for the different selection schemes when the eavesdropper is close to the destination.

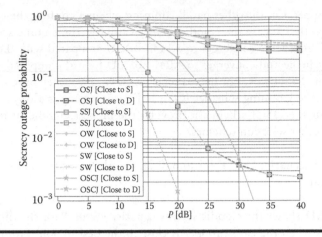

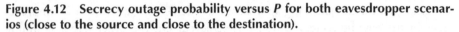

Figure 4.12 Secrecy outage probability versus *P* for both eavesdropper scenarios (close to the source and close to the destination).

From Figures 4.10 and 4.11, it is concluded that

■ The worst results of the ergodic secrecy capacity for all selection schemes are obtained when the eavesdropper is close to the source.
■ When the eavesdropper is close to the source or close to the destination, the nonjamming schemes are inefficient.
■ There is no need to apply the hybrid switching schemes, as they follow the jamming schemes' behavior.
■ The OSCJ scheme provides the highest ergodic secrecy capacity than the other selection schemes for both cases.

Comparing Figures 4.4, 4.10, and 4.11, it is found that as long as the eavesdropper is far from the source and the destination, an improvement in the ergodic secrecy capacity is achieved for both jamming and nonjamming schemes.

Figure 4.12 shows the secrecy outage probability comparison for both eavesdropper scenarios: when the eavesdropper is close to the source and close to the destination. It is clear that the best results of the secrecy outage probability for the OSCJ, OSJ, and OW selection schemes are obtained when the eavesdropper is close to the destination while the SSJ and SW selection schemes show a slight degradation for $P > 25$ dB.

4.4.3 Impact of the Presence of Multiple Eavesdroppers

The simulation environment follows the model presented in Figure 4.2. In the following, two scenarios with different numbers of eavesdroppers are considered.

First scenario: Number of eavesdroppers $M = 2$ and their locations are fixed at

$$\{x_{E_i}, y_{E_i}\}_{i=1}^2 = \{(0,1),(1,1)\}$$

Second scenario: Number of eavesdroppers $M = 3$ and their locations are fixed at

$$\{x_{E_i}, y_{E_i}\}_{i=1}^3 = \{(0,1),(0.5,0.5),(1,1)\}$$

From Figures 4.13 and 4.14, it is clear that by increasing the number of eavesdroppers in the system,

- The performance of different selection schemes is degraded.
- The OS scheme becomes inefficient and should not be used in these systems.
- The OSJ scheme is preferred in these systems due to the ability of jamming nodes to confuse eavesdroppers and significantly increase the ergodic secrecy capacity.
- The OSCJ scheme achieves the best performance due to the ability of the destination node to decode the jamming signals.

Comparing Figures 4.4, 4.13, and 4.14, it is concluded that the ergodic secrecy capacity is decreased when the number of eavesdroppers is increased for different relay selection schemes.

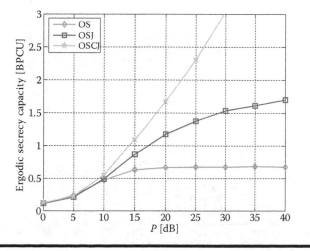

Figure 4.13 Ergodic secrecy capacity versus *P* for *N* = 4 and *M* = 2.

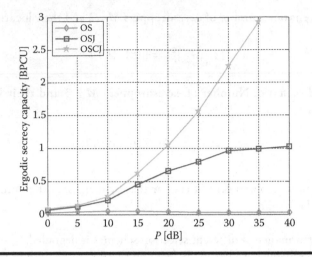

Figure 4.14   Ergodic secrecy capacity versus *P* for *N* = 4 and *M* = 3.

- For $M = 1$: $C_{s\_OS} = 1.3702$, $C_{s\_OSJ} = 2.6087$, and $C_{s\_OSCJ} = 5.7494$.
- For $M = 2$: $C_{s\_OS} = 0.6695$, $C_{s\_OSJ} = 1.6950$, and $C_{s\_OSCJ} = 4.6367$.
- For $M = 3$: $C_{s\_OS} = 0.0292$, $C_{s\_OSJ} = 1.0233$, and $C_{s\_OSCJ} = 3.7812$.

Figure 4.15 presents the secrecy outage probability comparison for two different numbers of eavesdroppers: when $M = 2$ and $M = 3$. It is clear that the secrecy outage probability increased by increasing the eavesdroppers' number for different relay selection schemes.

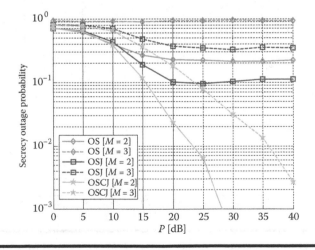

Figure 4.15   Secrecy outage probability versus *P* for *M* = 2 and *M* = 3.

4.5 Conclusion

This chapter has studied different relay and jammer selection schemes for one-way cooperative networks with physical layer secrecy consideration. The proposed schemes achieve an opportunistic selection of one conventional relay node and two jamming nodes to increase the security against eavesdroppers based on both instantaneous and average knowledge of the eavesdroppers' channels. Selection in the proposed schemes was made with the presence of direct links and the assumption that the broadcast phase was unsecured. The obtained results showed that jamming schemes such as OSJ and SSJ are effective for scenarios with strong eavesdropper links. In order to overcome jamming limitations for scenarios with weak eavesdropper links, a hybrid scheme for switching between jamming and nonjamming schemes was introduced, which further improves the system performance in terms of ergodic secrecy capacity and secrecy outage probability. The obtained results showed also that as long as the eavesdropper has comparable links with the source and the destination, the ergodic secrecy capacity and the secrecy outage probability are improved. Finally, increasing the eavesdropper nodes in the system degrades the system performance.

Chapter 5

Relay and Jammer Selection Schemes for Secure Two-Way Cooperative Networks

5.1 Introduction

Due to the broadcast nature of wireless communication networks, the adversarial nodes "eavesdroppers" can intercept transmissions in their coverage area and try to recover parts of the transmitted message. This issue makes security solutions quite challenging to implement in wireless communications. Recently, security technologies that are designed for the physical layer (PHY) have gained considerable attention [86]. The basic idea of these PHY-based approaches is to exploit the characteristics of wireless medium, like channel fading and inherent randomness of the noise, to allow legitimate nodes (source and destination) to communicate securely at a nonzero rate in the presence of eavesdroppers, provided that the source–eavesdropper channel is a degraded version of the main channel. The security is quantified by the secrecy capacity, which is defined as the maximum rate at which information is transmitted with a perfect secrecy from the source to the destination. In Csiszár and Körner [87] and Liang *et al.* [132], the secrecy capacity of the Gaussian wiretap channel was extended to signal transmission over the broadcast and wireless fading channels, respectively.

Several works have been done to increase the secrecy capacity of wireless networks with both cooperative relaying and cooperative jamming protocols

[66,91,102,103,120,121,124–128,131,133]. In [121] and [133], Dong *et al.* proposed effective decode-and-forward (DF) and amplify-and-forward (AF)–based cooperative relaying protocols for physical layer security, respectively. In [120], Liang *et al.* address the reliability, stability, and physical layer security for wireless broadcast networks. Furthermore, joint optimal power control and optimal scheduling schemes were proposed to enhance the secrecy rate of the intended receiver against cooperative and noncooperative eavesdroppers. In [124], Lai and El Gamal show that even if the source–destination rate is zero, a positive secrecy rate can be achieved if the relay is closer to the destination than the eavesdropper. Moreover, an efficient secrecy rate for networks with several potential relays and multiple eavesdroppers can be verified via relay selection by keeping the complexity relatively low. In Krikidis [125], two relay selection techniques have been proposed with different levels of feedback overhead. Krikidis *et al.* in [102] extended the work presented in [125] for cooperative networks with jamming protection without considering the direct links. In [103], Ibrahim *et al.* extended the work presented in [102] with the presence of direct links, the assumption that broadcast phase is unsecured, and when one or more eavesdroppers are present in the system. In Ibrahim *et al.* [66] and Beres and Adve [126], different strategies for relay selection were introduced for improving the secrecy rate in [125]. In [131], Al-nahari *et al.* proposed multiple relay selection schemes which improve the secrecy rate and enhance the outage performance for secrecy-constrained cooperative networks with multiple eavesdroppers. In Tekin and Yener [91] and Simeone and Popovski [127,128], the authors showed that the capacity of the eavesdropper link can be reduced via jamming schemes which produce an artificial interference at the eavesdropper node.

5.1.1 Related Work

The two-way relay channel has attracted much interest because of its bandwidth efficiency and potential application to cellular networks and peer-to-peer networks [134–138]. In [134] and [135], Rankov and Wittneben extended the one-way relay channels AF and DF protocols to the general full-duplex discrete two-way relay channel and half-duplex Gaussian two-way relay channel, respectively. In Chen *et al.* [136], joint relay and jammer selection schemes have been studied to ensure secure communication in DF two-way cooperative networks when there is no direct link between the two sources. Furthermore, the signal transmission consists of three phases, and the authors deal with the secrecy outage probability metric only. In Zhou *et al.* [137], different relay and jammer selection schemes in DF two-way relay networks have been investigated in terms of the ergodic secrecy rate metric only and with a perfect instantaneous knowledge of each link in the presence of one eavesdropper. In Chen *et al.* [138], several relay and jammer selection schemes in AF two-way cooperative networks with physical layer security consideration have been studied with the assumption that jamming signal is unknown at the other intermediate nodes and considering one eavesdropper network model.

5.1.2 Chapter Contributions

The main contribution of this chapter is to propose three different categories of relay and jammer selection schemes to improve the physical layer security of two-way cooperative networks. These categories are selection schemes without jamming, selection schemes with conventional jamming (where the jamming signal is unknown at the destinations), and selection schemes with controlled jamming (where the jamming signal is known at the destinations). The considered network consists of two sources: multiple intermediate nodes and one or more eavesdroppers. The proposed schemes select three intermediate nodes during two communication phases. In the first phase, a friendly jammer is selected to create intentional interference at the eavesdroppers' nodes. In the second phase, two relay nodes are selected: one node is selected to operate as a conventional relay and assists the sources to deliver their data to the corresponding destinations via the DF strategy, while the other node behaves as a jammer node in order to confuse the eavesdroppers in this phase. The proposed schemes are analyzed with two different channel knowledge sets: a global instantaneous knowledge of all links and an average knowledge of the eavesdroppers' links.

The performance of the proposed schemes is analyzed in terms of ergodic secrecy rate and secrecy outage probability. The obtained results show that the selection schemes with jamming outperform the schemes without jamming when the intermediate nodes are distributed dispersedly between the sources and the eavesdropper. However, when the intermediate nodes cluster gets close to one of the sources, they are no longer superior due to the strong interference on the destination nodes. Therefore, a hybrid scheme which switches between jamming and nonjamming selection schemes is proposed to overcome jamming limitations and seems to be an efficient solution for practical applications with critical secrecy constraints. Moreover, the impact of changing both the eavesdroppers and the intermediate nodes location on the system performance is discussed in this chapter. Finally, we discuss the impact of the presence of multiple cooperating and noncooperating eavesdroppers on system performance metrics. The obtained results reveal that, despite the presence of cooperating eavesdroppers, the proposed selection schemes are still able to improve both the secrecy rate and the secrecy outage probability of the two-way cooperative networks.

The rest of this chapter is organized as follows. In Section 5.2, the network model is described and the problem is formulated. In Section 5.3, the different proposed selection schemes are introduced. Numerical results and discussion are shown in Section 5.4, and finally, the main conclusions are drawn in Section 5.5.

5.2 Network Model and Assumptions

5.2.1 Single Eavesdropper Model

5.2.1.1 Network Model

A simple network configuration consisting of two sources S_1 and S_2, one eavesdropper E, and an intermediate node set $S_{relay} = \{1, 2,..., N\}$ with N nodes, as shown in Figure 5.1. The intermediate nodes operate in half-duplex mode; therefore, they cannot transmit and receive simultaneously and the communication process is performed in two phases. During the first phase, S_1 and S_2 transmit their data to the intermediate nodes, and due to the broadcasting nature of the transmission, the eavesdropper overhears the transmitted information. In addition, according to the security protocol, one node J_1 is selected from the S_{relay} set to operate as a "jammer" and transmits intentional interference to degrade the sources–eavesdropper links in this phase. During the second phase, an intermediate node R is selected to operate as a conventional relay which forwards the source's messages to the corresponding destinations. Node R belongs to a decoding set C_d ($C_d \subseteq S_{relay}$) which includes the relays that can successfully decode the source's messages. A second jammer J_2 is selected from S_{relay}, for the same reason as in J_1. Note that the artificial interference from the jamming nodes is unknown at S_1 and S_2, and thus, they are not able to mitigate it; this is referring to applications with critical secrecy constraints.

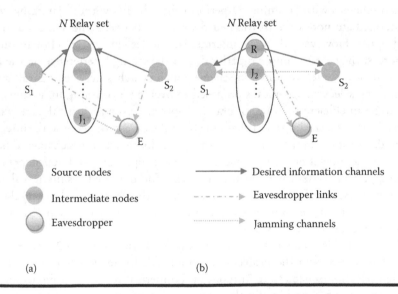

Figure 5.1   Network model with one eavesdropper. (a) First phase. (b) Second phase.

In this chapter the following assumptions are made:

- There is no direct link between the two sources.
- The jamming signal is known at the rest nodes of S_{relay}; the interference will not degrade the performance of the sources–relay links.
- Selection in the proposed schemes is made with the secrecy constraints.
- In both phases, a slow, flat, and block Rayleigh fading environment is assumed, i.e., the channel remains static for one coherence interval and changes independently in different coherence intervals with a variance $\sigma_{a,b}^2 = d_{a,b}^{-\beta}$, where $d_{a,b}$ is the Euclidean distance between node a and node b, and β is the path-loss exponent [130].
- Furthermore, additive white Gaussian noise (AWGN) is assumed with zero mean and unit variance.

Let $P^{(S)}$, $P^{(R)}$, and $P^{(J)}$ denote the transmitted power for the source nodes, the relay node, and the jamming nodes, respectively. In order to maximize the benefits of the proposed schemes and protect the destinations from the artificial interference, the jamming nodes transmit with a lower power than the relay node, and thus, their transmitted power is defined as $P^{(J)} = P^{(R)}/L'$ (with $P^{(S)} = P^{(R)}$), where $L' > 1$ denotes the power ratio of relay to jammer [102].

5.2.1.2 Problem Formulation

The instantaneous secrecy rate with the decoding set C_d for source S_i is given by Dong *et al.* [120]:

$$R_{S_i}^{|C_d|}(R, J_1, J_2) = \left[\frac{1}{2}\log_2\left(1 + \Gamma_i\right) - \frac{1}{2}\log_2\left(1 + \Gamma_{E_j}\right) \right]^+ \quad \textit{for } |C_d| > 0 \qquad (5.1)$$

where $[\chi]^+ \triangleq \max\{0, \chi\}$ and Γ_i and Γ_{E_j} denote the signals-to-interference-plus-noise ratios (SINRs) of link $S_j \to S_i$ ($i, j = 1, 2, i \neq j$) and link $S_j \to E$, respectively, and they are given by

$$\Gamma_i = \frac{\gamma_{R,S_i}}{\gamma_{J_2,S_i} + 1} \qquad (5.2)$$

$$\Gamma_{E_j} = \frac{\gamma_{S_j,E}}{\gamma_{S_i,E} + \gamma_{J_1,E} + 1} + \frac{\gamma_{R,E}}{\gamma_{J_2,E} + 1} \qquad (5.3)$$

where $\gamma_{a,b} \triangleq P^{(a)} |f_{a,b}|^2$ denotes the instantaneous signal-to-noise ratio (SNR) for the link $a \rightarrow b$ modeled as a zero-mean, independent, circularly symmetric complex Gaussian random variable with variance $\sigma_{a,b}^2$. The overall secrecy performance of the system is characterized by the ergodic secrecy rate, which is the expectation of the sum of the two sources' secrecy rates, $\mathrm{E}\left[R_S^{|C_d|}(R, J_1, J_2) \right]$, where

$$R_S^{|C_d|}(R, J_1, J_2) = R_{S_1}^{|C_d|}(R, J_1, J_2) + R_{S_2}^{|C_d|}(R, J_1, J_2) \tag{5.4}$$

Sometimes secrecy performance of the system is characterized by the secrecy outage probability, which is defined as the probability that the system secrecy rate is less than a target secrecy rate $R_T > 0$. Secrecy outage probability is written as

$$P_{out} = \sum_{n=1}^{N} P_r \left\{ R_S^n(R, J_1, J_2) < R_T \right\} P_r \left\{ |C_d| = n \right\} \tag{5.5}$$

The ultimate objective is to select appropriate nodes R, J_1, and J_2 in order to maximize the instantaneous secrecy rate for different types of channel feedback. The optimization problem can be formulated as [103]

$$(R^*, J_1^*, J_2^*) = \arg \max_{\substack{J_1 \in S_{relay} \\ R \in C_d \\ J_2 \in \{S_{relay} - R^*\}}} \left\{ R_S^{|C_d|}(R, J_1, J_2) \right\}, \quad s.t. \psi_u (for \quad u = 0,1) \tag{5.6}$$

where R^* and J_1^* and J_2^* denote the selected relay and jamming nodes, respectively. Note that the selected jammers J_1^* and J_2^* in the two phases may be the same node, which is determined by the instantaneous secrecy rate. ψ_0 denotes a global instantaneous knowledge for all the links, and ψ_1 denotes an average channel knowledge for the eavesdropper links.

5.2.2 Multiple Eavesdroppers Model

5.2.2.1 Network Model

Here, the presence of M eavesdroppers set $S_{eves} = \{1, 2,..., M\}$ is considered, as shown in Figure 5.2.

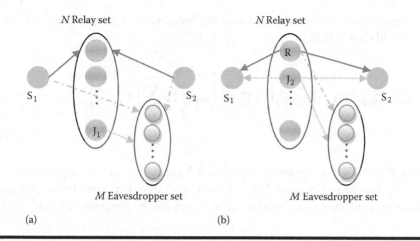

Figure 5.2   Network model with multiple eavesdroppers. (a) First phase. (b) Second phase.

5.2.2.2 Problem Formulation

In the network with multiple eavesdroppers, the eavesdroppers may cooperate or not cooperate with each other in two different scenarios as follows.

First scenario: When the eavesdroppers are noncooperative (i.e., each eavesdropper tries to decode the sources information individually)

The instantaneous secrecy rate with the decoding set C_d for source S_i is given by Al-nahari *et al.* [131]:

$$R_{S_i}^{|C_d|}(R, J_1, J_2) = \left[\frac{1}{2}\log_2\left(1+\Gamma_i\right) - \frac{1}{2}\log_2\left(1 + \max_{E_m \in S_{eves} \forall m}\left\{\Gamma_{E_{m_j}}\right\} \right) \right]^+ |C_d| > 0$$

(5.7)

where Γ_i is given by Equation 5.2 and $\Gamma_{E_{m_j}}$ can be expressed as follows:

$$\Gamma_{E_{m_j}} = \frac{\gamma_{S_j,E_m}}{\gamma_{S_i,E_m} + \gamma_{J_1,E_m} + 1} + \frac{\gamma_{R,E_m}}{\gamma_{J_2,E_m} + 1}$$

(5.8)

Note that Equation 5.7 represented the worst case, in which the eavesdropper can achieve the maximum rate. In other words, the secrecy rate achieved at the destination node is limited by the maximum rate achieved at the eavesdroppers.

Second scenario: When the eavesdroppers are cooperative (i.e., malicious eavesdroppers cooperate together in order to overhear the sources information)

The instantaneous secrecy rate with the decoding set C_d for source S_i is given by Lai and El Gamal [124]:

$$R_{S_i}^{|C_d|}(R, J_1, J_2) = \left[\frac{1}{2}\log_2\left(1 + \Gamma_i\right) - \frac{1}{2}\log_2\left(1 + \sum_{m=1}^{M}\left(\Gamma_{E_{m_j}}\right)\right) \right]^+ \quad for \; |C_d| > 0$$

(5.9)

where Γ_i and $\Gamma_{E_{m_j}}$ are given by Equation 5.2 and Equation 5.8, respectively.

It is clear from Equation 5.9 that the cooperation between the eavesdroppers adds more constraints on the achievable secrecy rate by the source nodes.

5.3 The Proposed Relay and Jammer Selection Schemes

Three different categories of relay and jammer selection schemes, namely, selection schemes without jamming, selection schemes with conventional jamming (where the jamming signal is unknown at the destinations), and selection schemes with controlled jamming (where the jamming signal is known at the destinations), will be discussed in the following subsections.

5.3.1 Selection Schemes in the Presence of One Eavesdropper

5.3.1.1 Selection Schemes without Jamming

In a conventional cooperative network, the relay scheme does not have a jamming process, and therefore, only one relay accesses the channel during the second phase of the protocol. The existing selections are summarized as follows:

■ **Conventional Selection (CS)**

This solution does not take the eavesdropper channels into account, and the relay node is selected according to the instantaneous SNR of the channel between node S_1 and node S_2 [125]. Therefore, the SINR given in Equation 5.2 can be written as follows:

$$\Gamma_i^{CS} = \gamma_{R,S_i}$$

(5.10)

Hence, the conventional selection scheme can be expressed as

$$
\begin{aligned}
R^* &= \arg \max_{R \in C_d} \left\{ R_{S_1}^{|C_d|}(R) + R_{S_2}^{|C_d|}(R) \right\} \\
&= \arg \max_{R \in C_d} \left\{ \frac{1}{2} \log_2 \left(1 + \Gamma_1^{CS} \right) + \frac{1}{2} \log_2 \left(1 + \Gamma_2^{CS} \right) \right\} \\
&= \arg \max_{R \in C_d} \left\{ \left(1 + \Gamma_1^{CS} \right) \cdot \left(1 + \Gamma_2^{CS} \right) \right\}
\end{aligned}
\tag{5.11}
$$

Although the selection in Equation 5.11 is an effective solution for non-eavesdropper environments, it cannot support the secrecy constraints for eavesdropper environments.

■ **Optimal Selection (OS)**

This solution takes the relay–eavesdropper link into account and selects the relay node according to the knowledge set ψ_0. The SINRs given in Equation 5.2 and Equation 5.3 can be rewritten as

$$
\Gamma_i^{OS} = \Gamma_i^{CS} = \gamma_{R,S_i}
\tag{5.12}
$$

$$
\Gamma_{E_j}^{OS} = \frac{\gamma_{S_j,E}}{\gamma_{S_i,E} + 1} + \gamma_{R,E}
\tag{5.13}
$$

The optimal selection scheme is given by Zhou *et al.* [117]:

$$
\begin{aligned}
R^* &= \arg \max_{R \in C_d} \left\{ R_{S_1}^{|C_d|}(R) + R_{S_2}^{|C_d|}(R) \right\} \\
&= \arg \max_{R \in C_d} \left\{ \frac{1}{2} \log_2 \left(1 + \Gamma_1^{OS} \right) - \frac{1}{2} \log_2 \left(1 + \Gamma_{E_2}^{OS} \right) \right. \\
&\qquad\qquad \left. + \frac{1}{2} \log_2 \left(1 + \Gamma_2^{OS} \right) - \frac{1}{2} \log_2 \left(1 + \Gamma_{E_1}^{OS} \right) \right\} \\
&= \arg \max_{R \in C_d} \left\{ \frac{1 + \Gamma_1^{OS}}{1 + \Gamma_{E_2}^{OS}} \cdot \frac{1 + \Gamma_2^{OS}}{1 + \Gamma_{E_1}^{OS}} \right\}
\end{aligned}
\tag{5.14}
$$

■ **Optimal Selection with Max–Min Instantaneous Secrecy Rate (OS-MMISR)**

It is common that the sum of the two sources secrecy rates, i.e., $\left\{ R_{S_1}^{|C_d|}(R) + R_{S_2}^{|C_d|}(R) \right\}$, may be driven down to a low level by the source with the lower secrecy rate. As a result, for a low complexity, the relay node, which maximizes the minimum secrecy rate of the two sources, can be selected to achieve a near-optimal performance. In addition, in some scenarios, the considered secrecy performance takes into account not only the total secrecy rate of both sources but also the individual secrecy rate of each one. If one source has a low secrecy rate, the whole system is regarded as secrecy inefficient. The OS-MMISR scheme maximizes the worse instantaneous secrecy rate of the two sources with the assumption of knowledge set ψ_0, and therefore, it is given by

$$
\begin{aligned}
R^* &= \arg \max_{R \in C_d} \ \min \left\{ R_{S_1}^{|C_d|}(R), R_{S_2}^{|C_d|}(R) \right\} \\
&= \arg \max_{R \in C_d} \ \min \left\{ \frac{1+\Gamma_1^{OS}}{1+\Gamma_{E_2}^{OS}}, \frac{1+\Gamma_2^{OS}}{1+\Gamma_{E_1}^{OS}} \right\}
\end{aligned}
\tag{5.15}
$$

where Γ_i^{OS} and $\Gamma_{E_j}^{OS}$ are given by Equation 5.12 and Equation 5.13, respectively.

■ **Suboptimal Selection (SS)**

This scheme avoids the OS scheme instantaneous estimation of the relay eavesdropper link by deciding the appropriate relay based on the knowledge set ψ_1. The suboptimal selection scheme can be expressed as Krikidis [125]:

$$
R^* = \arg \max_{R \in C_d} \left\{ \frac{1+\Gamma_1^{SS}}{1+\Gamma_{E_2}^{SS}} \cdot \frac{1+\Gamma_2^{SS}}{1+\Gamma_{E_1}^{SS}} \right\}
\tag{5.16}
$$

where

$$
\Gamma_i^{SS} = \Gamma_i^{OS} = \gamma_{R,S_i}
\tag{5.17}
$$

$$
\Gamma_{E_j}^{SS} = \frac{E[\gamma_{S_j,E}]}{E[\gamma_{S_i,E}]+1} + E[\gamma_{R,E}]
\tag{5.18}
$$

Comparing the OS scheme in Equation 5.14 and the SS scheme in Equation 5.16, the SS scheme is more useful in practice, as it depends on average channel knowledge for the eavesdropper links.

■ **Suboptimal Selection with Max–Min Instantaneous Secrecy Rate (SS-MMISR)**

The SS-MMISR scheme maximizes the worse instantaneous secrecy rate of the two sources with the assumption of the knowledge set ψ_1, and therefore, it can be expressed as

$$R^* = \arg \max_{R \in C_d} \min\left\{ R_{S_1}^{|C_d|}(R), R_{S_2}^{|C_d|}(R) \right\}$$

$$= \arg \max_{R \in C_d} \min\left\{ \frac{1+\Gamma_1^{SS}}{1+\Gamma_{E_2}^{SS}}, \frac{1+\Gamma_2^{SS}}{1+\Gamma_{E_1}^{SS}} \right\} \tag{5.19}$$

where Γ_i^{SS} and $\Gamma_{E_j}^{SS}$ are given by Equation 5.17 and Equation 5.18, respectively.

5.3.1.2 Selection Schemes with Conventional Jamming

In this subsection, several node selection schemes based on the optimization problem given by Equation 5.6 are presented in order to maximize the expectation of the sum of the two sources' secrecy rates.

■ **Optimal Selection with Jamming (OSJ)**

The optimal selection with jamming assumes the knowledge of the ψ_0 set and ensures a maximization of the sum of instantaneous secrecy rates of node S_1 and node S_2 given in Equation 5.4, which gives credit to

$$(R^*, J_1^*, J_2^*) = \arg \max_{\substack{J_1 \in S_{relay} \\ R \in C_d \\ J_2 \in \{S_{relay} - R^*\}}} \left\{ R_S^{|C_d|}(R, J_1, J_2) \right\}$$

$$= \arg \max_{\substack{J_1 \in S_{relay} \\ R \in C_d \\ J_2 \in \{S_{relay} - R^*\}}} \left\{ \frac{1+\Gamma_1}{1+\Gamma_{E_2}} \cdot \frac{1+\Gamma_2}{1+\Gamma_{E_1}} \right\} \tag{5.20}$$

where Γ_i and Γ_{E_j} are given by Equation 5.2 and Equation 5.3, respectively. The cooperative relay and jammer selection in Equation 5.20 tends to promote the system's secrecy performance by maximizing Γ_i, which promotes the assistance to the sources and minimizing Γ_{E_j}, which is equivalent to enhance the interference to the eavesdropper.

■ **Optimal Selection with Jamming with Max–Min Instantaneous Secrecy Rate (OSJ-MMISR)**

In the OSJ-MMISR scheme, the selected relay and jamming nodes aim to maximize the worse instantaneous secrecy rate of the two sources with the assumption of the knowledge set ψ_0, and therefore, it can be expressed as

$$
\begin{aligned}
(R^*, J_1^*, J_2^*) = \arg \quad & \max_{\substack{J_1 \in S_{relay} \\ R \in C_d \\ J_2 \in \{S_{relay} - R^*\}}} \quad \min\left\{ R_{S_1}^{|C_d|}(R, J_1, J_2), R_{S_2}^{|C_d|}(R, J_1, J_2) \right\} \\
= \arg \quad & \max_{\substack{J_1 \in S_{relay} \\ R \in C_d \\ J_2 \in \{S_{relay} - R^*\}}} \quad \min\left\{ \frac{1+\Gamma_1}{1+\Gamma_{E_2}}, \frac{1+\Gamma_2}{1+\Gamma_{E_1}} \right\}
\end{aligned}
\tag{5.21}
$$

where Γ_i and Γ_{E_j} are given by Equation 5.2 and Equation 5.3, respectively.

■ **Suboptimal Selection with Jamming (SSJ)**

In practice, an average knowledge of eavesdropper links available from long-term supervision of the eavesdropper transmission provides suboptimal selection metrics. The selection metric is modified as

$$
(R^*, J_1^*, J_2^*) = \arg \quad \max_{\substack{J_1 \in S_{relay} \\ R \in C_d \\ J_2 \in \{S_{relay} - R^*\}}} \quad \left\{ \frac{1+\Gamma_1}{1+\Gamma'_{E_2}} \cdot \frac{1+\Gamma_2}{1+\Gamma'_{E_1}} \right\}
\tag{5.22}
$$

where Γ_i is given by Equation 5.2 and Γ'_{E_j} can be calculated as follows:

$$
\Gamma'_{E_j} = \frac{E[\gamma_{S_j,E}]}{E[\gamma_{S_i,E}] + E[\gamma_{J_1,E}] + 1} + \frac{E[\gamma_{R,E}]}{E[\gamma_{J_2,E}] + 1}
\tag{5.23}
$$

■ **Suboptimal Selection with Jamming with Max–Min Instantaneous Secrecy Rate (SSJ-MMISR)**

In the SSJ-MMISR scheme, the selection policy maximizes the worse instantaneous secrecy rate of the two sources with the assumption of the knowledge set ψ_1. The selection metric is modified as

$$(R^*, J_1^*, J_2^*) = \arg \max_{\substack{J_1 \in S_{relay} \\ R \in C_d \\ J_2 \in \{S_{relay} - R^*\}}} \min \left\{ \frac{1+\Gamma_1}{1+\Gamma_{E_2}'}, \frac{1+\Gamma_2}{1+\Gamma_{E_1}'} \right\} \tag{5.24}$$

where Γ_i and Γ_{E_j}' are given by Equation 5.2 and Equation 5.23, respectively.

5.3.1.3 Selection Schemes with Controlled Jamming

In this subsection, an optimal selection with controlled jamming (OSCJ) scheme is proposed. Unlike the previous conventional jamming schemes, where the jamming signal is unknown at the destinations and the eavesdropper, the OSCJ scheme assumes that the jamming signal can be decoded at the destinations but not at the eavesdropper. In this case, the SINR of the link from S_j (for $j = 1, 2$) to E remains the same as given by Equation 5.3. The SINR of the link from S_j to S_i (for $i, j = 1, 2$, $i \neq j$) is modified as follows:

$$\Gamma_i^{OSCJ} = \gamma_{R, S_i} \tag{5.25}$$

5.3.1.4 Hybrid Selection Schemes

■ **Optimal Switching (OW)**

The original idea of using jamming nodes is to introduce interference on the eavesdropper links. However, based on the assumption that jamming signal is unknown at destinations, the continuous jamming from J_2 in the second phase may decrease the secrecy rate of both sources seriously, specifically when J_2 is close to one destination. In order to overcome this "negative jamming" effect which leads to excessive interference at destinations, an intelligent hybrid selection scheme which switches between optimal selection with jamming and optimal selection without jamming is proposed. The required condition for the participation of the jamming nodes is

$$\left\{ R_{S_1}^{|C_d|}(R, J_1, J_2) + R_{S_2}^{|C_d|}(R, J_1, J_2) \right\}_{OSJ} > \left\{ R_{S_1}^{|C_d|}(R) + R_{S_2}^{|C_d|}(R) \right\}_{OS}$$

i.e.,

$$\left\{ \frac{1+\Gamma_1}{1+\Gamma_{E_2}} \cdot \frac{1+\Gamma_2}{1+\Gamma_{E_1}} \right\} > \left\{ \frac{1+\Gamma_1^{OS}}{1+\Gamma_{E_2}^{OS}} \cdot \frac{1+\Gamma_2^{OS}}{1+\Gamma_{E_1}^{OS}} \right\} \tag{5.26}$$

where Γ_i, Γ_{E_j}, Γ_i^{OS}, and $\Gamma_{E_j}^{OS}$ are given by Equation 5.2, Equation 5.3, Equation 5.12, and Equation 5.13, respectively.

If the condition in Equation 5.26 was achieved, the OSJ scheme provides higher instantaneous secrecy rate than OS does and is preferred. Otherwise, the OS scheme is more efficient in promoting the system's secrecy performance and should be employed. Because of the uncertainty of the channel coefficient $h_{a,b}$ for each channel $a \rightarrow b$, OW should outperform either the continuous jamming scheme or the nonjamming one.

■ **Suboptimal Switching (SW)**

Given the fact that jamming is not always a positive process for the performance of the system, the suboptimal switching (SW) scheme uses the available knowledge set ψ_0 to make intelligent switching between SSJ and SS schemes. More specifically, the required condition for switching from SS to SSJ is

$$\left\{ R_{S_1}^{|C_d|}(R, J_1, J_2) + R_{S_2}^{|C_d|}(R, J_1, J_2) \right\}_{SSJ} > \left\{ R_{S_1}^{|C_d|}(R) + R_{S_2}^{|C_d|}(R) \right\}_{SS}$$

i.e.,

$$\left\{ \frac{1+\Gamma_1}{1+\Gamma_{E_2}'} \cdot \frac{1+\Gamma_2}{1+\Gamma_{E_1}'} \right\} > \left\{ \frac{1+\Gamma_1^{SS}}{1+\Gamma_{E_2}^{SS}} \cdot \frac{1+\Gamma_2^{SS}}{1+\Gamma_{E_1}^{SS}} \right\} \tag{5.27}$$

where Γ_i, Γ_{E_j}', Γ_i^{SS}, and $\Gamma_{E_j}^{SS}$ are given by Equation 5.2, Equation 5.23, Equation 5.17, and Equation 5.18, respectively.

5.3.2 Selection Schemes in the Presence of Multiple Eavesdroppers

5.3.2.1 Selection Schemes with Noncooperating Eavesdroppers

■ **Selection Schemes without Jamming**
As shown by Al-nahari *et al.* [131], the optimal relay selection is given by

$$R^* = \arg \max_{R \in C_d} \left\{ \frac{1+\Gamma_1^{OS}}{1+ \max\limits_{E_m \in S_{eves} \forall m} (\Gamma_{E_{m_2}}^{OS})} \cdot \frac{1+\Gamma_2^{OS}}{1+ \max\limits_{E_m \in S_{eves} \forall m} (\Gamma_{E_{m_1}}^{OS})} \right\} \tag{5.28}$$

where Γ_i^{OS} is given by Equation 5.12 and $\Gamma_{E_{m_j}}^{OS}$ can be expressed as

$$\Gamma_{E_{m_j}}^{OS} = \frac{\gamma_{S_j,E_m}}{\gamma_{S_i,E_m}+1} + \gamma_{R,E_m} \tag{5.29}$$

■ **Selection Schemes with Conventional Jamming**
As shown by Al-nahari *et al.* [131], the selection policy which maximizes the instantaneous secrecy rate given in Equation 5.7 and therefore maximizes the sum of the two sources secrecy rates given in Equation 5.4, assuming that $|C_d| > 0$, is given as

$$(R^*, J_1^*, J_2^*) = \arg \max_{\substack{J_1 \in S_{relay} \\ R \in C_d}} \left\{ \frac{1+\Gamma_1}{1+ \max\limits_{E_m \in S_{eves} \forall m} (\Gamma_{E_{m_2}})} \cdot \frac{1+\Gamma_1}{1+ \max\limits_{E_m \in S_{eves} \forall m} (\Gamma_{E_{m_1}})} \right\}$$

$$J_2 \in \{S_{relay} - R^*\} \tag{5.30}$$

where Γ_i and $\Gamma_{E_{m_j}}$ are given by Equation 5.2 and Equation 5.8, respectively.

■ **Selection Schemes with Controlled Jamming**
In this scheme, the selection policy which maximizes the instantaneous secrecy rate given in Equation 5.7 and therefore maximizes the sum of the

two sources secrecy rates given in Equation 5.4, assuming that $|C_d| > 0$, is given by Al-nahari *et al.* [131]:

$$(R^*, J_1^*, J_2^*) = \arg \max_{\substack{J_1 \in S_{relay} \\ R \in C_d \\ J_2 \in \{S_{relay} - R^*\}}} \left\{ \frac{1 + \Gamma_1^{OSCJ}}{1 + \max_{E_m \in S_{eves} \forall m} (\Gamma_{E_{m2}})} \cdot \frac{1 + \Gamma_2^{OSCJ}}{1 + \max_{E_m \in S_{eves} \forall m} (\Gamma_{E_{m1}})} \right\} \tag{5.31}$$

where Γ_i^{OSCJ} and $\Gamma_{E_{m_j}}$ are given by Equation 5.25 and Equation 5.8, respectively.

5.3.2.2 Selection Schemes with Cooperating Eavesdroppers

■ **Selection Schemes without Jamming**

As shown by Ibrahim *et al.* [103], the optimal relay selection is given by

$$R^* = \arg \max_{R \in C_d} \left\{ \frac{1 + \Gamma_1^{OS}}{1 + \sum_{m=1}^{M} (\Gamma_{E_{m2}}^{OS})} \cdot \frac{1 + \Gamma_2^{OS}}{1 + \sum_{m=1}^{M} (\Gamma_{E_{m1}}^{OS})} \right\} \tag{5.32}$$

where Γ_i^{OS} and $\Gamma_{E_{m_j}}^{OS}$ are given by Equation 5.12 and Equation 5.29, respectively.

■ **Selection Schemes with Conventional Jamming**

As described by Ibrahim *et al.* [103], the selection policy which maximizes the instantaneous secrecy rate given in Equation 5.9 and therefore maximizes the sum of the two sources secrecy rates given in Equation 5.4, assuming that $|C_d| > 0$, is given as

$$(R^*, J_1^*, J_2^*) = \arg \max_{\substack{J_1 \in S_{relay} \\ R \in C_d \\ J_2 \in \{S_{relay} - R^*\}}} \left\{ \frac{1 + \Gamma_1}{1 + \sum_{m=1}^{M} (\Gamma_{E_{m2}})} \cdot \frac{1 + \Gamma_2}{1 + \sum_{m=1}^{M} (\Gamma_{E_{m1}})} \right\} \tag{5.33}$$

where Γ_i and $\Gamma_{E_{m_j}}$ are given by Equation 5.2 and Equation 5.8, respectively.

■ **Selection Schemes with Controlled Jamming**

In this scheme, the selection policy which maximizes the instantaneous secrecy rate given in Equation 5.9 and therefore maximizes the sum of the two sources secrecy rates given in Equation 5.4, assuming that $|C_d| > 0$, is given by Ibrahim *et al.* [103]:

$$(R^*, J_1^*, J_2^*) = \arg \max_{\substack{J_1 \in S_{relay} \\ R \in C_d \\ J_2 \in \{S_{relay} - R^*\}}} \left\{ \frac{1 + \Gamma_1^{OSCJ}}{1 + \sum_{m=1}^{M}(\Gamma_{E_{m_2}})} \cdot \frac{1 + \Gamma_2^{OSCJ}}{1 + \sum_{m=1}^{M}(\Gamma_{E_{m_1}})} \right\} \quad (5.34)$$

where Γ_i^{OSCJ} and $\Gamma_{E_{m_j}}$ are given by Equation 5.25 and Equation 5.8, respectively.

5.4 Numerical Results and Discussion

In this section, some computer simulations are provided to show the effectiveness of the proposed selection schemes in improving the ergodic secrecy rate and secrecy outage probability of the two-way cooperative networks. For simplicity, the sources and the relay transmit with the same power, i.e., $P^{(S)} = P^{(R)}$. The relay and the jammers transmit with a relay–jammer power ratio of $L' = 100$. The transmission spectral efficiency is equal to $R_0 = 2$ bits per channel use (BPCU), and the target secrecy rate is equal to $R_T = 0.1$ BPCU. The secrecy performance is discussed for two different network models as follows:

5.4.1 Secrecy Performance for the Single Eavesdropper Model

The simulation environment of this model was presented in Figure 5.1 and consists of two sources S_1, S_2, one eavesdropper E, and an intermediate nodes cluster with $N = 4$ nodes. All the nodes are located in a 2-D square topology within a 1×1 unit square. The sources and the eavesdropper are fixed at $\{X_{S_1}, Y_{S_1}\} = \{0, 0\}$, $\{X_{S_2}, Y_{S_2}\} = \{1, 0\}$, and $\{X_E, Y_E\} = \{0.5, 1\}$, respectively, while the relays are located randomly in the 2-D space considered, and their exact location is given for each example separately.

5.4.1.1 Secrecy Performance When Changing the N-Relays Set Location in the Considered Area

In this subsection, three different topologies are handled as follows.

Topology 1: When the *N*-relays are distributed dispersedly between S_1, S_2, and E nodes

Figure 5.3 shows the considered topology, as well as the ergodic secrecy rate of the different selection schemes. It is clear that the selection schemes with jamming

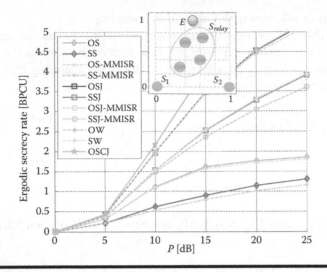

Figure 5.3 Ergodic secrecy rate versus transmitted power (*P*) for the different selection schemes when the relays are distributed dispersedly between S_1, S_2, and *E*.

outperform their nonjamming counterparts along the transmitted power range. For example, at $P = 20$ dB, the ergodic secrecy rate of OSJ is approximately higher than that of OS by 3 BPCU, and the ergodic secrecy rate of SSJ is approximately higher than that of SS by 2 BPCU. In addition, for this topology, the selection schemes which maximize the minimum instantaneous secrecy rates of the two sources (e.g., OS-MMISR, SS-MMISR, OSJ-MMISR, and SSJ-MMISR) have a slightly lower ergodic secrecy rate than the corresponding optimal ones (e.g., OS, SS, OSJ, and SSJ). Therefore, they can be efficiently used in order to get the appropriate relay and the jamming nodes. Regarding the hybrid schemes, it can be seen that the OW and SW schemes perform almost the same as the OSJ and SSJ schemes, respectively, as the jamming schemes are the best in promoting a system's secrecy performance compared with the nonjamming ones. It is also clear that the performance of the OSCJ scheme outperforms all the other selection schemes and achieves the highest ergodic secrecy rate when the transmitted power increases due to the ability of the destinations to decode the artificial interference in this scheme.

Figure 5.4 shows the secrecy outage probability metric of the considered selection schemes using the above topology. The presented results are in line with the above ergodic secrecy rate results and show that the integration of jamming provides lower secrecy outage probability for both the OS and SS schemes. Furthermore, all the near-optimal selection schemes have a slightly higher secrecy outage probability than the corresponding optimal ones along the transmitted power range, except the OSJ-MMISR scheme, which shows a considerable higher outage probability than OSJ for $P > 20$ dB. Regarding the hybrid schemes, OW follows OSJ's performance,

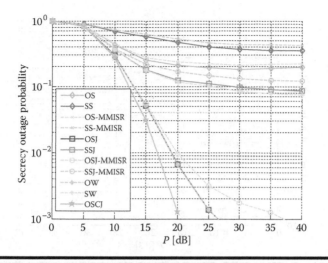

Figure 5.4 **Secrecy outage probability versus P for the different selection schemes with $R_T = 0.1$ BPCU and $R_0 = 2$ BPCU.**

and SW has a lower outage probability than SSJ when the transmitted power increases. For the OSCJ scheme, it gives the best performance because the effect of jamming signals is removed at the destination nodes.

Figures 5.5 and 5.6 present a comparison between the proposed two-way relay and jammer selection schemes and the one-way schemes presented by Ibrahim *et al.* [103] in terms of ergodic secrecy rate and secrecy outage probability, respectively. The obtained results show that when the relays are distributed dispersedly between

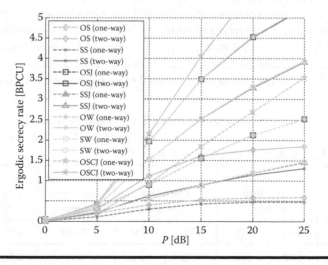

Figure 5.5 **Ergodic secrecy rate versus P for the two-way proposed selection schemes and the one-way schemes presented by Ibrahim *et al.* [103].**

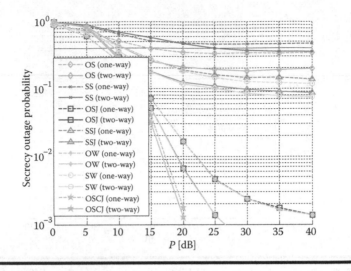

Figure 5.6 Secrecy outage probability versus *P* for the two-way proposed selection schemes and the one-way schemes presented by Ibrahim *et al.* [103].

S_1, S_2, and E all the proposed schemes outperform the schemes presented in [103], especially when the transmitted power is increased.

Tables 5.1 and 5.2 present the ergodic secrecy rate and secrecy outage probability comparison of the proposed two-way relay and jammer selection schemes and the one-way relay and jammer selection schemes presented by Ibrahim *et al.* [103] at *P* = 20 dB, respectively.

From the numerical results presented in Tables 5.1 and 5.2, it is observed that at *P* = 20 dB, all the proposed two-way relay and jammer selection schemes are more efficient than the one-way relay and jammer selection schemes presented in [103] in terms of both performance metrics, except the proposed two-way SS scheme, which shows its secrecy outage probability effectiveness for *P* > 20 dB.

Topology 2: When the *N*-relays are located close to the eavesdropper (*E*) node

Table 5.1 Ergodic Secrecy Rate Comparison of the Proposed Two-Way Selection Schemes and the One-Way Selection Schemes Presented by Ibrahim *et al.* [103] at *P* = 20 dB

Network Type	Without Jamming Schemes		Conventional Jamming Schemes		Controlled Jamming Schemes	Hybrid Schemes	
	OS	SS	OSJ	SSJ	OSCJ	OW	SW
One-way	0.58	0.47	2.13	1.2	2.7	2.13	1.21
Two-way	1.77	1.15	4.53	3.29	5.71	4.53	3.31

Table 5.2   Secrecy Outage Probability Comparison of the Proposed Two-Way Selection Schemes and the One-Way Selection Schemes Presented by Ibrahim *et al.* [103] at *P* = 20 dB

Network Type	Without Jamming Schemes		Conventional Jamming Schemes		Controlled Jamming Schemes	Hybrid Schemes	
	OS	*SS*	*OSJ*	*SSJ*	*OSCJ*	*OW*	*SW*
One-way	3.5e-1	4.7 e-1	1.6e-2	1.9e-1	1.8e-3	1.6e-2	1.8e-1
Two-way	2.1e-1	4.7e-1	6.6e-3	1.2e-1	1.3e-3	6.5e-3	1.2e-1

Figure 5.7 shows the considered topology, as well as the ergodic secrecy rate of the different selection schemes. It is clear that the performance of nonjamming approaches is very bad, as the relays have a strong link with the eavesdropper. On the other hand, the jamming schemes confuse the eavesdropper and significantly increase the ergodic secrecy rate. For this configuration, the OS-MMISR, SS-MMISR, OSJ-MMISR, and SSJ-MMISR selection schemes have almost the same performance as the OS, SS, OSJ and SSJ selection schemes, respectively. Furthermore, the hybrid schemes (OW, SW) have a similar performance to the jamming schemes (OSJ, SSJ), as jamming is always beneficial in this case. For the OSCJ scheme, it is noted that it gives the best performance because the effect of jamming signals is removed at the destination nodes.

Figure 5.8 shows the secrecy outage probability metric for the considered selection schemes using the above topology. The obtained results show that OSJ and SSJ provide a lower secrecy outage probability than the OS and SS selection schemes,

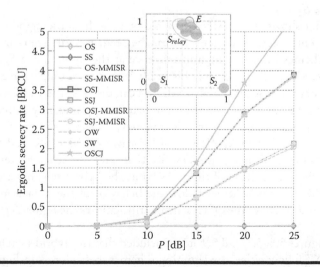

Figure 5.7   Ergodic secrecy rate versus *P* for the different selection schemes when the relays are located close to *E*.

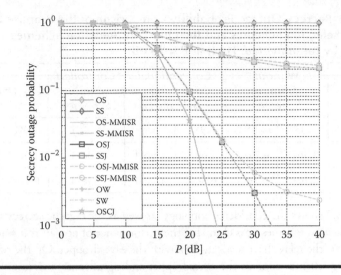

Figure 5.8 Secrecy outage probability versus *P* for the different selection schemes with R_T = 0.1 BPCU and R_0 = 2 BPCU.

respectively. The OSJ-MMISR and SSJ-MMISR schemes have a higher outage probability than the OSJ and SSJ schemes for *P* > 25 dB. Regarding the hybrid schemes, the OW and SW schemes follow the performance of jamming schemes OSJ and SSJ, respectively. For the OSCJ scheme, it achieves the lowest outage probability.

Topology 3: When the *N*-relays are located close to one of the source nodes, for example S_2

Figure 5.9 shows the considered topology, as well as the ergodic secrecy rate of the different selection schemes. As can be seen, for this scenario, continuous jamming schemes introduce high interference at the source node S_2 and become less efficient. The main reason for this result is that for strong source–relay links, jamming becomes stronger at high SNRs and can decrease the secrecy performance achieved. On the other hand, nonjamming schemes significantly increase the ergodic secrecy rate. It is clear that in this topology, there is a difference in the ergodic secrecy rate of the OS-MMISR, SS-MMISR, OSJ-MMISR, and SSJ-MMISR selection schemes and that of the OS, SS, OSJ, and SSJ schemes, respectively, so they cannot be used instead of them. As far as the hybrid schemes are concerned, both the OW and SW schemes follow the behavior of the nonjamming schemes OS and SS, respectively. The OSCJ scheme still achieves the highest ergodic secrecy rate.

From Figures 5.3, 5.7, and 5.9, it is concluded that the hybrid switching schemes OW and SW follow either the jamming schemes' behavior as in Figures 5.3 and 5.7 or the nonjamming schemes' behavior as in Figure 5.9 based on what either promotes as the system's secrecy performance.

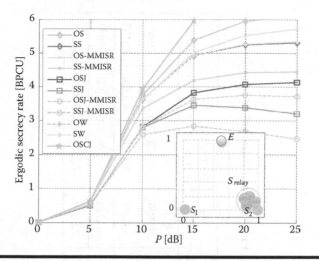

Figure 5.9 **Ergodic secrecy rate versus *P* for the different selection schemes when the relays are located close to S_2.**

Tables 5.3 and 5.4 present the ergodic secrecy rate and secrecy outage probability comparison of the proposed selection schemes for different topologies at $P = 20$ dB, respectively.

From Tables 5.3 and 5.4, it is concluded that

■ In Topology 1, where S_{relay} distributed dispersedly between S_1, S_2, and E nodes, the jamming schemes have a considerable effect on improving both system's secrecy metrics. The OS-MMISR, SS-MMISR, OSJ-MMISR, and SSJ-MMISR selection schemes show a slight degradation in system performance compared with the OS, SS, OSJ, and SSJ schemes, respectively. The hybrid switching schemes OW and SW almost follow the behavior of jamming schemes OSJ and SSJ, respectively, as jamming schemes are preferred in this case.

■ In Topology 2, where S_{relay} is close to the E node, the nonjamming schemes are inefficient, but the jamming schemes confuse the eavesdropper and significantly improve the secrecy performance. The OS-MMISR, SS-MMISR, OSJ-MMISR, and SSJ-MMISR selection schemes have almost the same performance as the OS, SS, OSJ, and SSJ schemes, respectively. For the hybrid schemes, they follow the jamming schemes' behavior.

■ In Topology 3, where S_{relay} is close to the S_2 source node, in this topology, the nonjamming schemes give better performance than the jamming schemes. The OS-MMISR, SS-MMISR, OSJ-MMISR, and SSJ-MMISR selection schemes show a noticeable degradation in secrecy performance compared with the OS, SS, OSJ, and SSJ schemes, respectively. For hybrid

Table 5.3 Ergodic Secrecy Rate Comparison of the Proposed Selection Schemes for the Three Different Topologies at P = 20 dB

Topologies	Schemes without Jamming				Conventional Jamming				Controlled Jamming	Hybrid Schemes	
	OS	SS	OS-MMISR	SS-MMISR	OSJ	SSJ	OSJ-MMISR	SSJ-MMISR	OSCJ	OW	SW
Topology 1	1.7688	1.1455	1.7239	1.0091	4.5282	3.2885	4.4652	3.0411	5.7078	4.5330	3.3135
Topology 2	0.0146	0.0010	0.0146	0.0014	2.8902	1.4822	2.8785	1.4514	3.6745	2.8902	1.4822
Topology 3	5.9340	5.2313	5.5106	4.4099	4.0674	3.3855	3.7527	2.6818	7.2704	5.9467	5.2521

Table 5.4 Secrecy Outage Probability Comparison of the Proposed Selection Schemes for the Three Different Topologies at $P = 20$ dB

| Topologies | Schemes without Jamming | | | | Conventional Jamming | | | | Controlled Jamming | Hybrid Schemes | |
	OS	SS	OS-MMISR	SS-MMISR	OSJ	SSJ	OSJ-MMISR	SSJ-MMISR	OSCJ	OW	SW
Topology 1	2.1e-1	4.7e-1	2.4e-1	5.3e-1	6.6e-3	1.2e-1	9.5e-3	1.7e-1	1.3e-3	6.5e-3	1.2e-1
Topology 2	9.9e-1	9.9e-1	9.9e-1	9.9e-1	9.4e-2	4.5e-1	9.4e-2	4.6e-1	3.5e-2	9.4e-2	4.5e-1
Topology 3	5e-4	6e-4	3.9e-3	1.3e-2	3.6e-3	4.7e-2	2.9e-2	1.4e-1	4e-4	5e-4	6e-4

schemes, they follow the nonjamming schemes' behavior due to their efficiency.

■ In the three topologies, the OSCJ scheme outperforms all the other selection schemes, as it provides the highest ergodic secrecy rate and the lowest secrecy outage probability.

5.4.1.2 Secrecy Performance When Changing the Eavesdropper Location with Respect to the Two Sources (S_1 and S_2)

Figure 5.10 shows the considered topology as well as the ergodic secrecy rate of the different selection schemes. It is clear that the nonjamming approaches are inefficient due to the strong link between the sources and the eavesdropper. On the other hand, jamming schemes confuse the eavesdropper and significantly increase the ergodic secrecy rate. For this configuration, the OS-MMISR, SS-MMISR, OSJ-MMISR, and SSJ-MMISR selection schemes have almost the same performance as the OS, SS, OSJ, and SSJ selection schemes, respectively. Furthermore, the hybrid schemes (OW, SW) have a similar performance to the jamming schemes (OSJ, SSJ), as jamming is always beneficial in this case. We also note that the OSCJ scheme gives the best performance because the effect of jamming signals is removed at the destination nodes.

From Figures 5.3 and 5.10, it is concluded that the performance of all the selection schemes is degraded when the eavesdropper node is approaching the S_1 and S_2 nodes.

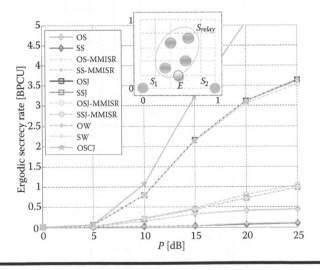

Figure 5.10 **Ergodic secrecy rate versus P for the different selection schemes when E is located close to S_1 and S_2.**

5.4.2 Secrecy Performance for the Multiple Eavesdroppers Model

The simulation environment for this model follows the model presented in Figure 5.2. Assuming the presence of two eavesdroppers, which are fixed at $\left\{x_{E_i}, y_{E_i}\right\}_{i-1}^{2} = \{(0.5,\ 1),(0,\ 0.5)\}$, Figures 5.11 and 5.12 present the ergodic secrecy rate and secrecy outage probability comparison for $N = 4$ and $M = 2$ cooperating

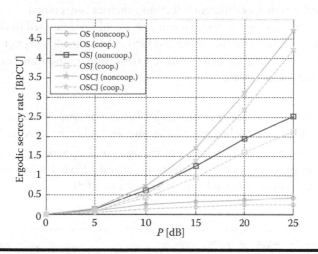

Figure 5.11 Ergodic secrecy rate versus P for $M = 2$ noncooperating and cooperating eavesdroppers.

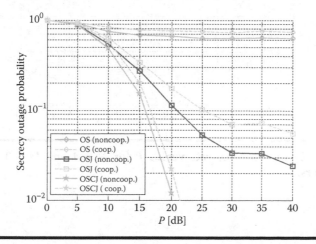

Figure 5.12 Secrecy outage probability versus P for $M = 2$ noncooperating and cooperating eavesdroppers.

and noncooperating eavesdroppers, respectively. The obtained results show that eavesdroppers' cooperation degrades both secrecy metrics.

Figure 5.13 shows the ergodic secrecy rate versus P for $N = 4$ and $M = 3$ cooperating eavesdroppers, which are fixed at $\left\{ x_{E_i}, y_{E_i} \right\}_{i=1}^{3} = \{(0.5, 1), (0, 0.5), (1, 0.5)\}$.

Comparing Figures 5.3, 5.11, and 5.13, it is concluded that increasing the number of cooperating eavesdroppers in the system results in the following:

- The performance of different selection schemes is degraded.
- The OS scheme becomes inefficient and should not be used in these systems.
- The demand increases for using the OSJ scheme in these systems due to the ability of jamming nodes to confuse eavesdroppers and significantly increase the ergodic secrecy rate.

Figure 5.14 shows the secrecy outage probability metric comparison of the proposed two-way selection schemes and the one-way schemes presented by Ibrahim *et al.* [83] in the presence of $M = 3$ cooperating eavesdroppers. The obtained results show that the proposed two-way selection schemes have a lower secrecy outage probability than the one-way schemes presented in [103].

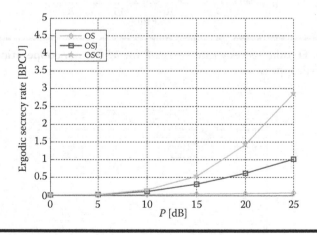

Figure 5.13 Ergodic secrecy rate versus *P* for *N* = 4 and *M* = 3 cooperating eavesdroppers.

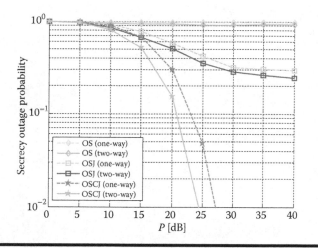

Figure 5.14 Secrecy outage probability versus *P* for the two-way proposed selection schemes and the one-way schemes presented by Ibrahim *et al.* [103] in the presence of *M* = 3 cooperating eavesdroppers.

5.5 Conclusion

Three categories of relay and jammer selection schemes were proposed in this chapter to improve the physical layer security of two-way cooperative networks. These categories are selection schemes without jamming, selection schemes with conventional jamming, and selection schemes with controlled jamming. Moreover, a hybrid scheme which switches between selection schemes with jamming and schemes without jamming was introduced to overcome the negative effects of interference. The obtained results showed the effectiveness of hybrid switching schemes (e.g., OW, SW), which follow the behavior of the OSJ and SSJ jamming schemes when *N*-relays are distributed dispersedly between system nodes and when they are located close to the eavesdropper, and follow the behavior of the OS and SS (without jamming) schemes when *N*-relays are located close to one of the source nodes. Despite the eavesdroppers' cooperation, which further degrades secrecy performance, the proposed selection schemes are still able to improve both the secrecy rate and the secrecy outage probability, especially the OSJ scheme. Finally, we showed the effectiveness of the proposed two-way relay selection schemes over the one-way relay selection schemes in promoting the system's secrecy performance.

SECURITY AND DATA RELIABILITY IN WIRELESS SENSOR NETWORKS

II

Dr. Emad S. Hassan
Dr. Amir S. Elsafrawey
Prof. Moawad I. Dessouky

Chapter 6

Overview on Sensor Networks

Sensor networks typically have little or even no infrastructure. It consists of a few tens to thousands of sensor nodes. These nodes are working together to monitor and observe a region under consideration and obtain data about the environment where they are deployed. These nodes are connected through a wireless channel, so they are called wireless sensor network (WSN).

6.1 Wireless Sensor Network

A schematic view of a WSN is depicted in Figure 6.1. Nowadays, the design and development of WSNs for various real applications, such as health monitoring, environmental monitoring, industrial process automation, battle fields surveillance, and seismic monitoring, have become possible owing to the rapid advances in both of wireless communications and sensor technology. This type of network is cost-effective and attractive to a wide range of application mission-critical situations, so they gain significant popularity compared with other networks [139].

A WSN is a collection of low-powered, physically tiny devices, smaller, cheaper, and intelligent, called sensor nodes. These sensors are equipped with many parts such as a power supply (battery), memory, processor, an actuator, and a radio transceiver interface so they can communicate with one another to form a network.

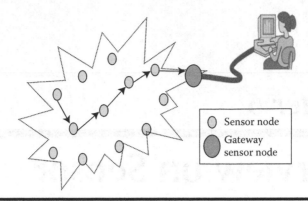

Figure 6.1   View of wireless sensor networks.

A radio is implemented for wireless communication to transfer the data to a base station. The battery is considered the main power supply in a sensor, with a secondary power supply in the form of solar panels that harvest power from the environment. An actuator is an electric mechanical device that can be used to control many components. It may actuate many sensing devices and can adjust sensor parameters. The actuator can also actuate the sensor to move and can monitor sensor power. There are a variety of sensing components which may be involved within the sensor node to measure the characteristic of the environment, such as thermal, chemical, mechanical, optical, biological, and magnetic sensors [139]. Furthermore, a WSN possesses a central gathering point called the sink (or base station), where all the collected data can be stored.

The major challenge in the design of a WSN is due to the constraints that are imposed on the sensing, storage, processing, and communication features of the sensor nodes. More precisely, the sensor nodes suffer from the constrained power supplies, which shortens the sensors' lifetime. It is very crucial to consider the design and implementation of this type of network for correct operation and longevity [140].

An advantage of WSNs is that the position of the sensor does not need to be predetermined. The WSN must have protocols that possess self-organizing capabilities. Another unique feature of these networks is the cooperation principal followed between sensor nodes. In addition, sensor nodes have an on-board processor, so they have the ability to locally perform some simple processing and transmit only the required data instead of sending all data to the nodes responsible for the fusion.

6.1.1 Types of WSNs

There are different classifications of WSNs; they can be classified according to the deployment or according to the environment.

6.1.1.1 Deployment Classification

According to the deployment, the WSN can be classified into structured and unstructured:

Unstructured WSN

It consists of a dense combination of sensors that are deployed in an ad hoc fashion in the region under consideration. In this type of deployment, sensor nodes are randomly placed. Once they are deployed, the network is kept to perform their reporting and monitoring functions.

Structured WSN

In this type, all or some of the nodes are placed in a preplanned fashion. The sensors have a known and fixed location. Since the nodes are placed at specific locations, they provide good coverage rather than the ad hoc deployment in an unstructured network, which may have uncovered regions [141].

6.1.1.2 Environment Classification

According to the environment, the WSN may be deployed on land, underground, or underwater, depending on the environment. A sensor network faces many challenges. There are five types of WSNs:

Terrestrial WSN

Terrestrial WSNs [142] typically consist of tens to thousands of low-cost nodes deployed in an area under consideration, either in a structured or unstructured manner. For unstructured, the sensors may be dropped from an airplane, so that they can be randomly placed into the area under consideration. For structured, there are grid placement [143], optimal placement [144], and 2D and 3D placement [145] models.

Underground WSN

Underground WSNs [146,147] consist of sensor nodes suppressed underground or in a cave that are used to monitor and measure underground conditions. On the other hand, a data sink is positioned above the ground to collect information from the underground nodes for sending to the distention base station. The underground environment causes a challenge in communication due to signal losses and attenuation, so energy consumption is a critical parameter in underground WSNs.

Underwater WSN

Underwater WSNs [148,149] consist of a number of sensor nodes and vehicles deployed underwater. Autonomous underwater vehicles are used for gathering data from nodes located underwater. Typical underwater wireless communication can be done through the transmission of acoustic waves. The main challenge in underwater communication is the propagation delay, limited bandwidth, and signal fading issue.

Multimedia WSN

Multimedia WSNs [150] are used to enable monitoring and tracking of events in the form of multimedia such as video, audio, and imaging. It consists of low-cost sensor nodes equipped with cameras and microphones. These sensor nodes are deployed in a preplanned manner. Challenges in this type of network are high-energy consumption, large bandwidth demand, and quality of service (QoS) provisioning.

Mobile WSN

Mobile WSNs consist of sensor nodes that have the ability to move, sense, compute, and communicate like static nodes. The main difference is that mobile nodes can do reposition and organize themselves. Data collected by a mobile node can be sent to another mobile node when they are within range. Another difference is data distribution. In a static WSN, data can be distributed using fixed routing, while in a mobile WSN, dynamic routing is used. Challenges in mobile WSNs include localization, deployment, navigation, self-organization, energy, control, coverage, maintenance, and data processing.

6.1.2 WSN Modes of Operation

After the network is set up, WSNs may operate in one of these modes of operation:

- The WSN can operate with periodic sensing, where sensors acquire data from the environment at specific time intervals and route it to the sink.
- The WSN can operate on demand; that is, sensors might be idle and wait for the sink to issue instructions.
- The WSN is an event-driven WSN where sensors only acquire data upon occurrence of specific events.

6.1.3 WSN Applications

Sensor networks may consist of many types of sensors such as seismic, thermal, magnetic, infrared, visual, acoustic, and radar, which are able to monitor a wide variety of ambient conditions such as the following [151]:

- Temperature
- Humidity
- Vehicular movement
- Lightning condition
- Pressure
- Soil makeup
- Noise levels

- The presence or absence of certain objects
- Mechanical stress
- Other characteristics such as speed, direction, and size of an object

Sensor nodes can be used in new application areas. We categorize the applications as follows:

6.1.3.1 Industrial Control and Monitoring

A large industrial facility has a small control room that has indicators and displays which describe the state of the plant, as well as input devices that control actuators in the physical plant (valves, heaters, etc.) that affect the observed state of the plant. Significant cost savings may be achieved if inexpensive wireless means were available to provide communication between the sensors and the control room. A WSN of many nodes, providing multiple message routing paths of multihop communication, can meet these requirements [4].

An example of wireless industrial control is the control of commercial lighting. A further example is the use of WSN for industrial safety applications to detect the presence of poisonous, noxious, or dangerous materials, providing early detection and identification of leaks or spills of chemicals or biological agents before serious damage can result [152].

6.1.3.2 Security and Military Sensing Applications

WSNs represent a part of military control, communications, intelligence, and surveillance systems. The rapid deployment and self-organization characteristics of WSNs make them very promising for military use [152]. Military sensing applications such as monitoring friendly forces, equipment and ammunition, battlefield surveillance, critical terrains, approach routes, and targeting.

6.1.3.3 Intelligent Agriculture and Environmental Sensing Applications

An example of the use of WSNs in agriculture is the rain gauge. Large farms and ranches may cover several square miles, and they may receive rain only sporadically and only on some portions of the farm. Irrigation is expensive, so it is important to know which fields have received rain, so that irrigation may be omitted, and which fields have not been and must be irrigated. Such an application is ideal for WSNs.

6.1.3.4 Health Monitoring Applications

The health applications include integrated patient monitoring, diagnostics, telemonitoring of human physiological data, drug administration, monitoring the internal processes of insects, and tracking and monitoring doctors and patients inside a hospital [153–158].

6.1.3.5 Home Automation and Consumer Electronics

As technology advances, smart sensor nodes and actuators can be buried in appliances such as vacuum cleaners, refrigerators, and microwave ovens. Smart sensors allow end users to manage home devices locally and remotely more easily [159,160].

6.1.3.6 Other Commercial Applications

Other commercial applications such as monitoring material fatigue, managing inventory, constructing smart office spaces, monitoring product quality, environmental control, robot control, interactive toys, interactive museums, monitoring disaster areas, factory process control and automation, machine diagnosis, transportation, and instrumentation of semiconductor processing chambers, rotating machinery, wind tunnels, and anechoic chambers [154,161–163].

6.1.4 Factors Influencing Sensor Network Design

WSN design is influenced by different factors such as fault tolerance, scalability, production costs, and hardware constraints [139]. They are explained as follows:

6.1.4.1 Fault Tolerance

Fault tolerance is defined as the ability to sustain sensor network functionalities without an interruption due to sensor failures. Some sensors may fail due to physical damage, power, or environmental interference. The network should not be affected by this failure [164].

6.1.4.2 Scalability

Depending on the application, the number of sensors ranges from hundreds to thousands or millions. Schemes and protocols must have the ability to work with this large number. In addition, the node density, depending on the application, ranges from few to few hundred sensor nodes in the area where they are deployed. For machine diagnosis applications, the node density is around 300 sensor nodes in a 5×5 m^2 region, and the density for the vehicle tracking application is around 10 sensor nodes per region. In general, the density can be as high as 20 sensor nodes/m^2 [163].

6.1.4.3 Production Costs

The cost of a node is an important parameter in controlling the total cost of the networks. We must guarantee that the cost of each sensor node has to be kept low. State-of-the-art technology allows a Bluetooth radio system to be less than $10. As a result, the cost of a sensor node is a very challenging issue given the limited amount of functionalities that come with a price of much less than a dollar [165].

6.1.4.4 Hardware Constraints

The node is formed from four basic parts, as shown in Figure 6.2: a sensing unit, a transceiver unit, a processing unit, and a power unit. The node may have additional components depending on the application, for example, a location position system, a mobilizer, and a secondary power supply. Sensing units consist of two subunits: sensors and ADCs. The processing unit consists of a small storage unit and manages the processes that make the sensor work. A transceiver unit connects the sensor to the others. The most important unit is the power unit, which supplies all the units with the required power [139].

Node power is a great challenge; it is a scarce resource due to the limited size of the node. It is possible to increase the lifetime by using solar cells, which is an example of a technique used for energy scavenging.

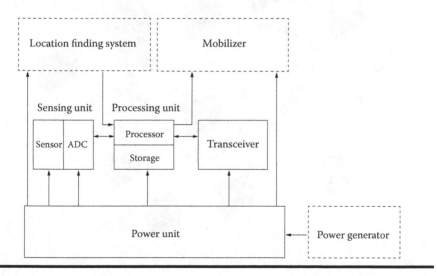

Figure 6.2   Sensor node parts.

The transceiver unit may be an optical device or a radio frequency (RF) device. RF communications require modulation, band pass, filtering, demodulation, and multiplexing circuitry, which make them more complex and expensive [161].

Most of the sensor network requires the knowledge of position. It is assumed that the sensor will have a global positioning system (GPS) unit that has at least 5 m accuracy [166,167]. It is considered that equipping all nodes with a GPS is not viable, so one approach would be to use a smaller number of nodes that have GPS which would help the other nodes find their locations [168].

6.2 Unattended WSN

Unattended WSN (UWSN) is a class of WSN. Both WSN and UWSN have the same features except that the sink does not exist all the time in UWSN. The concept of UWSNs with a mobile sink, as shown in Figure 6.3, is very suitable for some environments and applications. In some applications, the region under

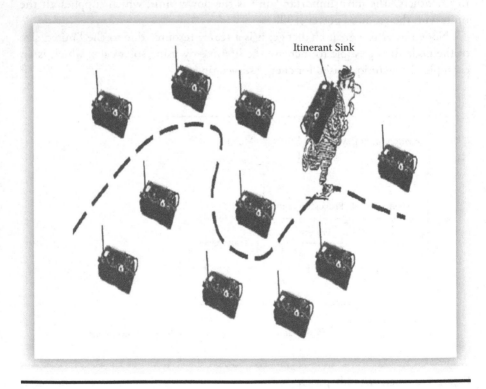

Figure 6.3   Unattended wireless sensor network with mobile sink.

consideration may be too far from the base station (destination), so sending data through multihops may cause a problem in security or increase the energy consumption of the nodes close to the base station, thereby reducing the lifetime of the network. In this case, a mobile sink is the preferred solution [8].

In general, the security requirements in UWSNs include integrity, confidentiality, authenticity, and availability of generated data. Encryption is one straight way of achieving data security, but it comes with computational cost. Public key cryptography solves most of the data security issues, and it is considered as computationally expensive in the past. Thus, for a long time, research had been focused on symmetric key cryptography until only recently [169–171] when it was proven that public key encryption is feasible using low-cost sensors. According to the results in [172] and [173], wireless communication in sensors causes more energy loss than performing computations for encryption.

A UWSN has distinctive characteristics in that it operates without the supervision of a sink and stores accumulated data in local sensors' storage for a long time. This has become attractive to a new kind of adversary: a mobile one [174,175] that can stealthily learn a network's strategies, compromise sensor nodes, and ultimately take control of them. Moreover, it is able to roam the network from node to node, corrupting and taking over one and then moving on to another. Research into UWSN strategies to counteract the effects of adversaries is as urgent as it is timely.

6.2.1 Advantages of UWSN

There are many advantages of UWSNs that justify their utilization in all these different aspects of life. Following is a list of the advantages of UWSNs:

6.2.1.1 Robustness/Ability to Withstand Rough Environmental Conditions

Because of their shrinking size, their ability to communicate through a lot of materials, and the possibility to cover the particular nodes in robust cases, UWSNs can be used in a wide variety of environments. They are designed to defy harsh weather conditions. That is one of the reason they are already used for things like forest fire detection or seismic monitoring.

6.2.1.2 Ability to Cover Wide and Dangerous Areas

In many areas, infrastructural issues and economic considerations prevent wired networks from being used. For example, setting up a wired network on a battlefield would obviously be useless. UWSNs can fill this gap because of their lack of infrastructure and their low setup costs.

6.2.1.3 Self-Organizing

With the abilities of network discovery and multihop broadcast, UWSNs are able to self-organize in small amounts of time when set up. This is interesting, as we only need to turn on the system and the rest is organized by the network itself.

6.2.1.4 Ability to Master Node Failures

UWSNs are able to overcome node failures (dead or destroyed). This can be done by changing the routing path. If, for example, during war, an enemy destroys some sensor nodes, this will not affect the whole network.

6.2.1.5 Mobility of Nodes

Node mobility has many advantages in the UWSN field. For instance, node mobility can be used in vehicle tracking applications, self-organizing, and energy harvesting. Modern UWSN protocols and architectures are able to handle node mobility in term of shifting in position and to maintain routing.

6.2.1.6 Dynamic Network Topology

UWSNs are able to have a dynamic network topology, which means that the topology is variable and determines the neighbor relationships to be maintained by the nodes. For example, if a cluster head in the topology drops out, another sensor can jump in and take the place of that cluster head, which leads to a change of the topology.

6.2.1.7 Heterogeneity of Nodes

The fact that the monitored data of the sensors is first converted into digital signals and then transmitted benefits the situation that a special UWSN can contain a variety of different sensors in one network. Every node can also have multiple different sensors implemented on it.

6.2.1.8 Multihop Communication

Since UWSN has a large number of densely deployed sensor nodes, the nodes are very close; multihop is expected to save more power than a single hop communication. Therefore, that transmission power can be kept small. Multihop communication can also effectively overcome some of the signal propagation effects experienced in long-distance wireless communication.

6.2.1.9 Unattended Operation

Designed and configured correctly, UWSNs are able to work unattended. This saves working time and minimizes the effort that has to be made to administer these systems. This advantage is probably very interesting for home applications, where nontrained customers want to benefit from the system with minimal effort.

6.2.2 Disadvantages of UWSN

On the other hand, the UWSNs have some disadvantages:

6.2.2.1 Limited Energy Resources

With the absence of a fixed infrastructure, wireless sensor nodes are forced to manage the small amounts of battery-provided power they have carefully. This limits their computational power and memory size and prevents them from using the full bandwidth due to higher energy costs. Working only on battery power also means that after a certain life span, a sensor node will die because the battery is empty.

6.2.2.2 Lower Data Rates

One of the biggest problems of wireless networks in general is the low data rates. The amount of data that can be transmitted in a period of time depends on the frequency that is used. A higher frequency results in higher data rates but at the same time causes more interference issues. This leads to the fact that wireless networks cannot be as quick as their wired brothers.

6.2.2.3 Communication Failures

Wireless networks have a higher error rate than their wired counterparts. They use electromagnetic waves to transmit packets, and these waves can be affected by phenomena like reflection, refraction, diffraction, or scattering. These phenomena can fragment or garble the package and, that way, produce error in transmission.

6.2.2.4 Security Issues

Wireless networks in general are much easier to attack from the outside than wired systems are. The wireless channel is accessible to unwanted listeners, and several passive and active attacks can be conducted. Methods like encryption are also limited by the energy resources that tend to be small in WSNs, which strengthens the problems.

6.2.3 Applications of UWSNs

Sensor nodes are densely deployed either very close or directly inside the phenomenon to be observed. Thus, they work unattended in locations such as the following [8]:

- Busy intersections
- Inside a large machine
- The bottom of an ocean
- Inside a twister
- At the ocean's surface during a tornado
- Chemically or biologically contaminated field
- Battlefield beyond the enemy lines
- Home or a large building
- Large warehouse
- Attached to animals
- Fixed on fast moving vehicles
- Drain or river moving with current

From the abovementioned applications, it is clear that the sensor nodes can work in harsh environments, under high pressure, under extreme heat or cold, and in a noisy environment. Applications of UWSNs as mentioned in [9] and [12] include the following:

- Network of nuclear emission sensors deployed to monitor any potential nuclear activity
- Underground sensor to monitor the troop movements
- Airborne sensor network tracking fluctuations in air turbulence and pressure to detect enemy aircrafts
- Monitoring system to detect poaching in a national park
- Submarine sensor network to track the movements of animals
- Monitoring underground or underwater oil pipelines to detect the leakage of oil
- A WSN mounted on a tree to monitor firearm discharge or illegal cutting of the trees
- A sensor network deployed in the international border to monitor illegal crossings
- A new DARPA-initiated program called LANdroids [11], in which nodes are located in hostile environments to collect critical information until the soldiers approach the network

In general, UWSNs are needed in places where nodes are deployed in hostile or inaccessible environments, as shown in Figure 6.4. Sensors may be deployed in

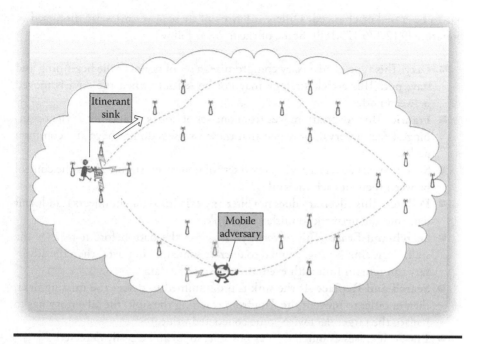

Figure 6.4 Unattended wireless sensor network with mobile sink and adversary compromising nodes.

hostile environments where the sink is not secure; if the sink gets compromised, then potentially valuable data can be lost. In environments where the number of sensors deployed is large, one sink may not be enough to cover the entire area. In some environments, the sink may be physically present in the network but needs to conserve its energy by switching itself off periodically [12]. The size of the deployment area, inaccessibility, and difficulty for the sink to be hidden motivates the need for UWSNs.

6.2.4 Mobile Adversary

Initial investigation on mobile adversary was done in [176]. This adversary can compromise a fixed number of nodes in a round and migrate to a new set of nodes in the next round. Given enough rounds, the adversary can compromise all the nodes in the network for which the security of the network will be undermined. Proactive cryptography techniques can provide security against a mobile adversary [177,178].

In UWSNs, adversaries are the ones that try to modify or destroy the data. Adversaries are mostly classified into two types: proactive and reactive. Proactive adversary starts compromising the nodes before it identifies the target node. Reactive adversary remains inactive until it locates the target node where the data has to be erased or changed. Different flavors of these adversaries are discussed in literature [9,12,176,179,180]. Some of them are as follows:

- **Lazy.** This type of adversary compromises a set of nodes at the beginning and stays put. This attack strategy may not be sensible when the data is moved between nodes.
- **Frantic.** This adversary moves from one set of nodes to another at the beginning of each interval, provided that these two sets do not have any common nodes.
- **Smart.** Smart adversary selects two sets of nodes and simply alters the control between them at each interval.
- **Polluter.** This adversary does not alter any existing data but injects fraudulent data into the network to mislead the sink.
- **Search-and-Erase.** This adversary can erase the data before it reaches the sink. If the sink is programmed to tolerate some missing data, then the adversary can remain in stealth even after erasing the data.
- **Search-and-Replace.** If the sink is programmed to detect the missing data, then in order to prevent the data from reaching the sink, the adversary has to replace the target data with some concocted value.
- **Eraser.** It erases as much data as possible, so it can be easily detected.
- **Curious.** This aims to learn as much data as possible. It is a read-only-type adversary; it does not erase the data. This type of adversary may be interested in learning some specific data which might be of high importance.

Most of the adversaries remain in the stealth mode. In a few cases, based on the goals of the adversaries, they choose not to remain undetected. However, the adversary tries to remain undetected all the time. Since the network is unattended most of the time, it is also possible for the adversary to physically destroy the sensors.

6.2.5 Security Goals

While dealing with security in UWSNs, we mainly focus on achieving some or all of the following security goals [8]:

- **Forward Security.** This is a critical parameter in UWSNs, so that even if the adversary can compromise the current key, it is infeasible for it to generate the previous keys using the current key. The adversary cannot even forge the authentication tags for the data generated and authenticated before compromise.

- **Backward Security**. The compromise of a secret key at any time must not lead to the compromise of the secret key in the future, so that if the adversary gets the current secret key of a sensor, the adversary cannot decrypt the data generated after compromise.
- **Data Confidentiality**. Confidentiality emphasizes that the data must be inaccessible and unreachable for an unauthorized user. In UWSNs, data in the sensors should be encrypted in a way that it can only be read by the sink.
- **Data Integrity**. Data integrity is a defense against unauthorized alteration of the data. Data integrity can be achieved if the network can detect any manipulations done to the data by unauthorized parties, i.e., insertion, substitution, and deletion. Data integrity protects against the "search and replace" and "search and erase" types of adversaries.
- **Data Authentication**. Authentication applies to both nodes and data. It ensures the identity of the node with which it is communicating, i.e., the two communicating parties can identify each other. Information delivered through the network should be authenticated with respect to the generation time, date of origin, origin etc.

6.2.6 Security Challenges

For all the security schemes, we need to consider resource constraints such as the storage and communication overhead incurred by the schemes on the nodes. In UWSNs, since the nodes are deployed in hostile and unattended environments, they become an easy target for an adversary. As mentioned in [181], the first question to be asked while designing a security scheme for UWSNs would be "How can a disconnected sensor network protect itself from an adversary?" The nodes in the network should be able to prevent the adversary from learning the sensed data even though the adversary can break into the node and learn all its secrets. In general, the greatest challenge in UWSNs is to secure the data long enough until the arrival of the sink.

Tamper-resistant hardware [182] can be used to avoid many attacks in the network, but they add significant cost to the sensor hardware and are thus not a practical approach. Current message authentication code schemes and signature schemes are based either on a secure hardware or on a trusted online third party (sink); these techniques are not feasible in the case of UWSNs.

Therefore, the security schemes must be inexpensive (in terms of both hardware and computation), simple, and efficient. The authentication schemes designed should hide the data origin, data content, and time of data collection [11]. Another challenge with the UWSNs would be finding a way to combine lightweight cryptographic schemes involving node collaboration with bandwidth-conscious data migration methods [174].

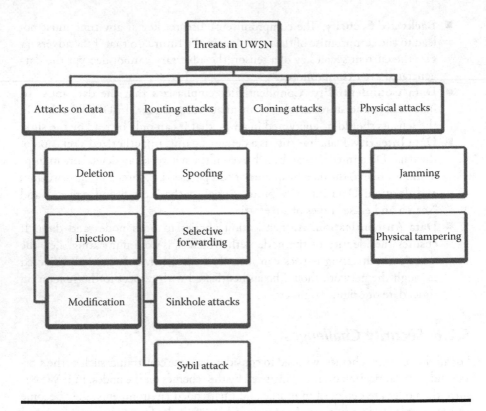

Figure 6.5   Threats in unattended wireless sensor networks.

6.2.7 Possible Attacks on Nodes

To appreciate the challenge of securing a network, all the possible threats should be considered. Threats in UWSNs [12,183,184] can be broadly classified into four categories, as shown in Figure 6.5.

Chapter 7

Cooperative Hybrid Self-Healing Randomized Distributed Scheme for UWSN Security

7.1 Introduction

Unattended wireless sensor networks (UWSNs) are operated in hostile environments without continuous supervision by a trusted data sink, so these networks face the risk of being compromised by threats and adversaries. This chapter proposes a cooperative hybrid self-healing randomized distributed (CHSHRD) scheme, where there is a new mechanism to enhance the confidentiality of the data collected by the network. It uses both types of peers, reactive and proactive, to enhance data reliability and secrecy. CHSHRD helps the unattended compromised sensors, such that they can self-heal and restore their backward secrecy. The compromised sensors will ask for help from their best-qualified neighbors. After getting help, the sensors will generate a new secret key unknown to the adversary, and they will regain their secrecy. In addition, the sick sensor will dispense the parts of its data between its qualified neighbors, which will keep the data protected from eavesdropping and enhance data reliability.

This chapter also presents a powerful, realistic, and agile adversary model and shows how the CHSHRD scheme can result in the sensor regaining secrecy and achieving high data reliability, in spite of the adversary's efforts to do the contrary. The estimation of the CHSHRD relies on both theoretical analysis and simulation.

123

The results showed that the hybrid scheme provides better guard than other schemes that use either proactive peers or reactive peers.

The rest of this chapter is organized as follows: In Section 2, we will give an overview of UWSN security challenges. Section 3 presents the UWSN system model. The proposed CHSHRD scheme is described in Section 4. Section 5 presents numerical results and discussions. A summary of this chapter is presented in Section 6.

7.2 Overview of UWSN Security Challenges

UWSNs have become the subject of attention in the security research community [9,10,185,186]. Unattended sensors are deployed in a hostile and harsh environment. In this environment, the sensors are easy targets for compromise, and the data collected can be erased, altered, substituted, or even damaged. Sensor compromising is a common threat because sensors are often mass-produced commodity devices without any secure hardware or tamper-resistant components [10,17]. UWSNs are appropriate for a wide range of applications such as introducing a military UWSN application for border surveillance, target acquisition, and situational awareness [187,188]. UWSNs pose many challenges in security; a mobile adversary, which roams, can compromise and release sick sensors behind. Another challenge is data survival; data "survives" if they are stored locally within the sensor until they are transmitted to the data sink at its next visit [9].

In UWSNs, the data sink is absent most of the time. Thus, the sensors must collect the data and store it, and wait to off-load these data to the data sink. The incapability of off-loading data in such real-time fashion will expose the potentially sensitive accumulated data to certain risks. When the data sink visits the network, it will collect stored data from all sensors. To achieve the UWSNs feature, a data sink must be moveable and sporadically visit the network. However, in the time between two visits of the sink, the adversary can step by step compromise sets of sensors within its range and subvert the data security and integrity [9]. The adversaries can be divided into reactive and proactive adversaries [11,12,16]. The reactive adversary stays close to the network, waiting for communications between sensors. As soon as it notices any sensor trying to send data to another, it reaches the source of the data and then tries to attack and compromise it. The proactive adversary compromises all sensors lying in its range, whether they send data or not.

The sensor can be compromised from two different points of view: physical and logical [9,189,190]. For physical compromise, the adversary will physically plug into the sensors. After that, it will try to hack the sensor's circuits, so that it can cause a memory dump. The adversary can also use the sensor to act as its agent in the sensor network. This chapter does not focus on this kind of compromising, but it assumes that the sensor is enclosed in a reliable and self-destructive envelop that physically protects it. It is assumed that sensors are logically compromised. Once a sensor is logically compromised, the adversary picks up keys and cracks the sensor functions, so that it can decrypt the collected data of this sensor [9,12,190]. The adversary can compromise a set of sensors within its range and within a particular interval that may be shorter than the time between two consecutive visits of the data sink. Therefore, the adversary can subvert the whole unattended network as it moves between sets of sensors before the next visit of the data sink. This type of adversary is known in cryptographic literature as the mobile adversary [176].

Zooming into the problem of data secrecy, consider that a sensor is compromised for a certain duration. The data collected by this sensor can be partitioned into three groups: (1) before, (2) during, and (3) after the compromise. Obviously, nothing can be done about the secrecy of the data for Group 2 since the sensor is already under the control of the adversary. There are two challenges: For Group 1, it is how to ensure that the precompromise data will not be exposed once a sensor is compromised. This is called forward secrecy (FSe). This can be obtained by periodic key updates. For Group 3, it is how to keep the postcompromise data from being exposed. This is called backward secrecy (BSe). This means that the data must remain a secret even though a compromise has occurred. The following text focuses on the confidentiality of the data collected in Groups 1 and 3. Another challenge is data reliability, i.e., how to achieve data survival.

The related works that deal with the aforementioned challenges are as follows: In [181], the authors propose a technique that allows unattended sensors to recover by asking for help from the reactive peers. In [17], the authors built a self-healing scheme that allows sensors to recover by using proactive peers. In [12] and [191–193], the authors propose a data replication strategy, wherein multiple copies of sensed data are sent to multiple sensors until the sink receives a copy. However, this strategy increases the overhead communication. In [19], the authors assume a more resource-efficient scheme; they move the data from one sensor to another. The authors in [194] present an adversary model which traces back an event source by sequentially moving to the intermediate senders of the messages corresponding to the event source. In [195], the authors discuss location privacy in sensor networks in a situation involving a strong and global eavesdropper.

None of these works considered a hybrid scheme which selects the best neighbors (sponsors) from reactive and proactive peers that have the lowest compromising probability. Therefore, this chapter proposes a cooperative hybrid self-healing randomized distributed (CHSHRD) scheme that combines the benefits of both reactive and proactive peers. The selected sponsors are asked to do two functions. The first function is to dispense data parts among them. In the second function, the sick sensor asks the sponsors for a random contribution to generate a new and unknown secret key; the sick sensor will use this new key to self-heal and restore its backward secrecy. Moreover, a powerful, realistic, and agile adversary model is proposed in this chapter. Several results are carried out to validate the outcomes of the proposed CHSHRD in terms of probability of BSe to be compromised, compromising probability, and the probability of data reliability. Results show that the proposed hybrid scheme provides better data protection than other schemes that use either proactive or reactive peers alone.

7.3 UWSN System Model

7.3.1 Network Model

The proposed network model shares some characteristics as in [12] and [16]. A homogeneous UWSN contains N sensors; they are uniformly distributed over a region under consideration. A UWSN can be formulated as an undirected graph $G(V, E)$, where V refers to the sensors (vertices of the graph) = $\{v_1, v_2, v_3,...., v_N\}$ that are connected by undirected edge. The undirected edge is an edge that has no arrow; there is no head or tail, and E refers to the edge set (edges of the graph) = $\{e_1, e_2,...,e_M\}$, as shown in Figure 7.1 [196]. Each sensor has a unique identifier ID and is capable of sensing an area around itself called the sensing region. Each sensor s_i has a neighbor set N_{Bi}.

The data sink is itinerant. It periodically visits the network for backing up the data stored, clearing the memories, and updating the cryptographic material of each sensor. The sink could be an airplane flying close to the ground or an army troop approaching the network. The time between two visits, (T), is divided into rounds. In each round r, the sensor s_i generates data denoted as d_i^r. It stores this data locally and waits for an authorized mobile sink to off-load its data. Each sensor can perform one-way cryptographic hashing, symmetric key encryption based on a master symmetric key, and pseudo-random number generation initialized with a unique secret seed shared with the sink. It is assumed that the itinerant sink is a trusted party that cannot be compromised. In addition, it will reinitialize the secret keys and reset the round counters when it visits the network [197].

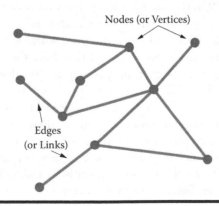

Nodes (or Vertices)

Edges (or Links)

Figure 7.1   Undirected graph.

7.3.2 Adversary Model

The UWSN may be attacked by different types of adversaries. In general, the adversary may be read-only or read–write. The former compromises and reads the memory and leaves no evidence behind. In contrast, a read–write adversary can read, delete, or modify and/or present its own fake data. Additionally, a more powerful adversary, referred to as a mobile adversary, can roam the network from one set of sensors to another, making such attacks more difficult to detect and prevent. This type of threat could be overcome by tamper-resistant devices [182], but this solution adds significant costs to the sensor hardware [198,199]. This chapter focuses on a powerful read-only mobile adversary that prefers roaming in the UWSN and compromising k sensors at each round. During an attack round, the k sensors under attack are unresponsive for a period of time (round attack period); during this time, the adversary can perform its attacking functions. It aims to remain stealthy, so it can attack many times. Our adversary model resembles that in [9] and [181] and has the following capabilities:

■ **Compromise Power**. The adversary can compromise up to $k < N$ sensors.
■ **Minimal Disruption**. The adversary is skillful; it can leave the sensor node in the same state as before the attack without disruption of regular node operation, so it will be undetected.
■ **Network Knowledge**. The adversary knows the topology of the network.
■ **Key-Centric**. The adversary is interested in learning the secret keys and cracking the sensors' functions, so it can decrypt the data.
■ **Periodic Operation**. At each round, the adversary picks up a set of k sensors to compromise.
■ **Local Eavesdropping**. The adversary is unable to make a global attack; it can only eavesdrop on communications in currently compromised sensors.

In this chapter, the adversary has the following characteristics:

■ The adversary moves randomly with a constant velocity (AV).
■ The adversary has an adversary range (AR).
■ Velocity, direction, and range will control the number of selected sensors to be compromised.
■ Its next movements in the next rounds are unpredictable and untraceable, so it is infeasible to predict its next attack location, but because of the unresponsive period, one can detect its previous path.

7.3.3 Data Secrecy

In sensor networks, the sensors use secret keys generated by their pseudo-random number generator (PRNG) in addition to the encryption functions to encrypt the generated data at the transmitting node and vice versa at the destination node. If an adversary compromises a sensor, the sensor becomes sick. The adversary can learn the sensor's secret key and functions, and it can decrypt the compromised the sensor's data. If the sensor computes a new and unknown secret key, it can self-heal. The adversary cannot decrypt the intercepted sensor's data even if it can crack the sensor's decryption function.

7.3.4 Sensor States

In the following, we summarize the sensor states that are considered in this thesis.

- **Occupied Sensor (*Os*).** The sensor is occupied by an adversary. The sensor is unresponsive; current and next-round keys are all eavesdropped on.
- **Sick Sensor (*Ss*).** The sensor is sick if the adversary has just released it. It is still compromised, but it becomes responsive again. The current and next keys are still known by the adversary.
- **Healthy Sensor (*Hs*).** The sensor is healthy if it is unknown to the adversary. Either it was not compromised or it was healed by the proposed scheme.

Figure 7.2 shows the sensor state transition diagram. A healthy sensor remains healthy until an adversary compromises it. The occupied sensor cannot become healthy without becoming sick first. The sick sensor can become healthy if and only if it has at least one contribution from a healthy sensor. The abbreviations used in this chapter are summarized in Table 7.1.

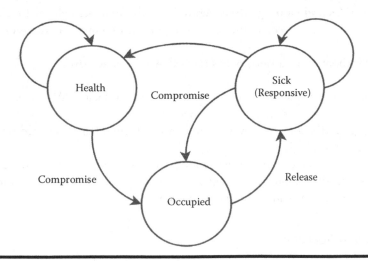

Figure 7.2   Sensor state transition diagram.

Table 7.1   List of Abbreviations

N	Total number of sensors in the network
s_i	Sensor i
r	Round index
t	No. of cooperative peers for each sensor
t_1, t_2	No. of selected reactive and proactive peers, respectively
k_i^r	Secret key of s_i at round r
k	Adversary capability
$H(.)$	One-way collision resistant hash function
v	Number of rounds between successive sink visits

7.4 CHSHRD Scheme

CHSHRD is used to enhance data confidentiality; it employs both proactive peers and reactive peers to enhance backward secrecy and data reliability. In the following, the adversary occupies and releases a sensor s_i at rounds r_1 and r_2, respectively ($r_2 > r_1$).

A simple way to provide FSe for s_i is to update its zero round secret key k_i^0 by applying one-way collision-resistant hash function $H(.)$. This function can be used to produce the current round secret key k_i^r from the previous round secret key k_i^{r-1} as follows:

$$k_i^r = H\left(k_i^{r-1}\right).k_i^0 \tag{7.1}$$

The benefits of hash function is its way of operation; it cannot work in the reverse direction, so the adversary cannot derive the previous round's secret keys before compromising. Therefore, FSe is guaranteed for any round ($r < r_1$).

With respect to BSe, the adversary can hold all secret keys $k_i^r \; \forall \; r \in [r_1, r_2]$ and crack all the encryption and decryption functions. Therefore, for any round r' ($r' > r_2$), the adversary can drive all future secret keys $k_i^{r'}$ and decrypt all future data generated; this means that the BSe is not guaranteed. Therefore, the encryption can only guarantee FSe, but it is not enough to achieve both BSe and data reliability. The proposed CHSHRD scheme uniquely differs from the previously mentioned related works in the following ways:

■ It employs both types of peers, proactive and reactive, to ensure both backward secrecy and data reliability rather than use a single type of peer as in [17] and [181].

■ Each sensor in the network has a compromising probability function of its physical conditions, such as the sensor's location, adversary's location, and adversary's range rather than the use of predefined, fixed, and nonreal compromising probability as in [197].

■ It also reduces the overhead of memory storage and communication by using the minimum number of peers provided that the required security level is achieved rather than using a fixed number of six peers as in [181].

7.4.1 CHSHRD Scheme Steps

The proposed scheme can be summarized using the following steps:

Step 1: Network Initialization. First, the sink picks a secure hash function $H(.)$ and a symmetric master encryption key k_m based on a PRNG initialized with a unique secret seed. Second, the sink preloads each sensor s_i by a hash function $H(.)$ and an initial secret key computed by the sink as $k_i^0 = H(k_{m\|i})$. At the end of each round, each sensor can update its secret key using $k_i^r = H(k_i^{r-1}).k_i^0$.

An important matter is how to securely manage the keys among the sink and the sensors when it leaves and comes again to the network after many rounds of its absence. Hence, the sink preloads all the sensors in the network with a single and unique secret symmetric master key K_m and with a PRNG, which initializes with a unique secret seed shared with the sink. Therefore, the sink can compute any secret key for any sensor i at any round r by using $H(.)$, K_m, i, and r.

Step 2: Encrypted Data Generation. At each round r, sensor s_i will sense and generate data unit d_i^r through three successive stages. The first stage is to generate a message authentication code (MAC) as follows:

$$MAC_i^r = H\left(d_i^t \big\| k_i^r\right) \tag{7.2}$$

The second stage is to generate plain text data:

$$PLtext_i^r = \left\{ d_i^r \| MAC_i^r \| r \| s_i \right\} \tag{7.3}$$

The third stage is to encrypt the plain text data:

$$ENtext_i^r = Enc(k_i^r, PLtext_i^r) = Enc\left(k_i^r, \left\{ d_i^r \| MAC_i^r \| r \| S_i \right\}\right) \tag{7.4}$$

where $ENtext_i^r$ is the data generated by s_i during round r to fully guarantee FSe.

Step 3: Data Part Generation. In this step, an erasure code A(m, n) with maximum security without redundancy ($n = m$) is used to encode the encrypted data into n fragments, each having a size of $1/m$ of the original encrypted data.

An example of such erasure coding scheme is Reed Salomon (RS) codes [200], which are used to encode $ENtext_i^r$ into n parts, $p_i^r = \left\{ p_{i,1}^r, p_{i,2}^r \cdots p_{i,n}^r \right\}$.

Step 4: Sensor Vulnerable Level Calculation. In this step, it is desired to calculate the vulnerable level for each sensor. Our calculation is based on the fact that for the adversary to attack some sensor nodes during the attack round, it requires the removal of these sensor nodes from the network for some time to perform attacking functions. During this time, the sensor nodes are unresponsive, and the removal of these sensor nodes can be observed by their neighbors [201,202]. These observations are based on the fact that, for any two neighbor nodes a and b, they must communicate (meet) periodically within time λ which is shorter than the round time. If this time λ expires and nodes a and b did not re-meet, then node a will observe the absence of node b and send an alarm message to its neighbors. If node b does not prove its presence within time δ, node b is considered unresponsive and node a will consider that node b was captured. After the adversary leaves the captured sensor nodes, they become responsive again (hence, the adversary has minimal disruption, and it removes the sensor node after the attack in the same state it was before the attack), and they can perform any measurements. However, the capture detection algorithm considers these sensor nodes as sick sensors.

The capture detection algorithm makes use of these observations and floods the network with an informative message, such that it asks node a to inform node b and its neighbors that it is a sick node and that they must revoke it from the network for security reasons [202]. In addition, the capture detection algorithm distinguishes, identifies, and clusters these revoked sick sensors. It provides information about their locations and identifications, so the coordinates that represent the boundaries of the square surrounding the cluster of the revoked sensors can be found, as shown in Figure 7.3.

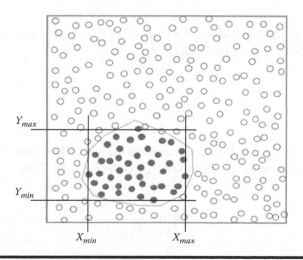

Figure 7.3   Boundaries of the square surrounding the cluster of the revoked sensors.

Using the above information, the coordinates of the cluster center (X_c, Y_c) and the radius of the cluster R_c are computed as follows:

$$X_c = \frac{X_{max} + X_{min}}{2} \tag{7.5}$$

$$Y_c = \frac{Y_{max} + Y_{min}}{2} \tag{7.6}$$

$$R_c = \frac{(X_{max} - X_c) + (Y_{max} - Y_c)}{2} \tag{7.7}$$

where (X_{min}, X_{max}) are the maximum and minimum coordinates of the revoked sensors along the X-axis and (Y_{min}, Y_{max}) are those for the Y-axis.

The coordinates of the cluster center represent the previous location of the adversary, and the radius the adversary's range. According to the past adversary location and its range, the sensors can be divided into two groups: the first group is the sensors within the cluster (i.e., adversary range) and they are called sick. The second group is the sensors outside the cluster, and they are healthy or not revoked sensors. Now the sensor vulnerable level will depend on the distance between the sensor's location (X_i, Y_i) and the adversary's location (X_c, Y_c). This distance can be computed as follows:

$$D = \sqrt{(X_i - X_c)^2 + (Y_i - Y_c)^2} \tag{7.8}$$

If $D < R_c$, the sensor is sick and exists within the cluster area, and its compromising probability is given by Equation 7.9:

$$P_i = 1 - \frac{D}{R_c} + \frac{P_{th} * D}{R_c} \tag{7.9}$$

On the other hand, if $D < R_c$, the sensor is healthy and exists outside the cluster area, and its compromising probability is given by Equation 7.10:

$$P_i = P_{th} * \left(1 - \frac{D - R_c}{D_{max} - R_c}\right) \tag{7.10}$$

where $P_{th} = 0.3$ is the threshold compromising probability, which is used to separate between sick and health states [197], and D_{max} is the maximum expected distance between sensor location and adversary location, which can be computed as

$$D_{max} = \sqrt{\left(x_{max} - x_{min}\right)^2 + \left(y_{max} - y_{min}\right)^2} \qquad (7.11)$$

Step 5: Node Selection Protocol. In this step, s_i will select t sponsors $(t_1 + t_2)$ from the best of its neighbors to share with them the data parts and ask them for random contributions for secret key evaluation. In [189] and [197], the authors assumed that each s_i has a compromising probability vector, $CPV_i = \{P_{i,1}, P_{i,2}, \cdots, P_{i,NBi}\}$, which reflects the security levels of the neighbors of s_i and assigned manually and independently on real network. Here, the CPV_i will depend on the sensor's location, adversary's location, and adversary's range. According to the proposed scheme, each s_i will transmit a message with its computed compromising probability P_i and its ID. The message contents depend on the value of P_i with respect to P_{th}. If s_i is sick, it has $P_i > P_{th}$ and message 1 (MS1) with the content "I need help" is sent. If s_i is healthy, it has $P_i < P_{th}$, and message 2 (MS2) with the content "I can help" is sent. Sensors will hear the messages of each other, and the compromised s_i will have two types of sponsors:

■ **Proactive Peers (Pro$_{peers}$).** The sick s_i will hear MS2 from Pro$_{peers}$ who offer their help. The interaction is initialized by Pro$_{peers}$ and then s_i will interact and choose the best t_1 peers from Pro$_{peers}$ and form the $CPV_{i,t_1}^{Pro} = \{P_{i,1}, P_{i,2}, \cdots, P_{i,t1}\}$. As shown in Figure 7.4a, healthy s_1 and s_2 transmit MS2 while the sick sensor s_i receives these messages, so they are considered as the Pro$_{peers}$ of s_i.
■ **Reactive Peers (Re$_{peers}$).** The sick s_i transmits MS1 to ask Re$_{peers}$ for help. The interaction is initialized by s_i who will choose t_2 peers from the Re$_{peers}$ and form the $CPV_{i,t_2}^{Re} = \{P_{i,1}, P_{i,2}, \cdots, P_{i,t2}\}$. As shown in Figure 7.4b, the sick s_i will transmit MS1 while s_3 and s_4 receive this message, so they are considered as the Re$_{peers}$ of s_i.

During the self-healing process, s_i will get the best qualified t sponsors $(t_1 + t_2)$ that have the lowest compromising probability. A required security level of 0.1% must be ensured to guarantee that the selected sponsors can achieve the healing task and keep the recovering probability of the original data at the required security level [197].

The sponsors are selected according to the algorithm presented below. In this algorithm, Pr_{recov} is the probability that the original data can be detected by the adversary. It must achieve the above required security level. It is clear that the higher the number of qualified neighbor peers, the lower the Pr_{recov}.

ALGORITHM USED TO SELECT THE SPONSORS

do;
select Pro_{peers};
select Re_{peers};
if length $(Pro_{peers}) \geq t1$;
$t_1 = minimum\ (Pro_{peers})$;
$CPV_{i,t1}^{Pro} = \{P_{i,1}, P_{i,2}, \cdots, P_{i,t1}\}$;
else if;
if length $(Re_{peers}) \geq t2$;
$t_2 = minimum\ (Re_{peers})$;
$CPV_{i,t2}^{Re} = \{P_{i,1}, P_{i,2}, \cdots, P_{i,t2}\}$;
$t = t_1 + t_2$
$CPV_{i,t}^{r} = CPV_{i,t1}^{Pro} + CPV_{i,t2}^{Re}$
end

$$Pr_{recov} = \prod_{j=1}^{t1} P_{i,j} \cdot \prod_{k=1}^{t2} P_{i,k}\ \forall (P_{i,j}\ \dot{E} P_{i,k}\ )\hat{I} CPV_{i,t}^{r}$$

While $(Pr_{recov} > 0.001)$

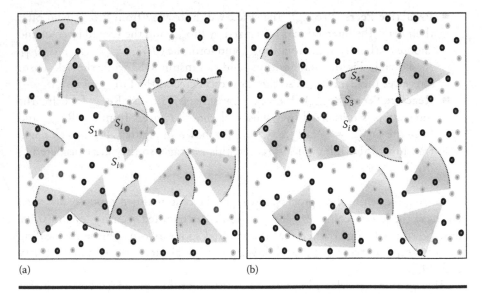

(a)   (b)

Figure 7.4   Sponsor types: (a) proactive peers, where the healthy sensors transmit MS2, and (b) reactive peers, where the sick sensors transmit MS1.

Step 6: Data Distribution. After selecting the qualified t sponsors, s_i will send distinct data parts to t sponsors. After that, the original data will be erased from s_i. This will enhance the data reliability by reducing the adversary's recovering probability; hence, the adversary must compromise all parts of data to know the encrypted data.

Step 7: Security Regaining (Self-Healing). Our goal is for a sick s_i to regain its BSe, recover from its prior compromising, and compute a new secret key. At least one contribution of secure randomness from a peer whose secret key is not compromised will achieve exactly the above goal. Using the proposed CHSHRD scheme, a sick s_i can get random contributions of secure randomness from Pro_{peers} and/or Re_{peers}. These contributions are derived by the peers using PRNG. The CHSHRD scheme gives more priority to Pro_{peers} where these peers initiate interaction to s_i, so the adversary has no information about them. On the other hand, the Re_{peers} are asked by the sick s_i for help, so the adversary may eavesdrop on their contributions. Therefore, s_i will use contributions from Pro_{peers} more than from Re_{peers}. To update the s_i secret key at the end of round r, s_i computes as follows [17]:

$$k_i^{r+1} = H(k_i^r \,\|\, c_{i,1}^r \,\|\, c_{i,2}^r \cdots \,\|\, c_{i,j}^r) \tag{7.12}$$

where $c_{i,j}^r$ is a contribution from peer j to s_i during the current round r. If a single contribution exists, we get the minimum requirement to achieve the proposed scheme goal of maintaining BSe:

$$k_i^{r+1} = H(k_i^r \,\|\, c_{i,1}^r) \tag{7.13}$$

At this point, s_i regains its security. It has a new and unknown secret key with respect to the adversary. Even if the adversary can crack the encryption and decryption functions, they become useless without knowing the new secret key. If the adversary wants to decrypt the data generated by s_i, it must occupy s_i again and learn the unknown secret key. After security is regained, there are two important issues: The first is how the sink can know which nodes contributed. The second is how the sink syncs with the healed sensor after that sensor performs rekeying during the self-healing process. Firstly, the sensor forms CPV_{i,t_1}^{Pro} and CPV_{i,t_2}^{Re} and gets the best qualified t sponsors $(t_1 + t_2)$ from them according to the "Algorithm Used to Select the Sponsors." We can thus say that the sick sensor knows well the complete information about the selected sponsors who make random contributions and it can tell the sink this information, including the sponsor IDs. Secondly, the sensor performs rekeying by using random contributions derived from the peers' PRNGs which are initialized with a unique secret seed shared with the sink. Therefore, the sink is aware of these random contributions used in the rekeying process, and it can be synced to any sensor. It can also decrypt any data stored in any sensor at any round after rekeying. The sink also has the power, memory, computational capabilities, and the same unique secret seeds used in the peers' PRNGs, so it can compute and derive any random values (contributions) generated by any peer's PRNG without the need of knowing the sensor nodes that contributed.

7.4.2 Analytical Model of CHSHRD Scheme

This subsection presents the analytical model that describes the proposed CHSHRD scheme. We follow the model presented in [190]. s_i can regain its BSe and protect its data by explicitly imploring cooperation with Re_{peers} and/or $\text{Pro}_{\text{peers}}$. Despite the similarity between the two types of peers, they are inherently different in behavior and compromising probability, as will be described in this subsection. First of all, let us define different probabilities, as given in Table 7.2.

7.4.2.1 Proactive Peers

Proactive peers, $\text{Pro}_{\text{peers}}$, will initialize and offer their help for a sick s_i. Now, assume a sick s_i at round r that wants to regain its health at the next round. According to the proposed scheme, there are only two events that can prevent it from being healed. In event (A_p), s_i does not receive MS2 from a healthy peer; in event (B_p), s_i receives MS2 from a healthy peer, but this contribution is intercepted by the adversary. Thus,

$$P(\text{Pro}_{\text{peers}}) = \frac{t_1}{n-1} \tag{7.14}$$

Table 7.2 Definition of Probabilities

Probability	Definition
p	Adversary probability of compromising
p_{comp}^r	Probability of s_i being compromised at round r and still compromised for the next rounds, $\forall\ s_i \in [Os^r \cup Ss^r]$
p_{health}^r	Probability of s_i being healthy at round r, $\forall\ s_i \in Hs^r$
p_{occu}^r	Probability of s_i being occupied at round r, $\forall\ s_i \in Os^r$
p_{sick}^r	Probability of s_i being sick at round r, $\forall\ s_i \in Ss^r$
$p_{sick}^{r,r+1}$	Probability of s_i being sick at round r and still sick in the next round

$$P(A_p) = 1 - P(\mathrm{Pro}_{\mathrm{peers}}) = 1 - \frac{t_1}{n-1} \tag{7.15}$$

$$P(B_p) = P(\mathrm{Pro}_{\mathrm{peers}}) \cdot p = \frac{t_1}{n-1} \cdot p \tag{7.16}$$

$$p_{sick.Pro}^{r,r+1} = (P(A_p) + P(B_p))^{Hs^r} \tag{7.17}$$

$$p_{sick.Pro}^{r,r+1} = \left(1 - \frac{t_1}{n-1} + \frac{t_1}{n-1} \cdot p\right)^{n \cdot p_{health}^{r-1} - k} = \left(1 - t_1 \cdot \frac{1-p}{n-1}\right)^{n \cdot p_{health}^{r-1} - k} \tag{7.18}$$

$$p_{sick}^{r} = (p_{occu}^{r-1} + p_{sick}^{r-1}) \cdot p_{sick.Pro}^{r,r+1} \tag{7.19}$$

$$p_{sick}^{r} = \left(p_{occu}^{r-1} + p_{sick}^{r-1}\right) \cdot \left(1 - t_1 \cdot \frac{1-p}{n-1}\right)^{n \cdot p_{health}^{r-1} - k} \tag{7.20}$$

$$p_{occu}^{r} = \frac{k}{n \cdot p_{health}^{r-1} + k} \tag{7.21}$$

The probability of s_i being compromised at round r and still compromised for the next rounds will be

$$p_{comp.Pro}^{r} = p_{sick}^{r} + p_{occu}^{r} \tag{7.22}$$

$$p_{comp.Pro}^{r} = \left(p_{occur}^{r-1} + p_{sick}^{r-1}\right) \cdot \left(1 - t_1 \cdot \frac{1-p}{n-1}\right)^{n \cdot p_{health}^{r-1} \, k} + \frac{k}{n \cdot p_{health}^{r-1} + k} \tag{7.23}$$

7.4.2.2 Reactive Peers

Reactive peers, $\mathrm{Re}_{\mathrm{peers}}$, are asked for help by a sick s_i. Now assume a sick s_i at round r that wants to regain its health in the next round. According to the proposed scheme, there are only two events that can prevent it from being healed. In event (A_R), there are no healthy peers from which to select t_2; in event (B_R), the selected t_2 peers' contributions are intercepted by the adversary.

Computing for the probability of event (A_R) can be done by getting the ratio of the combinations of t_2 from the sick and occupied sensors and the combinations of t_2 from all sensors:

$$P(A_R) = \frac{\binom{n.p_{comp}^{r-1} + k - 1}{t2}}{\binom{n-1}{t2}} \tag{7.24}$$

The probability of event (B_R) can be found by summing up the multiplication of the combinations of i healthy peers and the probability that the reminder (t_2-i) peers are not healthy and the adversary compromising probability of power i.

$$P(B_R) = \sum_{i=1}^{t_2} p^i \cdot \frac{\binom{n.\left(1 - p_{comp}^{r-1}\right) - k}{i}\binom{n.p_{comp}^{r-1} + k - 1}{t2 - i}}{\binom{n-1}{t2}} \tag{7.25}$$

The sum of the probabilities of two events will give the probability that the sick s_i will remain sick for the next round [190]:

$$p_{sick.Re}^{r,r+1} = \frac{\binom{n.p_{comp.Re}^{r-1} + k - 1}{t2}}{\binom{n-1}{t_2}} + \sum_{i=1}^{t_2} p^i \cdot \frac{\binom{n.\left(1 - p_{comp.Re}^{r-1}\right) - k}{i}\binom{n.p_{comp.Re}^{r-1} + k - 1}{t2 - i}}{\binom{n-1}{t2}} \tag{7.26}$$

The probability of s_i being sick will be

$$p_{sick.Re}^{r} = \left(p_{occu}^{r-1} + p_{sick}^{r-1}\right) \cdot p_{sick.Re}^{r,r+1} \tag{7.27}$$

The probability of s_i being occupied will be

$$p_{occu}^{r} = \frac{k}{n.p_{health}^{r-1} + k} \tag{7.28}$$

The probability of s_i being compromised at round r and still compromised for the next rounds will be

$$p^r_{comp.Re} = p^r_{sick} + p^r_{occu} \tag{7.29}$$

$$p^r_{comp.Re} = \left(p^{r-1}_{occu} + p^{r-1}_{sick}\right) \cdot p^{r,r+1}_{sick.Re} + \frac{k}{n \cdot p^{r-1}_{health} + k} \tag{7.30}$$

CHSHRD will use a combination of two types of peers, so we can get the advantages of both types of peers and circumvent the disadvantages. The Re_{peers} have the benefit that s_i can select and guarantee a specific number of t_2 sponsors because the initialization was done by the s_i, but it has the drawback that any action initialized by a sick s_i may be eavesdropped on by the adversary. This drawback can be settled by using the Pro_{peers} that will initialize the cooperation with the sick s_i, which will select t_1 sponsors from them, so the adversary cannot eavesdrop on this type of peers. The drawback, however, is that there is no guarantee that a sick s_i can receive any contribution from them, so s_i does not depend completely on the Re_{peers} or Pro_{peers} because all contributions from the former may be eavesdropped on, while there is no guarantee on the presence of the latter. If one of the peers does not exist, the scheme to be used is the cooperative self-healing randomized distributed (CSHRD) scheme. It is comparable to the proactive–reactive or push–pull schemes proposed in [17], [181], and [190]. According to the values of t_1 and t_2, we can define the parameter as $ratio = \dfrac{t_1}{t_2}$, which controls the choice between the CHSHRD and CSHRD schemes. The relation between the selected peers and the proposed scheme used is given in Table 7.3.

Table 7.3 Relation between Selected Peers and Proposed Scheme

Proactive Peers	Reactive Peers	Selected Peers	Relation	Ratio = t_1/t_2	Scheme
t_1	t_2	$t = t_1 + t_2$	$t_1 = t_2$	Ratio = 1	CHSHRD
t_1	t_2	$t = t_1 + t_2$	$t_1 > t_2$	Ratio > 1	CHSHRD
t_1	t_2	$t = t_1 + t_2$	$t_1 < t_2$	Ratio < 1	CHSHRD
$t_1 = 0$	t_2	$t = t_2$	–	Ratio = 0	CSHRD
t_1	$t_2 = 0$	$t = t_1$	–	Ratio undefined	CSHRD

The net compromising probability will differ according to the used scheme. For the CHSHRD scheme, it will be

$$p^r_{comp} = p^r_{comp.Pro} \cdot p^r_{comp.Re}$$

$$p^r_{comp} = \left\{ \left(p^{r-1}_{occu} + p^{r-1}_{sick} \right) \cdot \left(1 - t1 \cdot \frac{1-p}{n-1-t2} \right)^{n.p^{r-1}_{health}-k-t2} + \frac{k}{n.p^{r-1}_{health}+k} \right\}$$

$$\cdot \left\{ \left(p^{r-1}_{occu} + p^{r-1}_{sick} \right) \frac{\left(\begin{array}{c} n.p^{r-1}_{comp}+k-1 \\ t2 \end{array} \right)}{\left(\begin{array}{c} n-1-t1 \\ t2 \end{array} \right)} + \right.$$

$$\left. \sum_{i=1}^{t2} p^i \cdot \frac{\left(\begin{array}{c} n.\left(1-p^{r-1}_{comp}\right)-k-t1 \\ i \end{array} \right)\left(\begin{array}{c} n.p^{r-1}_{comp}+k-1 \\ t2-i \end{array} \right)}{\left(\begin{array}{c} n-1-t1 \\ t2 \end{array} \right)} + \frac{k}{n.p^{r-1}_{health}+k-t1} \right\}$$

$$(7.31)$$

For the CSHRD scheme, the compromising probability will be

$$p^r_{comp} = p^r_{comp.Pro}(t_2 = 0)$$

$$p^r_{comp} = \left\{ \left(p^{r-1}_{occu} + p^{r-1}_{sick} \right) \cdot \left(1 - t1 \cdot \frac{1-p}{n-1} \right)^{n.p^{r-1}_{health}-k} + \frac{k}{n.p^{r-1}_{health}+k} \right\} \quad (7.32)$$

Or,

$$p^r_{comp} = p^r_{comp.Re}(t_1 = 0)$$

$$p^r_{comp} = \left(p^{r-1}_{occu} + p^{r-1}_{sick} \right) \left[\frac{\left(\begin{array}{c} n.p^{r-1}_{comp}+k-1 \\ t2 \end{array} \right)}{\left(\begin{array}{c} n-1 \\ t2 \end{array} \right)} + \right.$$

$$\left. \sum_{i=1}^{t2} p^i \cdot \frac{\left(\begin{array}{c} n.\left(1-p^{r-1}_{comp}\right)-k \\ i \end{array} \right)\left(\begin{array}{c} n.p^{r-1}_{comp}+k-1 \\ t2-i \end{array} \right)}{\left(\begin{array}{c} n-1 \\ t2 \end{array} \right)} + \frac{k}{n.p^{r-1}_{health}+k} \right]. \quad (7.33)$$

7.5 Numerical Results and Discussions

In this section, a theoretical and a simulation model are used to validate CHSHRD and compare its results with the results of the schemes proposed in [179].

7.5.1 Theoretical Results

This subsection presents the theoretical results according to the above-mentioned analytical model. These results are obtained using N = 500, p = 0.2, and t = 6. Figure 7.5 shows the probability of compromised sensors versus the number of rounds using k = 100 and three different values of ratio, 2, 0.5, and 1. It is clear that during the first rounds, the compromising probability increases rapidly until the CHSHRD wins the game against the adversary attacks; the compromising probability is nearly fixed from round 5. It is clear also that the best result occurs at ratio = 1 because a balance between both types of peers is achieved. This is great evidence that both reactive and proactive peers are complementary in their performance.

Figure 7.6 shows the probability of compromised sensors versus the number of rounds using k = 100 and ratio = 1 for the proposed CHSHRD scheme and the stand-alone reactive and proactive schemes. The results illustrate that the CHSHRD achieves a compromised probability of 0.07 compared with 0.21 and 0.22 for the stand-alone reactive and proactive schemes, respectively.

Figure 7.7 presents the probability of compromised sensors versus the adversary capability per round using ratio = 1 for the proposed CHSHRD scheme and the

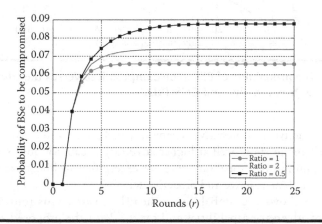

Figure 7.5 Probability of BSe to be compromised against rounds for different ratios.

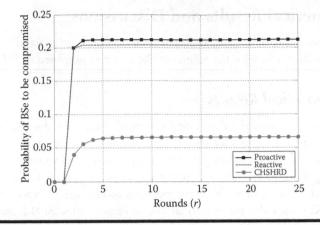

Figure 7.6 Probability of BSe to be compromised against rounds for the proactive, reactive, and CHSHRD schemes.

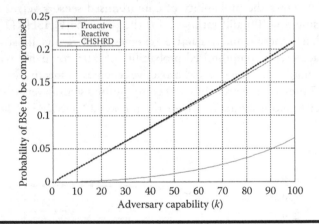

Figure 7.7 The probability of BSe to be compromised versus the adversary capability for proactive, reactive and CHSHRD.

stand-alone reactive and proactive schemes. This figure ensures the ability of the proposed scheme to stand well at higher adversary capability; for example, at $k = 100$, the compromising probability does not exceed 6.5% while it exceeds 20% for both the stand-alone reactive and proactive schemes.

Figure 7.8 shows the probability of health sensor versus rounds for different values of adversary capabilities and ratio = 1 for the proposed scheme. This figure shows how the CHSHRD scheme will interact with different values of

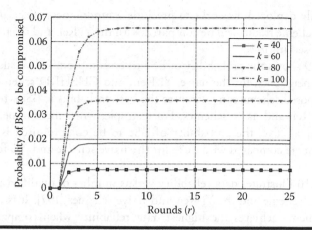

Figure 7.8 Probability of BSe to be compromised against number of rounds for different values of adversary capability.

adversary capabilities; as shown for the first few rounds, the adversary will degrade the performance and reduce the probability of the sensor to be healthy. The CHSHRD scheme will work to heal and alleviate the effect of the adversary, causing stabilization in the probability of healthy sensors to be about 93.5% at $k = 100$.

7.5.2 Simulation Results

In this subsection, simulation experiments are carried out to validate the proposed scheme. UWSN consists of 500 sensors; they are uniformly distributed in an area with a width of 500 m and a length of 500 m. Each sensor has a transmission range TR = 60 m and a mobile adversary of range AR and velocity AV = 2*AR m/round. The adversary is assumed to move within the region in random directions; all sensors that lay in this range will be occupied by the adversary, while sensors outside the adversary range will have a different compromising probability depending on their location according to adversary position. At the beginning of each round, the adversary moves in a random direction, leaving a set of sick sensors and occupying another set of sensors. The remaining sensors will have a different compromising probability according to the new position of the adversary. As soon as the sick s_i is released, it will get its sponsors from its neighbors according to the proposed scheme. To reduce the overhead of memory storage and communication, the sick s_i can reselect again the fewer number of the best qualified neighbors provided that the required security level must be satisfied. The simulation results are averaged over

200 randomly deployed networks to preclude a randomness effect. The proposed CHSHRD scheme results will be compared with the Yi Ren and the naive schemes' results proposed in [197].

Figure 7.9 shows the probability of BSe to be compromised versus the number of qualified neighbors, for the naive, Yi Ren, and CHSHRD schemes. It is clear that the proposed scheme can guarantee the best probabilistic BSe when compared with the Yi Ren and naive schemes. For example, when the number of qualified neighbors equals five, the probability of BSe to be compromised is zero for the proposed scheme compared with 30% for the naive scheme and 2% for the Yi Ren scheme.

Figure 7.10 illustrates data reliability versus number of qualified neighbors for the proposed scheme and the Yi Ren and naive schemes [197]. It is clear that the proposed scheme achieves the highest data reliability when compared with the other schemes. For example, when the number of qualified neighbors equals five, the probability of data reliability is about 96% in the proposed scheme compared with 70% for the naive scheme and 65% for the Yi Ren scheme.

Figure 7.11a, b, and c shows the average compromising probability versus sensor indices, the number of qualified neighbors, and rounds, respectively, for different values of adversary ranges (AR, 1.5*AR, 2*AR). In the simulation model, the adversary capability is determined by the adversary range (AR); hence, as the AR increases, the adversary capability will also increase. This figure shows that the proposed scheme will behave well even at high adversary capability. As shown in

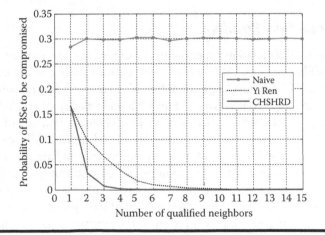

Figure 7.9 Probability of BSe to be compromised versus the number of qualified neighbors for the naive, Yi Ren, and CHSHRD schemes.

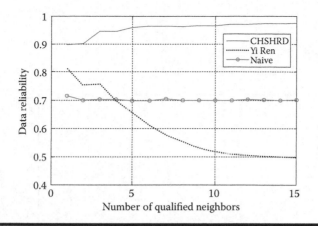

Figure 7.10   Data reliability versus the number of qualified neighbors for the naive, Yi Ren, and CHSHRD schemes.

Figure 7.11a, the probability of BSe for all sensors at a high adversary value on average does not exceed 0.27, which is less than the threshold value (0.3) between the health and the sick states.

Figure 7.11b emphasizes the above findings and ensures the fact that the self-healing will dominate the network using a lower number of qualified neighbors. Therefore, the proposed scheme reduces the storage overhead communication by reducing the number of sponsors to as little as possible.

Figure 7.11c emphasizes that, irrespective of the adversary capability, the probability of BSe is nearly fixed starting from round 4, achieving a match between the analytical and simulation models.

Figure 7.12 presents the probability of BSe against the number of rounds for the proposed scheme using both analytical and simulation models. The results have nearly the same performance. Both results have a sharp increase in the first few rounds, and then they begin to saturate. From round 4, the self-healing process can overcome the attacking caused by the adversary.

The main difference is in the saturation level, which can be described as follows: In the analytical model, the adversary occupies each round a subset of k sensors from the set of healthy sensors. In simulation, according to the adversary's mobility, it can occupy a subset of health sensors or a subset of both health and sick sensors; this will cause a reduction in the compromising probability of the simulation rather than the analytical model. Another reason is that in simulation, the adversary may attack in a network border or corner, so it may attack a lower number of sensors.

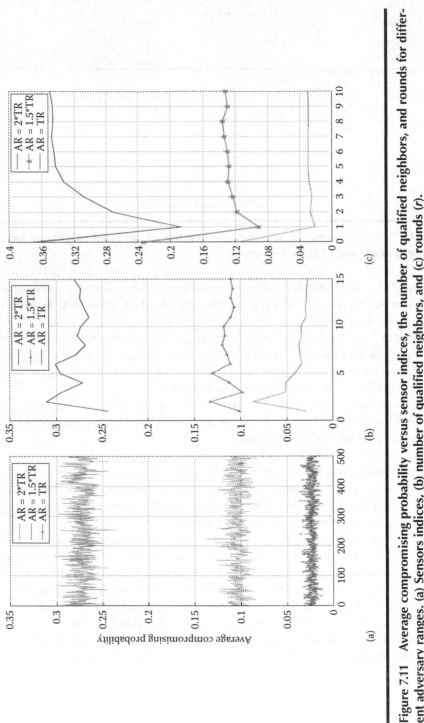

Figure 7.11 Average compromising probability versus sensor indices, the number of qualified neighbors, and rounds for different adversary ranges. (a) Sensors indices, (b) number of qualified neighbors, and (c) rounds (r).

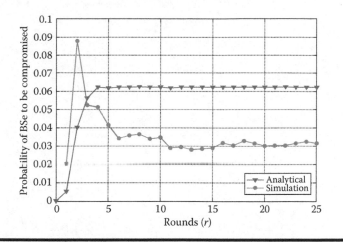

Figure 7.12   Probability of BSe to be compromised versus the number of rounds for the proposed scheme using analytical and simulation models.

7.6 Summary

In this work, a new CHSHRD scheme for a sensor to regain secrecy and high data reliability in UWSNs was proposed. The CHSHRD employs both proactive and reactive peers to ensure data secrecy and guarantee probabilistic BSe and compromised probability. Both theoretical and simulation results were used to evaluate the proposed scheme and compare its performance with other protection schemes. The results illustrated that the hybrid scheme offers better protection than others that use stand-alone proactive or reactive peers. The proposed scheme also reduces the storage overhead communication by reducing the number of sponsors as possible. Compared with other schemes, CHSHRD was more powerful and outperformed the Yi Ren and naive schemes. The results present that the CHSHRD achieves a compromised probability of 0.07 compared with 0.21 and 0.22 for stand-alone reactive and proactive schemes, respectively, when the number of qualified neighbors equals five. Moreover, the data reliability is about 96% in the CHSHRD compared with 70% for the naive scheme and 65% for the Yi Ren scheme.

Chapter 8

Self-Healing Cluster Controlled Mobility Scheme for Self-Healing Enhancement

8.1 Introduction

As explained in Chapter 7, UWSNs [203–205] contain resource-limited sensors deployed to sense data in an inimical or harsh environment. The sink does not exist all the time; it visits the network in a periodic manner to collect the data stored. There is no real-time off-loading of the data stored; thus, the stored data are exposed to adversary risks. An adversary has the ability to compromise sensors lying in its range, learn their secret keys, and alter, steal, or spoil the stored data. The most puzzling problem is how to reduce the risk of compromise. It is a formidable challenge, and there are many existing solutions to mitigate this risk.

The first solution depends on the existence of an online trusted sink; this solution is not suitable for a UWSN because the sink is not present all the time. The second solution works on the availability of a true random number generator (TRNG) on each sensor [10]; this solution is not suitable for large-scale UWSNs. The third solution considers that there is secure or tamper-resistant hardware on each sensor [17], but this is also not suitable for a low-cost sensor commodity. The fourth solution depends on the cooperation between sensors to perform self-healing. Sensor cooperation has been shown to be an effective solution in UWSNs [206]. Finally, it is the mobility that can be used to help and improve the chance of finding healthy neighbors and hence self-healing.

The authors in [206] assert that the CHSHRD scheme can overcome the performance of the other previously mentioned schemes. If a sick sensor gets at least one contribution of a secure randomness from a healthy peer, it can move to a healthy state and regain its security. However, if a sick sensor has a few or no healthy peers, it will be sick for the next rounds.

In this chapter, a new proposal called self-healing cluster controlled mobility (SH-CCM) scheme is presented. This is based on hybrid cooperation between proactive and reactive peers and the sick sensors at both network and cluster levels, to guarantee the security in UWSNs. The SH-CCM scheme uses the mobility of sensors inside a cluster of sick sensors beside the hybrid cooperation principal. Sensor mobility will enable a new set of possibilities in UWSNs; it can be considered a complementary solution to solve the problem of healthy peer leakage. The change in sensor location can be used to mitigate/solve many of the design challenges or network problems [207,208].

In this chapter, the mobility will be leveraged to improve cooperation, self-healing, and hence data security in UWSNs, thus avoiding expensive routing operations in static UWSNs [18]. The proposed SH-CCM scheme uses the mobility of sick and healed sensors inside a cluster of sick sensors; this will cause an infusion between sensors and thus will increase the chance of finding healthy neighbors. As a result, it enhances both data security and self-healing probability inside the network. Thus, we can say that SH-CCM will help the sick sensors to perform self-healing and restoring backward secrecy faster and better than other schemes that do not depend on mobility.

There are different types of mobility models used in network simulations [209–211]. One of the mobility models that give the best performance against a mobile adversary is the random jump mobility (RJM) model [12]. The mobility model considered in this chapter is called the random jump controlled mobility (RJCM) model, and it differs from the RJM model as it supports controlled mobility. The controlling parameters are the direction and distance of moving, sensor residual energy, and the number of moving sensors.

RJCM enables a controlled mobility for some of the cluster sensors. Cluster formation is based on the idea that sensor tampering by the adversary will require the sensor to be removed from the network for a nonnegligible time. In particular, this time ranges from 5 minutes for a short attack to about 30 minutes for medium attacks, and it may take some hours for a long attack [201]. Therefore, we can use a sensor capturing detection algorithm to distinguish a compromised sensor from healthy sensors [201].

Some previous work has considered the aforementioned challenges. In [203], the authors depend on the hybrid cooperation principal; sensor cooperation has been shown to be an effective solution against the aforementioned challenge. The solution in [204] depends on the presence of an online trusted sink which is not suitable for unattended networks and large scale networks. In [10], the solution depends on the existence of a TRNG installed in each sensor, but this is not suitable for cost-effective and large-scale UWSNs. In [17], the solution is based on each sensor containing tamper-resistant hardware, but this is not suitable for low-cost sensor products. In [212] and [213], the authors consider mobility besides hybrid cooperation to perform self-healing; sensor mobility enables a new set of possibilities in sensor networks, and it is the complementary solution for the leakage of healthy peers. The change in sensor position can be used to solve many of the network challenges [207,208,214,215]. In [216] and [217], the authors leveraged sensor mobility and collaboration in mobile UWSNs in order to obtain intrusion resilience with low overhead; they assumed a spherical deployment surface and a stationary adversary. Iida *et al.* [218] improved the schemes proposed in [217] from the viewpoint of both security and efficiency. The importance of mobility to improve the network coverage has been studied in [219]. Wang *et al.* in [220] explored the motion capability to relocate sensors to deal with sensor failure or respond to new events. In [221], Dutta *et al.* proposed a new self-healing process that depends on a key distribution scheme with revocation capability. This scheme requires constant storage of personal keys for each user, but this scheme requires sink presence, which is unsuitable for UWSNs. In [201], the authors used the sensor mobility to improve the sensor capture detection.

However, none of these works have examined the combination of controlled mobility within a cluster of healthy and sick sensors with hybrid cooperation between sick and healthy sensors, which is the main contribution of this chapter. A set of comprehensive simulation and analytic analysis is carried out to demonstrate the effectiveness of the proposed SH-CCM scheme in the presence of a powerful, realistic adversary. The performance is measured through different parameters such as the probability of BSe to be compromised, compromising probability, and data reliability. The results emphasizes that SH-CCM has a better performance over the other schemes that do not depend on controlled mobility.

The remainder of the chapter is ordered as follows: Section 8.2 presents the network model and system assumptions. The proposed SH-CCM simulation analysis is explained in Section 8.3. Then, Section 8.4 presents the analytical analysis of the proposed SH-CCM scheme. Section 8.5 introduces the simulation and analytical results and discussion. Section 8.6 gives the summary.

8.2 Network Model and Assumptions

8.2.1 Network Model

A homogeneous mobility enabled static UWSN consists of N sensors is assumed. Each sensor can sense a circular area of radius called the sensor range (SR). Each sensor has a neighbor set N_{Bi}. The basic security mechanism is the pairwise keys to secure the sensor to its neighbors and the sink to sensor communication [201,202]. We assumed an itinerant data sink to be a trusted party that cannot be compromised by an adversary [197]. It visits the network to collect the stored data, clear the sensors' memories, and finally update the sensor's cryptographic material. The sink could be an airplane or an army troop approaching the network. The time interval between two successive visits of the sink *T* is divided into rounds, and all sensor clocks are loosely synchronized [196]. During round time, the sensor can sense and generate data. This data is stored locally and waits until an authorized mobile sink off-loads them.

Adversaries can be divided into different classifications [196,222]. In this chapter, an adversary that is proactive, centralized, mobile, not logically focused, and trying to compromise a number of sensors logically within its range is assumed. Once a sensor is compromised, the adversary can pick up secrecy keys and collect the data even if the data was encrypted [202,222]. The adversary can undermine the entire network when it moves between sets of sensors before the next visit of the data sink. The adversary model used resembles that proposed in [206] and has the following exclusive capabilities:

- The adversary appears randomly once in each round within the network.
- Range and network density will control the number of compromised sensors.
- Due to random appearance, the adversary is unpredictable and untraceable.

8.2.2 Sensor States

The sensor may exist in one of the following four states: occupied, sick, healthy, and mobile. Figure 8.1 shows the four states and the transition between them.

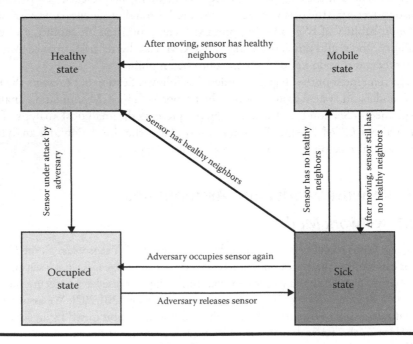

Figure 8.1   Sensor states' transition diagram.

8.2.3 Compromising and Data Secrecy

There are different types of adversaries [11,190,197,222–224]. A logical adversary is assumed. It can attack the sensors lying in its range logically. Also, an adversary has a proactive, centralized, mobile, and nonfocused feature; appears randomly once in each round; and is unpredictable and untraceable, so it is a strong threat.

If an adversary occupies the sensor at round r_1 and releases it at round r_2 $(r_2 > r_1)$, the stored data can be partitioned into three categories:

1. Before r_1, before occupation
2. Between r_1 and r_2, during occupation
3. After r_2, after occupation

Nothing about secrecy of data falls into category 2 since the adversary can learn and hold all keys $k_i^r \forall r \in [r_1, r_2]$ and drive all future keys $k_i^r \forall (r > r_2)$. The challenges become twofold: First, for category 1, the precompromise data must not be revealed if a sensor is compromised, and it is called FSe. It is guaranteed by updating its zero round secret key k_i^0 using one way collision-resistant hash function $H(.)$ and PRNG of this sensor to produce the current round secret key k_i^r from the previous one k_i^{r-1}. The benefit of the hash function is that it cannot work in the reverse direction. Second, for category 3, the challenge is how to guarantee that the postcompromise data is not exposed; this is called backward secrecy (BSe). Hence, the adversary drives all future keys $k_i^r \forall (r > r_2)$. Therefore, the BSe probability is not guaranteed, and data reliability is difficult to achieve. Thus, the encryption technique can only guarantee FSe but it is not enough to achieve BSe. The analysis will concentrate on the confidentiality and reliability of data collected in category 3.

8.3 SH-CCM Simulation Analysis

The proposed scheme can be summarized using the following steps:

Step 1: Adversary Attack. The adversary appears each round in a random location. It is unpredictable and untraceable. It occupies a set of sensors within its adversary range at round r_1. It can know secret keys $k_i^{r'}$, where $(r' \geq r_1)$, and crack the sensor encryption and decryption functions, so it can decrypt the data stored. At round r_2, the adversary occupies a new set of sensors. It leaves behind a set of sick sensors.

Step 2: Capturing Detection Algorithm. In each round, a capture detection algorithm [219] is used to distinguish, identify, revoke, and cluster the currently released sick sensors. In addition, the capture detection algorithm broadcasts information about the location of the revoked sensors to the entire cluster. It can find the coordinates of the sensors that have the maximum and minimum X, Y coordinates $(X_{min}, X_{max}, Y_{min}, Y_{max})$, which represent the borders of the square surrounding the cluster. From these coordinates we can find the coordinates of the cluster center (X_c, Y_c) and the radius of the cluster R_c as follows:

$$X_c = \frac{X_{max} + X_{min}}{2} \tag{8.1}$$

$$Y_c = \frac{Y_{max} + Y_{min}}{2} \tag{8.2}$$

$$R_c = \frac{(X_{max} - X_c) + (Y_{max} - Y_c)}{2} \tag{8.3}$$

Step 3: Node Selection Protocol after Formation of the Cluster. Sensors at the inner border of the cluster select t sponsors $(t_1 + t_2)$ from their peers at the outer border of the cluster. The protocol of selection of both t_1 and t_2 is similar to the one mentioned in [206], but it differs in such a way that the sick sensor sends a message MS1 "I need help-P_s," and the healthy sensor sends MS2 "I can help-P_h," where P_s and P_h are the compromising probabilities of sick and healthy sensors, respectively. It can be calculated as follows:

$$D = \sqrt{(X_i - X_c)^2 + (Y_i - Y_c)^2} \tag{8.4}$$

$$P_s = 1 - \frac{D}{R_c} + \frac{P_{th} * D}{R_c} \tag{8.5}$$

$$P_h = P_{th} * \boldsymbol{random} \tag{8.6}$$

where D is the distance between sensor location (X_i, Y_i) and the center of the cluster (X_c, Y_c), $P_{th} = 0.3$ is the threshold compromising probability [197], and $\boldsymbol{random}$ represents a random value between 0 and 1. Sensors at the inner and outer borders will hear the messages of each other, so the sick sensor forms compromising probability vectors:

$CPV_{i,t_1}^{Pro} = \left\{ P_{h,i,1}, P_{h,i,2}, P_{h,i,3} \ldots \ldots P_{h,i,t_1} \right\}$, from proactive peers

$CPV_{i,t_2}^{Re} = \left\{ P_{h,i,1}, P_{h,i,2}, P_{h,i,3}, \ldots P_{h,i,t_2} \right\}$, from reactive peers.

For a secure contribution and secret key generation, the selected sponsors must guarantee a security level of recovering probability of 0.1% [197]. Algorithm 1 shows the peer selection algorithm, where P_{recov} is the probability that the original data is recovered by the adversary. It is clear that the protocol gives priority to

Pro$_{\text{peers}}$ and reduces the number of sponsors to as little as possible provided that the required security level is achieved.

Step 4: Security Regaining without Mobility. Sick sensors at the cluster's border successfully get t sponsors from the healthy sensors and hence get a secure random contribution(s). Now cooperation and self-healing occur; sick sensors will regain their B_{se}, recover from the compromise, and become unknown to the adversary by computing a new secret, as in [17]:

$$k_i^{r+1} = H\left(k_i^r \left\| c_{i,1}^r \right\| c_{i,2}^r \ldots \right\| c_{i,t}^r\right) \tag{8.7}$$

where $c_{i,1}^r, c_{i,2}^r \ldots, c_{i,t}^r$ are the secure random contributions from the sponsors. The processes of cooperation and self-healing are repeated many times until all sensors become healthy. As shown in Figure 8.2, considering a single adversary attack,

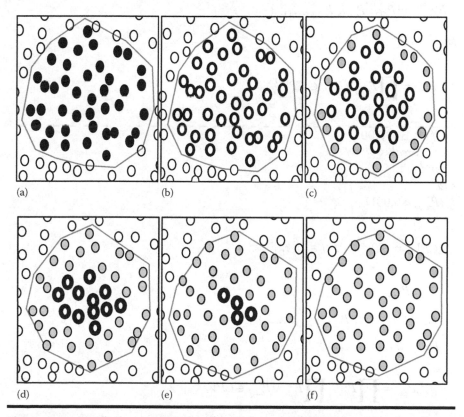

Figure 8.2 Healing processes for first five rounds after the adversary leaves (without mobility). (a) Occupied sensors. (b) Sick sensors. (c) First healing. (d) Second healing. (e) Third healing. (f) Fourth healing.

it is clear that about five rounds and four self-healing processes are required for the cluster to be completely healed.

Step 5: Node Selection and Security Regaining with Mobility. One round after the adversary leaves, the sick sensors at the cluster border successfully self-heals, while the other sick sensors that failed to self-heal due to the leakage of sponsors or the disability to achieve the required security level will remain compromised for the next rounds. Therefore, at the beginning of the second round, the CCM scheme is used to provide mobility to some sensors within the cluster.

Algorithm 1: Algorithm Used to Select the Sponsors

Sponsors selection protocol after attack
Select Pro_{peers}:

$$CPV_i^{Pro} = \{P_{h,i,1}, P_{h,i,2}, P_{h,i,3},\ldots\ldots\}$$

for $t_1=1:t/2$

$$P_{recov} = \prod_{j=1}^{t_1} P_{h,i,j} \; \forall P_{h,i,j} \in CPV_i^{Pro}$$

   If $(P_{recov} < 0.001)$
     break
  end
end
If $(P_{recov} < 0.001)$
  $t = t_1$

$$CPV_{i,t}^{r} = CPV_{i,j}^{Pro} \; \forall \; j \in \left[1, t_1\right]$$

else
  select Re_{peers}:

$$CPV_i^{Re} = \{P_{h,i,1}, P_{h,i,2}, P_{h,i,3},\ldots\ldots\}$$

  for $t_2=1:t/2$

$$P_{recov} = \prod_{j=1}^{t/2} P_{h,i,j} \cdot \prod_{k=1}^{t2} P_{h,i,k} \forall P_{h,i,j} \in CPV_i^{Pro},$$

$$P_{h,i,k} \in CPV_i^{Re}$$

If $(P_{recov} < 0.001)$
break
end
end
end

$$t = t_1 + t_2$$

$$CPV_{i,t}^{'} = \left\{ CPV_{i,j}^{D_u} \ \forall \ j \in \left[1, t_1\right] \cup CPV_{i,k}^{R} \ \forall \ k \in \left[1, t_2\right] \right\}$$

The selection of sensor is based on the amount of residual energy. Sensors with higher residual energy will perform some mobility. The direction of mobility will depend on the current location (X_i, Y_i) with respect to the cluster center (X_c, Y_c). It may be toward the cluster center or the cluster borders. If the sensor is near the cluster center $(D < R_s)$, then the direction of motion will be toward the borders, but if the sensor is near the cluster border $(D < R_s)$, the direction will be toward the center, as shown in Figure 8.3, R_s is the radius of the circle within

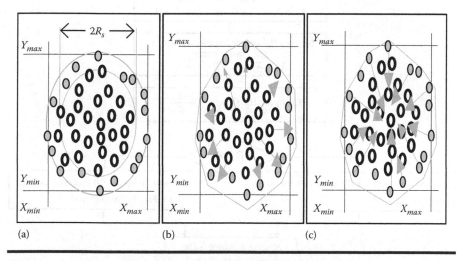

(a)   (b)   (c)

Figure 8.3 Moving directions of sick and healthy sensors inside a cluster. (a) Circle within the cluster containing sick sensors. (b) Some sick sensors moving toward the cluster border. (c) Some healthy sensors moving toward the cluster center.

the cluster that contains the sick sensors after one round of healing, which can be computed as

$$R_s = R_c \sqrt{1 - \frac{N_{hc}}{N_{tc}}} \tag{8.8}$$

where N_{tc}, N_{hc} are the number of total and healed sensors of the cluster, respectively. The distance of mobility will depend on the sensor location, the cluster center, and the cluster radius R_c.

The distance and direction of mobility will be calculated according to the flowchart shown in Figure 8.4.

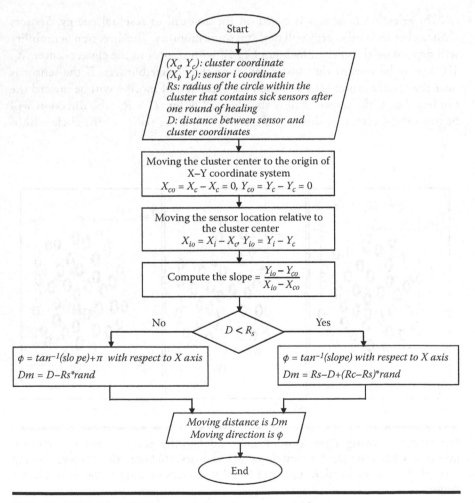

Figure 8.4 Moving distance and direction computation.

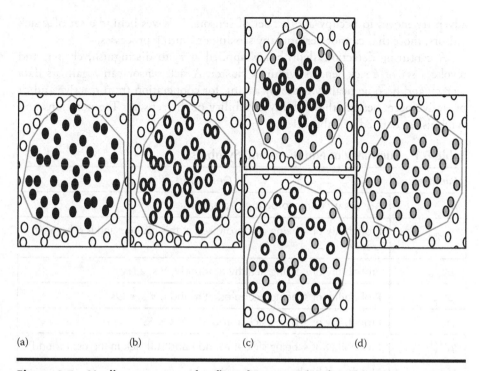

(a) (b) (c) (d)

Figure 8.5 Healing processes for first three rounds after the adversary leaves (with mobility). (a) Occupied sensors. (b) Sick sensors. (c) First healing and mobility. (d) Second healing.

After the motion, there is a fusion between the sick and healed sensors. At the end of the second round, the cooperation and self-healing processes start again, and Steps 3 and 4 are repeated. As shown in Figure 8.5, considering a single adversary attack, it is clear that about three rounds and two self-healing processes are required for the cluster to be completely healed. It is clear that a simple mobility model saves about 50% of the rounds, and the cooperation and self-healing processes are required for complete healing.

8.4 SH-CCM Scheme Analytical Analysis

Analytical analysis of the SH-CCM scheme involves initial steps such as network initialization, data generation and encryption, data segmentation, peer selection protocol, and data distribution. All of these steps are explained in detail in [203]. The analytical model will follow the model presented in [203], but with mobility synthesis and two types of peer consideration, it will be applied at both network level and cluster level. After the adversary occupies a set of k sensors at r_1, at r_2 the adversary moves to occupy another set of sensors. It leaves behind a set of k sick sensors, those that cannot perform any sensing or routing processes.

A capturing detection algorithm is applied at r_2 to distinguish, cluster, and revoke a set of k sick sensors within a cluster. A sick sensor can regain its data secrecy and become healthy again by asking for cooperation from reactive and/or proactive peers besides the mobility capability of the sensors. The notations and symbols are defined in Table 8.1.

Table 8.1 Summary of Notations and Symbols

Parameter	Definition
p	Adversary compromising probability
p_{comp}^r	Probability of s_i being compromised at this round r and for the next rounds
p_{health}^r	Probability of s_i being healthy at round r, $\forall\, s_i \in Hs^r$
p_{occu}^r	Probability of s_i being occupied at round r, $\forall\, s_i \in Os^r$
p_{sick}^r	Probability of s_i being sick at round r, $\forall\, s_i \in Ss^r$
$p_{sick}^{r,r+1}$	Probability of s_i being sick at round r and still sick in the next round
Hs^r	Number of healthy sensors at round r
Os^r	Number of occupied sensors at round r
Ss^r	Number of sick sensors at round r
r_i	Round i
t	Total number of peers
t_1	Number of proactive peers
t_2	Number of reactive peers
N	Number of sensor nodes in the network
k	Adversary capability
A	Event where there are no peers
B	Event where the peers' contributions are intercepted
A_N	Area of the network
A_C	Area of the cluster
N_{tc}	Number of sensors within the cluster
N_{hc}	Number of healed sensors within the cluster

8.4.1 Network Level Analysis

This analysis is during r_2. After the adversary leaves behind a set of sick sensors, the self-healing process will start depending on both types of peers.

Pro$_{peers}$ **Contributions**. Pro_{peers} will initialize the interaction and offer their help (random contributions) for the cluster of sick sensors. The sick s_i will use these contributions to self-heal, but there are two events (A_P, B_P) that can prevent the sick s_i to self-heal:

Event A_P represents the event that there are no Pro_{peers} from which to select t_1, with probability $P(A_P)$ given as

$$P(A_p) = 1 - P_{pro.\,peers} \tag{8.9}$$

where $P_{pro.peers}$ is the probability to select t_1 of Pro_{peers} and is given as

$$P_{pro.\,peers} = \frac{t_1}{n - k} \tag{8.10}$$

Event B_P represents the event that the adversary compromises the contribution of Pro_{peers}, with probability $P(B_P)$ given as

$$P(B_p) = P_{pro.\,peers} \cdot p = \frac{t_1}{n - k} \cdot p \tag{8.11}$$

The probability that s_i is still sick at round $r + 1$ is

$$p_{sick.Pro}^{r+1} = (\,P(A_P) + P(B_P)\,)^{Hs^r} \tag{8.12}$$

$$Hs^r = n.p_{health}^{r-1} - 2 * k \tag{8.13}$$

$$p_{sick.Pro}^{r+1} = \left(1 - \frac{t_1}{n-k} + \frac{t_1}{n-k} \cdot p\right)^{n.p_{health}^{r-1} - 2*k}$$

$$= \left(1 - t_1 . \frac{1-p}{n-k}\right)^{n.p_{health}^{r-1} - 2*k} \tag{8.14}$$

The probability of s_i being sick at round r will be

$$p^r_{sick.Pro} = \left(p^{r-1}_{occu.Pro} + p^{r-1}_{sick.Pro} \right) \cdot p^{r+1}_{sick.Pro} \tag{8.15}$$

$$p^r_{sick.Pro} = \left(p^{r-1}_{occu.Pro} + p^{r-1}_{sick.pro} \right) \cdot \left(1 - t_1 \cdot \frac{1-p}{n-k} \right)^{n \cdot p^{r-1}_{health} - 2 * k} \tag{8.16}$$

where

$$p^r_{occu.Pro} = \frac{k}{n \cdot p^{r-1}_{health} + k} \tag{8.17}$$

is the probability of s_i being occupied.

The probability of s_i to be compromised at round r will be

$$p^r_{comp.Pro} = p^r_{sick.Pro} + p^r_{occu.Pro}$$

$$p^r_{comp.Pro} = \left(p^{r-1}_{occu.Pro} + p^{r-1}_{sick.Pro} \right) \cdot \left(1 - t_1 \cdot \frac{1-p}{n-k} \right)^{n \cdot p^{r-1}_{health} - 2k} + \frac{k}{n \cdot p^{r-1}_{health} + k} \tag{8.18}$$

***Re*_{peers} Contributions.** The sick s_i will initialize the interaction and ask for help (random contributions) from the Re_{peers}. The sick s_i will use these contributions to generate a new secret key to self-heal, but there are two events (A_R, B_R) that can prevent the sick s_i to self-heal:

Event A_R represents the event that there are no Re_{peers} from which to select t_2, with probability $P(A_R)$ given as

$$P(A_R) = \frac{\left(\begin{array}{c} n \cdot p^{r-1}_{comp} \\ t2 \end{array} \right)}{\left(\begin{array}{c} n - k \\ t2 \end{array} \right)} \tag{8.19}$$

Event B_R represents the event that the adversary compromises the contribution of Re_{peers}, with probability $P(B_R)$ given as

$$P(B_R) = \sum_{i=1}^{t_2} p^i \cdot \frac{\binom{n.\left(1 - p_{comp}^{r-1}\right) - 2k}{i}\binom{n.p_{comp}^{r-1}}{t2 - i}}{\binom{n - k}{t2}} \quad (8.20)$$

The probability that s_i is still sick at round $r + 1$ is

$$P_{sick.Re}^{r+1} = \frac{\binom{n.p_{comp}^{r-1}}{t2}}{\binom{n - k}{t_2}} + \sum_{i=1}^{t_2} p^i \cdot \frac{\binom{n.\left(1 - p_{comp}^{r-1}\right) - 2k}{i}\binom{n.p_{comp}^{r-1}}{t2 - i}}{\binom{n - k}{t2}} \quad (8.21)$$

The probability of s_i being sick at round r will be

$$p_{sick.Re}^{r} = \left(P_{occu.Re}^{r-1} + P_{sick.Re}^{r-1}\right) \cdot p_{sick.Re}^{r+1} \quad (8.22)$$

The probability of s_i to be compromised at round r will be

$$p_{comp.Re}^{r} = p_{sick.Re}^{r} + p_{occur.Re}^{r} \quad (8.23)$$

$$p_{comp.Re}^{r} = \left(p_{occu.Re}^{r-1} + p_{sick.Re}^{r-1}\right) \cdot p_{sick.Re}^{r+1} + \frac{k}{n.p_{health}^{r-1} + k} \quad (8.24)$$

$$p_{comp.Re}^r = \left(p_{occu.Re}^{r-1} + p_{sick.Re}^{r-1}\right) \cdot \left[\frac{\binom{n.p_{comp}^{r-1}}{t_2}}{\binom{n-k}{t_2}} + \sum_{i=1}^{t_2} \frac{\binom{n.\left(1-p_{comp}^{r-1}\right)-2k}{i}\binom{n.p_{comp}^{r-1}}{t2-i}}{\binom{n-k}{t2}}\right]$$

$$+ \frac{k}{n.p_{health}^{r-1}+k}$$

The net probability of s_i to be compromised at round r will be

$$p_{comp}^r = p_{comp.Pro}^r \cdot p_{comp.Re}^r$$

$$p_{comp}^r = \left\{\left(p_{occu.Pro}^{r-1} + p_{sick.Pro}^{r-1}\right) \cdot \left(1-t1 \cdot \frac{1-p}{n-k-t2}\right)^{n.p_{health}^{r-1}-k-t2} + \frac{k}{n.p_{health+k}^{r-1}}\right\}$$

$$\cdot \left\{\left(p_{occu.Re}^{r-1} + p_{sick.Re}^{r-1}\right) \cdot \left[\frac{\binom{n.p_{comp}^{r-1}}{t2}}{\binom{n-k-t1}{t2}}\right]\right.$$

$$+ \sum_{i=1}^{t2} p^i \cdot \frac{\binom{n.\left(1-p_{comp}^{r-1}\right)-2k-t1}{i}\binom{n.p_{comp}^{r-1}}{t2-i}}{\binom{n-k-t1}{t2}} + \left.\frac{k}{n.p_{health}^{r-1}+k}\right\}$$

$$(8.26)$$

8.4.2 Cluster Level Analysis

At the first round of self-healing after the adversary leaves, the sick sensors near the cluster's border try to self-heal. Starting from the second round of self-healing, the analysis is transferred from the network level to the cluster level. The cluster now contains sick sensors inside it and healed sensors at its border. The RJCM scheme is applied within the cluster. A selected number of sick and healthy sensors are chosen to move, and a fusion between sick and healed sensors occurs within the cluster.

Now all sick sensors will have healthy neighbors. Therefore, the self-healing will start again, but it will be within the cluster. Also, we will consider both proactive and reactive peers.

***Pro**_{peers}* **Contributions.** *Pro*_{peers} will initialize the interaction and offer their help (random contributions) to the cluster of sick sensor. The sick s_i will use these contributions to self-heal, but there are two events (A_P, B_P) that can prevent the sick s_i to self-heal:

Event A_P represents the event that there are no *Pro*_{peers} from which to select t_1, with probability $P(A_P)$ given as

$$P(A_p) = 1 - P_{pro.\,peers} \qquad (8.27)$$

where $P_{pro.peers}$ is the probability to select t_1 of *Pro*_{peers} and is given as

$$P_{pro.\,peers} = \frac{t_1}{N_{hc}} \qquad (8.28)$$

where N_{hc} is the number of healed sensors within the cluster and is given as

$$N_{hc}^r = \begin{cases} k.p_{health}^{r-1} & r > 1 \\ 0 & r = 1 \end{cases} \qquad (8.29)$$

Event B_P represents the event that the adversary compromises the contribution of *Pro*_{peers}, with probability $P(B_P)$ given as

$$P(B_p) = P_{pro.\,peers} \cdot p = \frac{t_1}{N_{hc}} \cdot p \qquad (8.30)$$

The probability that s_i is still sick at round $r + 1$ is

$$p_{sick.Pro}^{r+1} = (P(A_P) + P(B_P))^{N_{hc}^r} \qquad (8.31)$$

$$p_{sick.Pro}^{r+1} = \left(1 - \frac{t_1}{N_{hc}} + \frac{t_1}{N_{hc}} \cdot p\right)^{k.p_{health}^{r-1}} = \left(1 - t_1 \cdot \frac{1-p}{N_{hc}}\right)^{k.p_{health}^{r-1}} \qquad (8.32)$$

The probability of s_i being sick at round r will be

$$p^r_{sick.Pro} = \left(p^{r-1}_{occu.Pro} + p^{r-1}_{sick.Pro} \right) \cdot p^{r,r+1}_{sick.Pro} \tag{8.33}$$

$$p^r_{sick.Pro} = \left(p^{r-1}_{occu.Pro} + p^{r-1}_{sick.pro} \right) \cdot \left(1 - t_1 \cdot \frac{1-p}{N_{hc}} \right)^{k \cdot p^{r-1}_{health}} \tag{8.34}$$

where

$$p^r_{occu.Pro} = \left(\frac{k}{n \cdot p^{r-1}_{health} + k} \right) \cdot \left(\frac{A_c}{A_n} \right) \tag{8.35}$$

is the probability of s_i being occupied, and A_c and A_n are the cluster and network areas, respectively. The net probability of s_i to be compromised at round r will be

$$p^r_{comp.Pro} = p^r_{sick.Pro} + p^r_{occu.Pro} \tag{8.36}$$

$$p^r_{comp.Pro} = \left(p^{r-1}_{occu.Pro} + p^{r-1}_{sick.Pro} \right) \cdot \left(1 - t_1 \cdot \frac{1-p}{N_{hc}} \right)^{k \cdot p^{r-1}_{health}}$$
$$+ \left(\frac{k}{n \cdot p^{r-1}_{health} + k} \right) \cdot \left(\frac{k}{n} \right) \tag{8.37}$$

where the ratio A_c/A_n is equal to k/n

Re_{peers} Contributions. As stated before, the sick s_i will initialize the interaction and ask for help from the Re_{peers}, so the sick s_i will try to self-heal, but there are two events (A_R, B_R) that can prevent the sick s_i to self-heal:

Event A_R represents the event that there are no Re_{peers} from which to select t_2, with probability $P(A_R)$ given as

$$P(A_R) = \frac{\left(\begin{array}{c} N_{tc} \cdot p^{r-1}_{comp} \\ t2 \end{array} \right)}{\left(\begin{array}{c} N_{hc} \\ t2 \end{array} \right)} \tag{8.38}$$

Event B_R represents the event that the adversary compromises the contribution of Re_{peers}, with probability $P(B_R)$ given as

$$P(B_R) = \sum_{i=1}^{t_2} p^i \cdot \frac{\binom{N_{tc} \cdot \left(1 - p_{comp}^{r-1}\right)}{i}\binom{N_{tc} \cdot p_{comp}^{r-1}}{t2-i}}{\binom{N_{hc}}{t2}}$$

(8.39)

The probability that s_i is still sick at round $r + 1$ is

$$p_{sick.Re}^{r+1} = \frac{\binom{N_{tc} \cdot p_{comp}^{r-1}}{t2}}{\binom{N_{hc}}{t_2}} + \sum_{i=1}^{t_2} p^i \cdot \frac{\binom{N_{tc} \cdot \left(1 - p_{comp}^{r-1}\right)}{i}\binom{N_{tc} \cdot p_{comp}^{r-1}}{t2-i}}{\binom{N_{hc}}{t2}}$$

(8.40)

The probability of s_i being sick at round r will be

$$p_{sick.Re}^{r} = \left(p_{occu.Re}^{r-1} + p_{sick.Re}^{r-1}\right) \cdot p_{sick.Re}^{r+1}$$

(8.41)

The probability of s_i to be compromised at round r will be

$$p_{comp.Re}^{r} = p_{sick.Re}^{r} + p_{occu.Re}^{r}$$

(8.42)

$$p_{comp.Re}^{r} = \left(p_{occu.Re}^{r-1} + p_{sick.Re}^{r-1}\right) \cdot p_{sick.Re}^{r+1} + \frac{k}{n.p_{health}^{r-1} + k} \cdot \left(\frac{k}{n}\right)$$

(8.43)

$$p_{comp.Re}^{r} = \left(p_{occu.Re}^{r-1} + p_{sick.Re}^{r-1}\right) \cdot \left[\frac{\binom{N_{tc} \cdot p_{comp}^{r-1}}{t2}}{\binom{N_{hc}}{t_2}} \right.$$

$$\left. + \sum_{i=1}^{t_2} p^i \cdot \frac{\binom{N_{tc} \cdot \left(1 - p_{comp}^{r-1}\right)}{i}\binom{N_{tc} \cdot p_{comp}^{r-1}}{t2-i}}{\binom{N_{hc}}{t2}} \right] + \frac{k}{n.p_{health}^{r-1} + k}\left(\frac{k}{n}\right)$$

(8.44)

The net probability of s_i to be compromised at round r will be

$$p_{comp}^r = p_{comp.Pro}^r \cdot p_{comp.Re}^r$$

$$p_{comp}^r = \left\{ \left(p_{occu.Pro}^{r-1} + p_{sick.Pro}^{r-1} \right) \cdot \left(1 - t_1 \cdot \frac{1-p}{N_{hc} - t2} \right)^{k \cdot p_{health}^{r-1}} \right.$$

$$+ \left(\frac{k}{n \cdot p_{health}^{r-1} + k} \right) \cdot \left(\frac{k}{n} \right) \right\} \cdot \left\{ \left(p_{occu.Re}^{r-1} + p_{sick.Re}^{r-1} \right) \cdot \left[\frac{\binom{N_{tc} \cdot p_{comp}^{r-1}}{t2}}{\binom{N_{hc} - t_1}{t2}} \right. \right.$$

$$\left. + \sum_{i=1}^{t_2} p^i \cdot \frac{\binom{N_{tc} \cdot \left(1 - p_{comp}^{r-1}\right) - t_1}{i} \binom{N_{tc} \cdot p_{comp}^{r-1}}{t2 - i}}{\binom{N_{hc} - t_1}{t2}} \right] + \frac{k}{n \cdot p_{health}^{r-1} + k} \cdot \left(\frac{k}{n} \right) \right\}$$

(8.45)

8.5 Results and Discussion

In this section, the theoretical expressions beside simulations will be used to test the proposed SH-CCM scheme and compare it with other schemes.

8.5.1 Simulation Results

In this section, a simulation model is used to test the performance of the proposed SH-CCM scheme and compare its performance with the schemes proposed in [206]. A simulator package based on MATLAB® is used to validate the proposed scheme. A UWSN that contains 500 sensors was considered; these sensors are uniformly distributed in an area with a width of 500 m and a length of 500 m. Each sensor has a transmission range TR = 60 m and a mobile adversary of range AR. The adversary is assumed to appear randomly at each round. All sensors lying in this range will be occupied by the adversary, while sensors outside the adversary range will have a different compromising probability according to their location with respect to the adversary.

In the first round, the adversary occupies a set of sensors. At the beginning of the next round, the adversary releases the previously occupied set of sensors and occupies another set of sensors in a different location. It leaves behind a set of sick sensors. After the adversary leaves, a cluster of currently released sensors is formed to be distinguished and revoked from the network. In the first round after the adversary leaves, the sick sensors at the cluster border will get their sponsors from their neighbors outside the cluster and self-healing occurs. At the beginning of the second round, the RJCM scheme is applied to select the sick and healthy sensors that will move within the cluster. Infusion between sick and healthy sensors occurs. Sick sensors can reselect a lower number and the best qualified sponsors provided that the required security level is satisfied. The simulation results are averaged over 200 times to preclude randomness. The proposed SH-CCM scheme results will be compared with the results of the CHSHRD scheme proposed in [206].

Figure 8.6 shows the probability of BSe to be compromised versus the number of qualified neighbors for both the proposed SH-CCM scheme and the CHSHRD scheme. It is clear that the proposed scheme guarantees that the network will be completely healthy earlier by about two rounds. Figures 8.7 and 8.8 illustrate the data reliability versus rounds and qualified neighbors, respectively; an enhancement in data reliability appears as qualified neighbors increase.

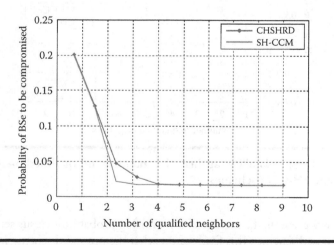

Figure 8.6 Probability of BSe to be compromised against the number of quali-fied neighbors for CHSHRD and SH-CCM schemes.

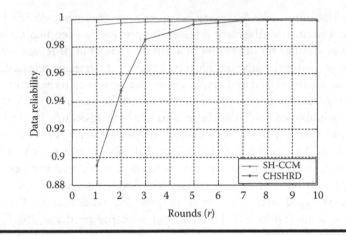

Figure 8.7 Data reliability against the number of rounds for CHSHRD and SH-CCM schemes.

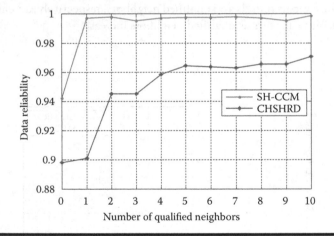

Figure 8.8 Data reliability against the number of qualified neighbors for CHSHRD and SH-CCM schemes.

Figure 8.9 presents the average compromising probability versus sensor indices for the proposed scheme. This figure shows that the proposed scheme will keep the average compromising probability of each sensor at the required security level of 0.001.

The following gives an evaluation of the proposed scheme against many attacks, such that two adversaries attack the network in parallel. Figure 8.10 shows another

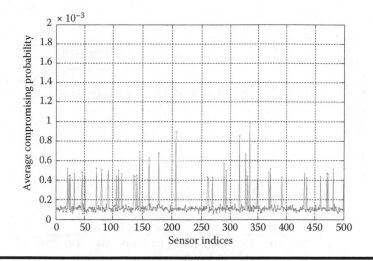

Figure 8.9 Average compromising probability against sensor indices.

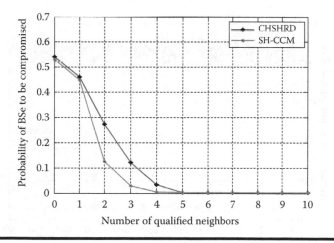

Figure 8.10 Probability of BSe to be compromised against the number of qualified neighbors (many attacks) for CHSHRD and SH-CCM schemes.

enhancement in the probability of BSe to be compromised against the number of qualified neighbors for the SH-CCM scheme with respect to CHSHRD. Figures 8.11 and 8.12 show the average data reliability and the average compromising probability, respectively, against sensor indices. Both figures show that the proposed scheme outperforms the CHSHRD scheme.

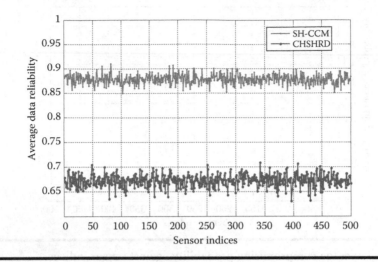

Figure 8.11   Average data reliability against sensor indices (many attacks) for CHSHRD and SH-CCM schemes.

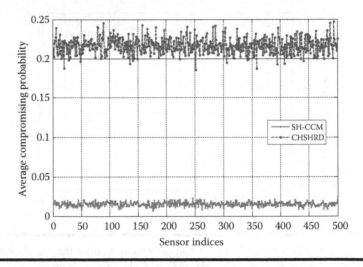

Figure 8.12   Average compromising probability against sensor indices (many attacks) for CHSHRD and SH-CCM schemes.

8.5.2 Analytical Results

The above theoretical expressions are used to validate the proposed SH-CCM scheme against schemes proposed in [203]. The analytical results are obtained using $N = 500$, $p = 0.2$, $k = 100$, and $t = 6$. Figures 8.13 and 8.14 show the probability of BSe to be compromised versus the number of rounds and the adversary capability for

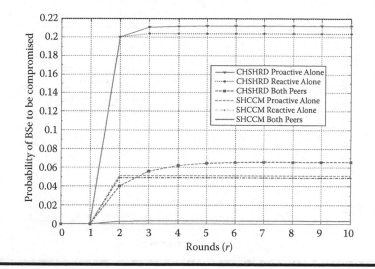

Figure 8.13 **Probability of BSe to be compromised against number of rounds for CHSHRD and SH-CCM schemes.**

the CHSHRD scheme proposed in [203] and the SH-CCM scheme for three cases: stand-alone proactive peers, stand-alone reactive peers, and both. It is clear that the SH-CCM scheme achieves two enhancements: In the first one, it has a better *BSe* over the other scheme. In the second one; compared with CHSHRD [203], the SH-CCM wins the game against the adversary early, as we see that the network reaches a steady state starting from round 2, compared with round 5 for the scheme in [203].

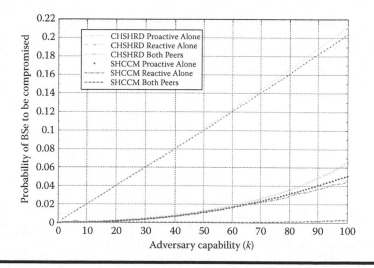

Figure 8.14 **Probability of BSe to be compromised against adversary capability for CHSHRD and SH-CCM schemes.**

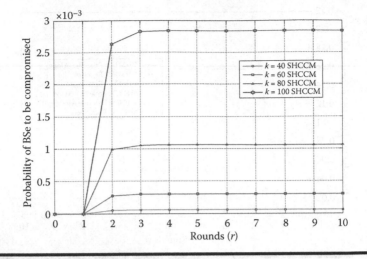

Figure 8.15 Probability of BSe to be compromised against number of rounds for different values of adversary capability.

Figure 8.15 presents the probability of BSe to be compromised versus the number of rounds for the SH-CCM scheme for different values of adversary capability. It is clear that the SH-CCM scheme stands well at different values of adversary capability.

Figures 8.16 and 8.17 show the probability of BSe to be compromised against rounds and adversary capability, respectively, for different values of t for SH-CCM

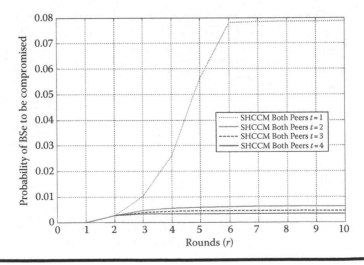

Figure 8.16 Probability of BSe to be compromised against number of rounds for different numbers of peers.

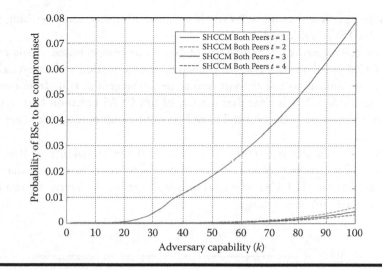

Figure 8.17   Probability of BSe to be compromised against adversary capability for different numbers of peers.

based on both types of peers. It is clear that as the number of peers increases, the performance is improved. Also, for $t < 3$, the performance is nearly the same, so we can say that three or four peers are sufficient to get the best performance. In contrast, the authors of [17] and [197] assumed that six peers were necessary.

Figure 8.18 shows the probability of BSe to be compromised against rounds for different values of p. It is clear that the SH-CCM scheme based on both proactive

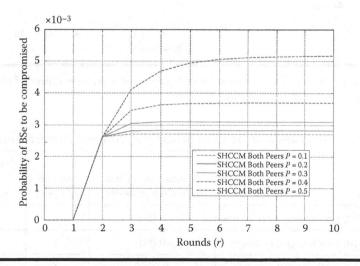

Figure 8.18   Probability of BSe to be compromised against number of rounds for different values of adversary compromising probability.

and reactive peers will stand well even if a higher adversary compromising probability ($P = 0.5$) is considered.

Figure 8.19 presents the probability of BSe to be compromised versus rounds for SH-CCM based on both types of peers for both simulation and analytical. It is clear that the simulation and analytic results are convergent starting from round 2. Also, from Table 8.2, it is clear that the use of SH-CCM based on both types of peers can reduce greatly the probability of BSe to be compromised at a lower number of peers.

Also, from the previous results, it is clear that the probability of BSe to be compromised reaches saturation as the number of rounds increases more than 2, so we can say that the SH-CCM scheme stands well against the adversary, and it can provide a sustainable self-healing.

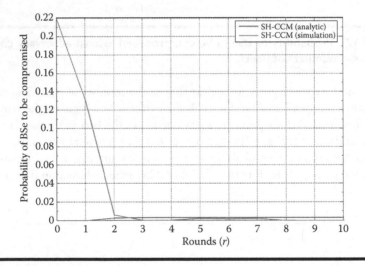

Figure 8.19 Probability of BSe to be compromised against number of rounds for simulation and analytical results.

Table 8.2 Comparison of SH-CCM Scheme at Different Peer Considerations

Scheme	Probability of BSe to Be Compromised	Number of Peers	Number of Rounds
SH-CCM, proactive peers alone	0.053	3	5
SH-CCM, reactive peers alone	0.051	3	5
SH-CCM, both proactive and reactive	0.0065	2	5

8.6 Summary

In this work, a new SH-CCM scheme for sensor regaining secrecy and high data reliability in UWSNs was proposed. The proposed scheme combines both cooperation and infusion of mobility with cooperative hybrid self-healing schemes to enhance the ability of security regaining and maximize both security level of data and guarantee probabilistic BSe and compromised probability. In addition to the above improvements, the proposed scheme will save about 50% of rounds of cooperation and self-healing processes required for complete self-healing; hence, it enhances the efficiency of the whole network. Analytical and simulation results showed approximately the same behavior, ensuring the efficiency of the proposed scheme. It also reduces the overhead communication by achieving the required security level using the minimum number of sponsors. Compared with other schemes, the proposed scheme was more powerful and outperforms the CHSHRD scheme [206].

Chapter 9

Self-Healing Single Flow Controlled Mobility within a Cluster Scheme for Energy Aware Self-Healing

9.1 Introduction

As stated in Chapter 8, the self-healing-based mobility scheme enhances the UWSN security better and faster. However, sensor mobility is considered as one of the major reasons for high energy consumption in the network. This chapter proposes a new self-healing single flow controlled mobility within a cluster (SH-SFCMC) scheme to optimize the trade-off among self-healing-based mobility and energy consumption in mobile UWSNs. The essential building elements of UWSNs are the mobile sensor devices that are usually energized by irreplaceable batteries with bounded power. These devices have diverse sensing, communication, and mobility capabilities.

In latest years, significant progress has been made in the field of mobile sensor networks; sensor mobility permits a novel set of prospects [208,214]. In [225], the authors presented the problem of energy consumption optimization associated with both motion and communication. They ensured that mobility can improve network communications; it can reduce the sum of all energy consumed and maximize the network lifetime. In [226], the authors investigated the influences of

179

sensor mobility on balancing and dissipation of energy in wireless networks. They showed that mobility can be used to exclude unfair energy dissipation patterns and gain a more energy-efficient process compared with immobile networks. However, extra mobility can lead to energy inefficiencies.

In [227], the authors utilized mobility to minimize the total communication and motion energy consumption of a robotic operation by co-optimizing the communication and motion strategies. They showed that, under some circumstances, it is more valued for the robot to spend energy on motion in order to move to a location that is better for communication. In [228], the authors made extensive simulations to evaluate the effectiveness of controlled mobility; they showed that there are many scenarios where controlled mobility can achieve substantial performance gains such as energy efficiency and improvement of communications. They also showed that mobility may be less effective due to various extenuating factors, including hardware limitations and traffic patterns. In [229], the authors showed that mobility can increase the network capacity, while in [230], the authors proved that extra mobility bounds the capacity. In [231] and [232], the authors showed that mobility can increase the network coverage and connectivity. In [233], the authors ensured that mobility could also help security in wireless networks by a method such that when nodes are more closer, they are more secure.

However, none of the above works have considered the benefits of mobility in self-healing and security regaining; hence, the changing in sensor location can be utilized to solve the problem of the leakage of healthy peers and improve cooperation and self-healing capabilities. This will enhance both data security and reliability in UWSNs [212,213]. In addition, the impact of mobility on various network aspects such as energy consumption is studied. Therefore, the objective of this chapter is to propose a new scheme called self-healing single flow controlled mobility within a cluster (SH-SFCMC) to enhance UWSNs' self-healing capability using controlled single flow mobility and to reduce the effects of sensor mobility on the performance characteristics of mobile UWSNs.

The proposed SH-SFCMC scheme is a modified version of the self-healing-cluster controlled mobility (SH-CCM) scheme proposed in [212] and [213] and presented in Chapter 8. The major modification is in the mobility subscheme. It will be a single flow under controlled mobility within a cluster, which differs from the cluster controlled mobility subscheme in the number, type, and position of the moved sensors. The proposed scheme also investigates the influences of sensor motion on the performance of various features of mobile UWSNs.

The obtained results show that using the proposed scheme, UWSNs can exploit controlled sensor mobility to enhance network capability in terms of self-healing and reduce communication-related energy consumption. In addition, the single flow controlled mobility of the proposed scheme does not disturb the number of neighbors per sensor and the network coverage.

The rest of this chapter is organized as follows: In Section 2, the system model and assumptions are presented. Section 3 shows the trade-off between mobility and other network aspects. The proposed SH-SFCCM scheme is described in Section 4. Section 5 presents the simulation setup and performance evaluation, while the summary is presented in Section 6.

9.2 System Model and Assumptions

This section introduces the network model, adversary model, mobility model, and energy models used in this chapter.

9.2.1 Network Model

A homogeneous mobility enabled static UWSN consisting of N sensors similar to the one explained in Chapter 8 is assumed.

9.2.2 Adversary Model

The adversary models used in this chapter is similar to the one explained in Chapter 8.

9.2.3 Mobility Model

The mobility model is planned to define the motion characteristics of the mobile sensors and show how their motions are varied over time. Sensor motion provides a novel set of prospects in sensor networks; the change in sensor position can be used to solve many of the network challenges [208,214]. The mobility is considered as the basic solution to the problem of the leakage of healthy peers, those required for self-healing process. There are diverse types of models used in network simulators [18,210,211]. One of the simplest and widely used to estimate the performance of mobile networks is the RJM model [18], which exhibits the best performance versus a moving adversary. In this model, each node moves independently from others; it selects a destination point at a round and moves directly until reaching the destination at the end of the same round.

The mobility model based mainly on the RJM model called the single flow controlled mobility within a cluster (SFCMC) model is presented. It differs from the RJM model in which it exhibits a controlling feature over the RJM model. This model is applied for a selected healed sensor within the cluster to reach a preassigned destination point. The controlling parameters are as follows:

- Sensor residual energy
- The connectivity and coverage
- The security level
- The shortest distance
- The destination point

The above constraints will control the selection and the motion of the selected sensor; they should be achieved as much as possible. For example, the selected sensor must have the highest residual energy, the lowest compromising probability, and the shortest distance of moving (nearest to the center of the cluster). Also, the motion of this sensor must have no effect on the network connectivity and coverage. Finally, the selected sensor must select the suitable speed to reach the destination point during a round.

9.2.4 Energy Models

These model the energy levels of the device during operation so that the total energy consumption can be calculated. In this chapter, two of the most energy-consuming cost functions are considered: communication consumption energy and mobility consumption energy.

9.2.4.1 Energy Communication Model

It is well known that the energy needed for successful data transmission is determined by the following parameters:

- Distance between nodes
- Noise level of the channel
- Electronic circuits
- Algorithms used for encoding, transmission, reception, and decoding

In this study, a transmission energy model of the network layer similar to the one used in [234] is adapted. Let $E_{tx}(d_c)$ be the energy needed for data transmission across a certain distance for 1 bit; then,

$$E_{tx}\left(d_c\right) = \left(d_c^{\alpha}.e_{tx} + e_{cct}\right) \tag{9.1}$$

where d_c is the communication distance between two sensor nodes, e_{tx} (Joule/bit) is the energy needed by the transceiver amplifier to send 1-bit data over a 1-m distance, and e_{cct} (Joule/bit) is the energy consumed in the circuits of the transceiver to send or receive a single bit. The value of e_{tx} ranges from some pico- to nano-Joule/bit per meter, and α is the path-loss exponent of the transmission medium that ranges $\alpha \in \{2, 6\}$ according to the environment, as shown in Table 9.1 [235].

The energy consumption for transmitting L data bits across distance d_c is

$$E_{tx}(L, d_c) = L.E_{tx}(d_c) \tag{9.2}$$

Table 9.1 Path-Loss Exponents for Several Environments

Environment	Path-Loss Exponent, α
Free space	2
Urban	2.7 to 3.5
Shadowed urban	3 to 5
In-building line of sight	4 to 6

On the other hand, the energy consumption for receiving L bits of data is defined as

$$E_{rx}(L) = L.e_{cct} \tag{9.3}$$

The energy consumption for receiving L bits of data is independent of the communicating distance between two nodes.

9.2.4.2 Energy Motion Model

Other than communication energy consumption, the energy consumption due to sensor mobility will be considered, or mobility cost. For mobility, it is assumed that every sensor is aware of its own location by using the localization technique [236] and does time synchronization prior to or during its motion [237]. The mobility energy cost depends on the parameters of locomotive module, mass of the mobile sensor, speed, acceleration, friction to the surface, transformation loss, and distance traveled. There are many motion energy models depending on some or all the aforementioned parameters [227,238]. For simplicity, a distance-proportional cost model is assumed, which is reasonable for wheeled vehicles similar to the one used in [225] and [235], so that, for traveling distance d, the energy required is

$$E_m(d) = M.d \tag{9.4}$$

where M is a constant called movement parameter, measured in Joule/meter; it depends on the environment and the mass of the mobile node. According to [239], a wheeled vehicle with rubber tires at 1 kg moving on concrete must expend 0.1 J/m to overcome 0.1 N force of dynamic friction.

9.3 Trade-Off between Mobility and Other Network Aspects

This chapter focuses on studying the impact of mobility on different network aspects. In literatures, there are different considerations about mobility; for example, in [227], the authors indicated that in several scenarios, it is beneficial for the sensor node to consume energy in mobility to move to a better place for communication. In [239], a case study has shown that the motion power is not the piece consuming the highest percentage of total energy. Instead, the wireless communications accounts for up to 65.3% of the total energy consumption. In other literatures, there are some trade-offs between mobility and other network issues such as security. Security and mobility seem to be at odds with each other; security is usually enforced by a static. But in [233], the authors proved that this intuition is wrong; they said that mobility can be useful to establish security between any two mobile nodes. Another trade-off is the impact of mobility on coverage and connectivity. There is an instinct that emphasizes the idea that mobility can harm both coverage and connectivity. But in [231] and [232], the authors showed that mobility increases the coverage and connectivity.

The most important trade-off is between mobility and communication. There is an important question: Considering a communication task, should the sensor node spend its energy on motion to move to a location with better communication quality? Or should it stay at its initial position and increase its transmission power? The answer to these questions differs in the literatures. In [235], the authors stated that the energy required for movements is much higher than that required for communications, if the communications do not involve high amounts of data.

9.4 Proposed SH-SFCCM Scheme

When an adversary of a compromising probability P attacks the unattended network at any round, it can occupy a set of sensors equal to $(P \times N)$, making them unresponsive during occupation time. After $1/P$ rounds and without self-healing process, the adversary wins the game and the network becomes fully inaccessible. The SH-SFCMC scheme proposed in this chapter is similar to the SH-CCM scheme [213,214]; both have self-healing and mobility subschemes. The self-healing mechanism is identical for both schemes; it depends mainly on both mobility and hybrid cooperation principal [203]. Mobility increases the chance of finding healthy peers while hybrid cooperation makes use of both proactive and reactive healthy peers, those that will cooperate with the sick sensor so that it can be healed. With respect to the mobility, an SFCMC subscheme is proposed that is similar to the CCM subscheme in the following aspects:

- Both of them depend on the RJM model.
- Both of them are applied inside a cluster.
- Both of them are applied during the second round.
- Both of them apply controlling parameters on the mobility within the cluster.
- Both of them will consider the selected sensors' energy.

The proposed SFCMC subscheme will have the following unique features:

- The mobility is applied to a single selected sensor (single flow) per cluster.
- The moving sensor is selected only from the healed sensors at the cluster border.
- The selected sensor must have the lowest compromising probability.
- The destination point is selected to be at the cluster center.

Considering that a single adversary attacks the UWSN at round ($r = 0$), it will occupy a cluster of sensors laying within its range, causing them to be unresponsive, as shown in Figure 9.1a. During the occupation round, the capture detection

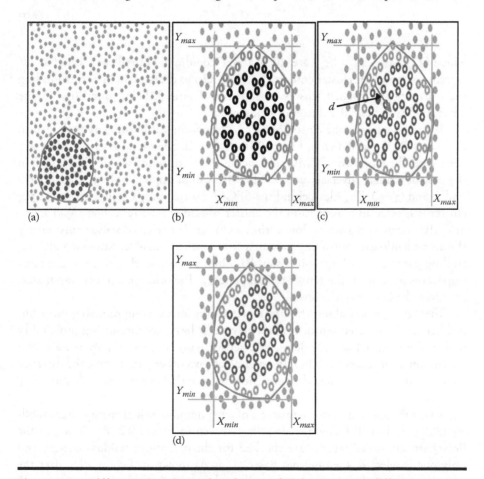

(a)   (b)   (c)

(d)

Figure 9.1   Different situations of a cluster of sick sensors at different rounds. (a) Set of occupied sensors at ($r = 0$). (b) Cluster of sick sensors, at round ($r = 1$), sick sensors at cluster boarder are self-healed. (c) At ($r = 2$), one of the healed sensor is selected, which has the maximum residual energy. (d) At ($r = 2$), the selected healed sensor is allowed to move.

algorithm (CDA) [203] functions to distinguish these unresponsive nodes. It detects and revokes the currently occupied sensors within a cluster and can provide information about the coordinates of these revoked sensors. By knowing the revoked sensors' coordinates, the coordinate of the cluster center and the radius of the cluster can be computed as follows:

$$X_c = \frac{X_{max} + X_{min}}{2} \tag{9.5}$$

$$Y_c = \frac{Y_{max} + Y_{min}}{2} \tag{9.6}$$

where (X_{min}, X_{max}, Y_{min}, Y_{max}) are the X and Y coordinates of sensors located at the extreme points of the cluster border; also, they refer to the straight lines surrounding the cluster of the revoked sensors, as shown in Figure 9.1b. At round (r = 1), the adversary leaves behind a cluster of sick sensors.

The sick sensors at the cluster border will perform self-healing, as explained in [203] and [213] and shown in Figure 9.1b. At (r = 2), one round after the adversary departure, the proposed SFCMC subscheme will be operated within the cluster; in the case of a single adversary attack, SFCMC will be applied only during the second round (r = 2). The idea behind the SFCMC is that only one healed sensor per cluster is selected to move within the cluster toward the cluster center (X_c, Y_c); i.e., SFCMC is based on a single flow within a cluster. The selected sensor must satisfy the aforementioned controlling parameters as much as possible. Satisfying all controlling parameters is impossible, so the highest priority level is given to the most important parameter: the sensor residual energy. The other remaining parameters are given the lower priority level.

There is no worry about the parameter of the lowest compromising probability; hence, the selected sensor is a healed one. It has a compromising probability within the required security level. The sensor residual energy must remain at a maximum after sensor mobility. It depends on two other parameters: the distance to the cluster center (it should be the shortest) and the sensor energy (it should be the largest).

To get the selected healed sensor (S_h) of a maximum residual energy after mobility (E_{res_max}), we will follow the flowchart shown in Figure 9.2. As shown in the flowchart, all healed sensors are checked for the maximum residual energy, and only one is selected, the one that has E_{res_max}. As shown in Figure 9.1c, after the selection process, the selected sensor is allowed to move toward the cluster center, as shown in Figure 9.1d; it will travel a distance d, which can be computed as follows:

$$d = \sqrt{(X_i - X_c)^2 + (Y_i - Y_c)^2} \tag{9.7}$$

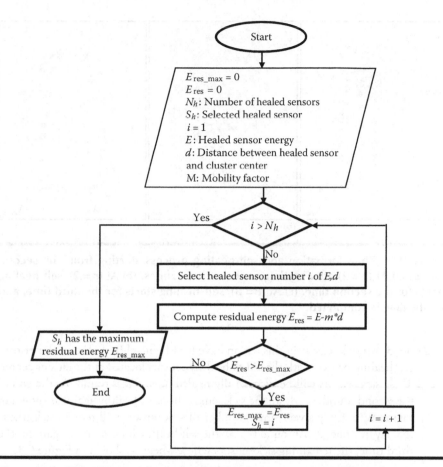

Figure 9.2   Flowchart used to select the healed sensor that will move.

where (X_i, Y_i) is the coordinate of the selected sensor, and it will consume energy due to mobility computed using Equation 9.4.

After this motion, there is a healthy sensor at the center of the cluster. The self-healing will start again during the second round ($r = 2$) for the second time. There are three zones inside the cluster, as illustrated in Figure 9.3a. They are treated as follows:

Zone_1. All sick sensors in this region have healthy neighbors. They will perform self-healing process with the aid of healthy sensors located at the inner border of the cluster, and these sick sensors will be successfully healed during this round.

Zone_2. All sick sensors in this region do not have any healthy neighbors; they are far from both the healthy sensors at the inner border of the cluster and the healthy sensor at the cluster center. These sick sensors cannot be healed during round ($r = 2$), and they will wait for the next round ($r = 3$).

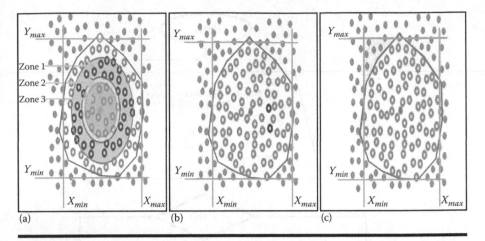

Figure 9.3 Zone formation and self-healing process starting from the second round. (a) At ($r = 2$) after motion, we have three zones. (b) At ($r = 2$), self-healing starts for the second time. (c) At ($r = 3$), self-healing starts for the third time, and self-healing is completed.

Zone_3. All sick sensors in this region have healthy neighbors. They will perform self-healing process with the aid of healthy sensors located at the cluster center. These sick sensors will be successfully healed during this round. At the end of the second round, nearly all the sick sensors become healthy again, as shown in Figure 9.3b. Only a very small number of sick sensors will remain sick (those that lay in Zone_2). At round ($r = 3$), the self-healing process starts again for the third time; the remaining sick sensors will self-heal, as shown in Figure 9.3c.

Compared with the case without mobility shown in Figure 9.4, it is clear that using the proposed SFCMC mobility subscheme will enhance the self-healing ability. Hence, the proposed scheme saves nearly half of the rounds required for complete self-healing, and this will save the power required for complete self-healing. In addition, the reduction of the rounds required for self-healing will cause the network to be accessible most of the time.

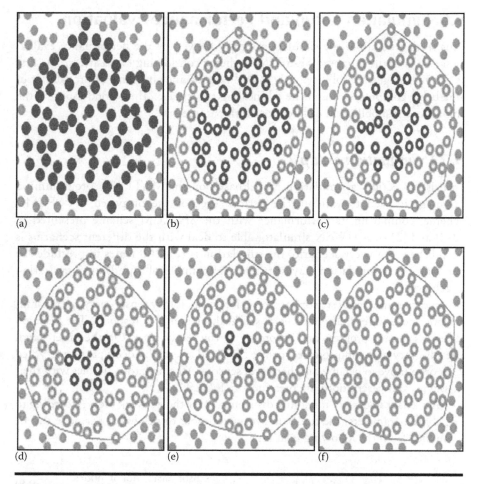

Figure 9.4 Self-healing process in the case without mobility. (a) At ($r = 0$): set of occupied sensors. (b) At ($r = 1$): cluster of sick sensors, self-healing for the first time. (c) At ($r = 2$): self-healing for the second time. (d) At ($r = 3$): self-healing for the third time. (e) At ($r = 4$): self-healing for the fourth time. (f) At ($r = 5$): completely healed cluster.

9.5 Simulation Setup and Performance Evaluation

This section describes the simulation setup followed by the performance evaluation results of the proposed SF-SFCMC scheme considering two aspects. The first is the advantages of mobility on UWSNs' performance in terms of probability of BSe being compromised and the self-healing percentage. The second aspect is the impact of the mobility on different network parameters.

9.5.1 Simulation Setup

In this subsection, a simulation model based on the MATLAB® programming language is used to validate the performance of the proposed SH-SFCMC scheme, comparing its performance with the SH-CCM scheme presented in [213] and [214]. A UWSN simulator able to deal with the different scenarios is established. For every scenario, the simulation is run 200 times to preclude the randomness. A simulation may be run for any number of rounds until saturation is achieved.

In each round, each sensor has a function to do; it may sense, encrypt, transmit, receive, move, offer help, need help, self-heal, or even not respond. The function differs as rounds progress and according to the sensor state. On the other hand, the adversary may compromise, acquire keys, and attempt to decrypt the encrypted data or to move to occupy another set of sensors. All simulator variables with their acronyms and descriptions are presented in Table 9.2.

The sensor nodes are randomly distributed over the network area. According to the above parameters, the resultant average number of neighbors per node can be computed from the following relation:

$$\text{Average no. of neighbors per node} = \frac{\text{sensor area} \times \text{no. of nodes}}{\text{network area}} \tag{9.8}$$

By substitution, the average number of neighbors is 15.7 neighbors/sensor. Some new metrics will be used to evaluate the performance of the proposed scheme. The percentage of self-healing is defined as the ratio of the number of sick sensors before self-healing and the number of sick sensors after a self-healing process during a round. The average communication distance is defined as the distance at which the data is transmitted or received. The average number of processes is defined as the average number of processes during the self-healing process during a round; these processes consist of transmission, reception, and computation. The network coverage is defined as the number of sensors covering each point in the network.

Table 9.2 Simulation Parameters

Parameter	Value
Number of nodes (*N*)	500 sensors
Network area (length × width)	500 m × 500 m
Sensor range (SR)	60 m
Number of rounds (*r*)	Until saturation is reached
e_{cct}	10^{-7} Joule/bit
e_{tx}	10^{-12} Joule/bit per meter
Path-loss exponent (α)	4
Data packet size	1 KB
Initial energy (E_0)	3.6-Volt lithium-ion battery rated at 850 mAh will maintain 11,016 Joules
Adversary compromising probability (*P*)	0.2
Threshold probability	0.3
Required security level	<0.001
Sensor mass	0.5 kg
Mobility factor (*M*)	0.1 Joule/m

9.5.2 Performance Evaluation

The simulation results and the analysis of the proposed scheme will concentrate on showing the effect of mobility on two important aspects. The first and the most important parameter in UWSNs is the effect of mobility energy consumption on the total energy consumption in the network. Second is its effect on UWSN parameters such as compromising probability, self-healing capability, coverage, connectivity, communication distance, and number of process.

9.5.2.1 Impact of Mobility Energy Consumption

Simulation results emphasize the benefits behind applying controlled mobility within a cluster and validating the proposed scheme. One of the most crucial network parameters is the energy consumption inside the network. As shown in Figure 9.5, at round 1, there is no energy consumption in mobility, and the sum of energy

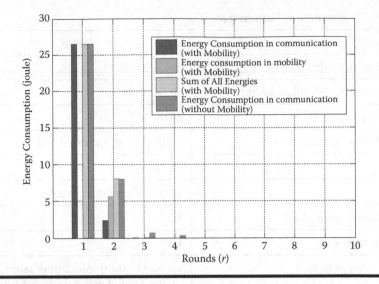

Figure 9.5 Energy consumption for both mobility and communication for the cases with and without mobility against rounds.

consumption without or with mobility is the same. At round 2, mobility is started, and we have the following results:

■ Mobility consumes energy nearly double that of the communication energy consumption.
■ Mobility can save the communication energy consumption to one third compared with that without mobility.
■ The sum of all energies (mobility energy and communication energy) consumption with mobility is nearly equal to the communication energy consumption without mobility.
■ At rounds 3 and 4, in the case without mobility, the network is still consuming some energy in communication. Figure 9.5 emphasizes that mobility can enhance system performance not only without any impact on system energy but also with energy saving.

9.5.2.2 Extensive Analysis of SH-SFCMC

After validation of the proposed scheme with respect to energy consumption, it is very important to study the cases with and without mobility to show the effect of mobility on other network parameters. Figure 9.6 shows the variation of the probability of BSe being compromised against the number of rounds for single attack adversary for three schemes, SH-SFCMC (proposed), SH-CCM [213,214], and CHSHRD [203]. This figure shows two facts. First, the schemes based on mobility exhibit an enhancement

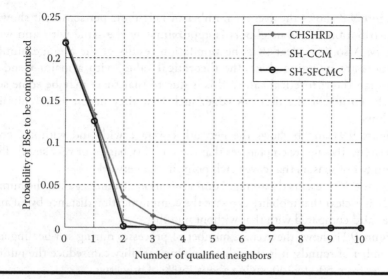

Figure 9.6 **Probability of BSe to be compromised versus number of qualified neighbors for CHSHRD, SH-CCM, and SH-SFCMC schemes.**

in BSe probabilities (smaller and faster) than the scheme that does not depend on mobility. Second, the SH-SFCMC exhibits the best performance over the other based on mobility scheme especially when mobility started at round ($r = 2$).

Figure 9.7 illustrates the effect of mobility on self-healing capability; it presents the percentage of self-healing against the number of rounds. As shown, the mobility can enhance the percentage of self-healing.

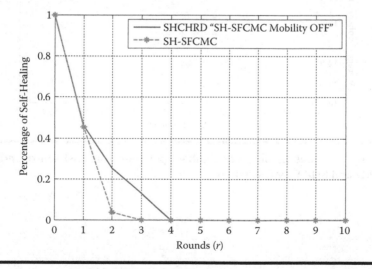

Figure 9.7 **Percentage of self-healing against rounds for SHCHRD and SH-SFCMC schemes.**

Figure 9.8 shows the average number of neighbors per sensor; it shows that the distribution of the neighbors is approximately the same with and without mobility. Also, it is clear that the simulation results of the average number of neighbors are swinging about the theoretical value, with a major trend to be lower than the theoretical value. This is due to that there may be some sensors near the corner or the network edge, such that they have a lower number of neighbors.

Figure 9.9a and b shows the network coverage with and without mobility, respectively. This figure emphasizes that the mobility has no or very small effect on the number of sensors that cover each point in the network.

Figure 9.10 presents the average communication distance against the number of rounds; it is clear that mobility can save the communication distance by an amount of one third compared with that without mobility.

Figure 9.11 shows the average number of processes during self-healing against the number of rounds; it is clear that applying mobility can reduce the number of processes from 500 to 200, saving about 150% of processes.

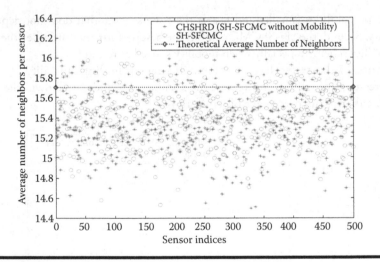

Figure 9.8 Average number of neighbors per sensor against sensor indices for CHSHRD and SH-SFCMC schemes.

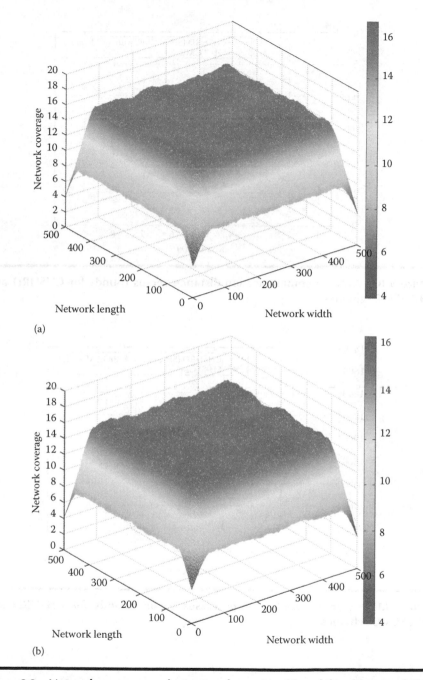

(a)

(b)

Figure 9.9 Network coverage against network area (a) with and (b) without mobility.

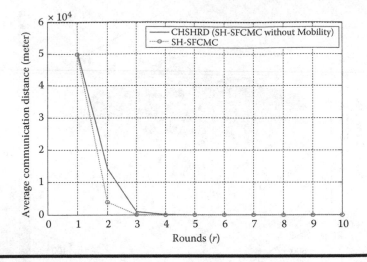

Figure 9.10 Average communication distance versus rounds for CHSHRD and SH-SFCMC schemes.

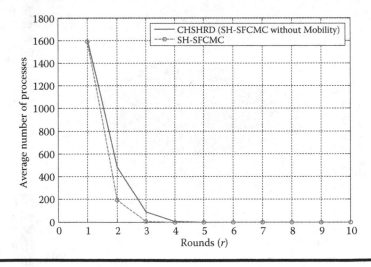

Figure 9.11 Average number of processes versus rounds for CHSHRD and SH-SFCMC schemes.

9.6 Summary

To the best of our knowledge, this is the first time to provide an extensive study of both the benefits and the impact of mobility for UWSNs. This chapter presented a mobility control scheme for enhancing security regaining in UWSNs, taking into consideration energy consumption. For the proposed SH-SFCMC scheme, a self-healing scheme based on a single flow controlled mobility within a cluster of sick sensors was presented. New metrics were defined, such as self-healing percentage, communication distance, and number of processes. All of these metrics can characterize the overall behavior of the UWSN. The proposed scheme provides solutions to the problem of self-healing associated in UWSN and the impact of energy consumption due to mobility. Simulation results have shown the feasibility of the proposed scheme that can provide significant performance enhancement on two aspects. First, mobility has no influence on energy consumption, which is one of the most important and critical network metrics. Hence, with the proposed scheme, the total energy consumption remains the same for the cases with and without mobility. In addition, the results showed that mobility does not disturb the number of neighbors per sensor and the network coverage. Second, the proposed scheme improves the UWSN security by enhancing both the probability of BSe being compromised and the self-healing percentage. In addition, the proposed scheme decreases the communication distance and the number of processes during self-healing.

Chapter 10

Conclusion and Future Work

10.1 Part One Conclusion

At the start of this study, the basic impairments of the wireless channel have been described. Then, the most popular diversity techniques which are used to address these impairments were illustrated. This is followed by a brief explanation for the MIMO system model as a spatial diversity-based model that aims to enhance the data rate and the transmission reliability. Then, the main problems of applying MIMO systems in small-size nodes, like those used in wireless cellular networks, are presented.

Next, the cooperative diversity has been studied as the most efficient virtual MIMO approach allowing the application of the MIMO system in small handheld mobiles without physically deploying antennas. Then, the relay channel concept, which is the basis for applying spatial diversity in cooperative communications, is illustrated. This is followed by a detailed overview of the cooperative communication working principle and its historical background. Then, different categories of cooperative communication protocols concerned with processing the source message at the relay node before forwarding it to the destination are described. Different relay selection metrics concerned with selecting the best relay among the available relays, which plays an important role in implementing cooperative diversity in distributed systems, were introduced. Then, various applications with pros and cons of cooperation were demonstrated.

In addition, the security in the wireless networks is handled to protect the transmitted messages against interception by unauthorized malicious receivers that lie in the coverage area of the transmission. At first, some issues concerned with the traditional security approaches and the main reasons which led us to recently study security approaches from a physical layer point of view have been briefly discussed. The basic idea behind the physical layer security was illustrated. This was followed by an overview about both key-based and keyless information-theoretic secrecy branches. Next, the multiple cooperative jamming techniques helping in achieving secrecy at the physical layer of a multiuser system were described. The idea of using cooperative jammers to improve security in multiple relay networks is also presented. Finally, at the end of this part, the interaction arising between cooperation and secrecy in the network with untrusted relays is studied.

In the next part of this thesis, various relay and jammer selection schemes have been proposed to improve the physical layer security in one-way cooperative networks. Firstly, the one-way network model that consists of one source, one destination, an eavesdropper, and multiple intermediate nodes is introduced. In the proposed schemes, an intermediate node is selected to operate as a DF relay that assists the source in delivering data to the corresponding destination. Meanwhile, another two intermediate nodes performed as jamming nodes are selected to transmit artificial interference for confusing eavesdroppers in different communication phases. The proposed schemes are analyzed for different complexity requirements based on global instantaneous knowledge of all links and average knowledge of the eavesdroppers' links.

The obtained results reveal that the proposed schemes with cooperative jamming can improve both the secrecy capacity and the secrecy outage probability metrics of the cooperative network. In addition to the investigation of these jamming-based selection schemes, it is observed that jamming is not always beneficial for security. Therefore, a hybrid scheme which switches between both jamming and nonjamming relay and jammer selection schemes is proposed to overcome jamming limitations. Moreover, the impact of changing the eavesdropper's location and the relay's location on the system performance metrics is also discussed. Finally, the effectiveness of the proposed selection schemes in improving the performance metrics in the presence of multiple eavesdroppers is shown.

At the end of this thesis, different joint relay and jammer selection schemes have been proposed to ensure secure communication in DF two-way cooperative networks. The two-way network model consisting of two sources, multiple intermediate nodes, and one or more eavesdroppers is firstly introduced. The proposed schemes select three intermediate nodes, one relay and two jammers during two communication phases. The performance of the proposed schemes is analyzed in terms of ergodic secrecy rate and secrecy outage probability.

The obtained results show the effectiveness of the jamming selection schemes over the nonjamming ones when the intermediate nodes are distributed dispersedly between the sources and the eavesdropper. However, when the intermediate nodes cluster gets close to one of the sources, they cause strong interference on the destination nodes. Therefore, a hybrid scheme is proposed to overcome jamming limitations and therefore seems to be an efficient solution for these critical secrecy constraints. Moreover, the impact of changing both the eavesdroppers and the intermediate nodes' location on the system's performance is discussed. Finally, the impact of the presence of multiple cooperating and noncooperating eavesdroppers on system performance metrics is discussed. The obtained results reveal that, despite the presence of multiple cooperating eavesdroppers, the proposed selection schemes are still able to improve both the secrecy rate and the secrecy outage probability of the two-way cooperative networks.

10.2 Part Two Conclusion

At the start of this study, the basic principles of the WSN have been described. This is followed by a detailed overview on a specific class of WSN called UWSN. As this type of networks is unattended and deployed in hostile environments, it faces security challenges. Different challenges of security and possible solutions have been studied. One of the best solutions for this type of network challenges is the self-healing based on cooperation process. Three proposals are introduced in this work for security regaining in UWSNs.

The first proposal is a CHSHRD scheme for sensor regaining secrecy and high data reliability in UWSNs. The CHSHRD employs both proactive peers and reactive peers to emphasize data secrecy and guarantee probabilistic BSe and compromised probability. Both theoretical and simulation results were used to evaluate the proposed scheme and compare its performance with the performance of other protection schemes. The results illustrated that the hybrid scheme provides better performance and higher protection than other schemes that use stand-alone proactive or reactive peers. The proposed scheme also reduces the storage overhead communication by reducing the number of sponsors to as little as possible. Compared with others, the CHSHRD scheme is more powerful and outperforms the Yi Ren and the naive schemes. The numerical results show that the CHSHRD scheme achieves a compromised probability of 0.07, compared with 0.21 and 0.22 for stand-alone reactive and proactive schemes, respectively, when the number of qualified neighbors equals 5. Moreover, the data reliability is about 96% for CHSHRD, compared with 70% for the naive scheme and 65% for the Yi Ren scheme.

The second proposed scheme is called SH-CCM; it combines both cooperation and infusion of mobility to enhance the ability of security regaining. SH-CCM guarantees probabilistic *BSe* to be compromised at a lower level compared with other schemes that have no mobility feature and that are based on a single type of peers. In addition, the SH-CCM scheme saves the number of rounds of self-healing processes until a steady state occurs. Also, the scheme needs a lower number of peers for self-healing. Due to a low number of rounds and peers, the SH-CCM scheme reduces the overhead communication, memory usage, and power consumption. The analytical analysis of the self-healing cluster controlled mobility scheme using both types of peers ensures better performance with respect to the backward secrecy being faster and lower than in the use of proactive peers alone. A set of analytical results are carried out to demonstrate the effectiveness of the proposed scheme in the presence of an adversary.

The third proposal presents a mobility control scheme for enhancing security regaining in UWSNs, taking into consideration energy consumption. It is called the SH-SFCMC scheme, a self-healing scheme based on a single flow controlled mobility within a cluster of sick sensors. New metrics were defined such as self-healing percentage, communication distance, and number of process. All of these metrics can characterize the overall behavior of the UWSN. The proposed scheme provides solutions to the problem of self-healing associated in UWSNs and the impact of energy consumption due to mobility. Simulation results have shown the feasibility of the proposed scheme that can provide significant performance enhancement on two aspects. Firstly, mobility has no effect on energy consumption, which is one of the most important and critical network metrics. Hence, with the proposed scheme, the total energy consumption remains the same for the cases with and without mobility. In addition, the results showed that mobility does not disturb the number of neighbors per sensor and the network coverage. Secondly, the proposed scheme improves the UWSN security by enhancing both the probability of BSe being compromised and the self-healing percentage. In addition, the proposed scheme improves the communication distance and the number of processes during self-healing.

10.3 Future Work

The work in this book can be extended to include the following points:

1. **Power Optimization Based on Link Condition**

   Wireless nodes generally have limited battery power. If relay based systems have some feedback mechanism, then power can be allocated based on link condition. Such dynamic allocation of the power may save battery power or boost the data transfer rate; hence, it is an optimization area to be investigated.

2. **Full-Duplex Operation of Relays**

   Relay operating in half-duplex mode creates a wide system bandwidth expansion. A full-duplex relay operating in single frequency can solve this problem. Therefore, the effect of full-duplex relay operation needs to be investigated.

3. **Complexity Performance Trade-Off**

   Relays can process the signal in a nonregenerative or regenerative mode depending on their functionality. The nonregenerative mode puts less processing burden on the relay as compared to the regenerative mode of operation;

hence, it is often preferred when complexity and/or latency is needed to be analyzed. Scope has been found to exist in the future for the nonregenerative mode of relay operation. Noise amplification is a major issue in this operation.

4. **Bidirectional User Cooperation**

   In this work, it is assumed that relay terminals do not have their own data and that they are just forwarding data received from the source. In the user cooperation technique, the user's terminal not only transmits its own data but also relays other users' data by sharing some of the resources. In this regard, it would be interesting to investigate bidirectional user cooperation extension.

5. **Base Station Cooperation**

   In a single cellular system, user terminals can communicate with the parent base station. Users near the outskirts of a cell can communicate with neighboring base stations and thus become a source for generating interfering signals. To overcome this, neighboring base stations can also cooperate with parent base stations and perform joint decoding of received signals. The work presented for relay cooperation can be extended for base station cooperation.

6. **Different Mobility Models**

   The mobility model used in this work is called RJCM, which depends mainly on the random jump mobility (RJM) model. There are other mobility models that can be used which differ in nature from RJM, such as the random waypoint model, the pathway mobility model, and the random walk model.

7. **Different Attackers**

   The adversary attacker considered here has the following characteristic: it is a proactive centralized logical adversary. Different characteristics for different types of attackers, such as the distributed denial of services (DDoS) attacker, can be applied, which has a different nature, compared with the adversary attack presented here. DDoS needs other techniques to be mitigated from the network, such as filtering and flooding techniques.

Appendix A: MATLAB® Simulation Codes for Chapter 4

```
%================================================================
              % First case "Eavesdropper near destination"
%================================================================

%relay selection techniques for implementing security at
%physical(PHY)layer
%direct links are available&unsecure broadcast phase is
assumed
%secrecy outage probability and secrecy ergodic capacity
versus
%(P)

  clear all
  clc
%================================================================
  %system parameters
%----------------------------------------------------------------
  N=4;              %number of relay nodes
  M=1;              %number of eavesdroppers
  Rs=.1;            %target secrecy rate (BPCU)-for outage
probability calculation
  R0=2;             %Transmission Rate-to check which relay
can decode correctly
  P_dB=0:5:40;      %Transmitted power (dB)
  Num_Trials=3000   %number of trials
  PL_exp=3;         %path loss exponent
  N0=1;             %noise variance
```

```
loc_s=[0 0];
loc_e=[.8 .2];
loc_d=[1 0];
loc_r=[.41 .58; .58 .44; .7 .35; .66 .7];

var_sd=(sqrt((loc_s(1)-loc_d(1)).^2+(loc_s(2)-
loc_d(2)).^2)).^(-PL_exp);
var_se=(sqrt((loc_s(1)-loc_e(1)).^2+(loc_s(2)-
loc_e(2)).^2)).^(-PL_exp);

z1=(loc_s(1)-loc_r(:,1)).^2;
z2=(loc_s(2)-loc_r(:,2)).^2;
z=sqrt(z1+z2);
var_sr=z.^(-PL_exp);

z1=(loc_d(1)-loc_r(:,1)).^2;
z2=(loc_d(2)-loc_r(:,2)).^2;
z=sqrt(z1+z2);
var_rd=z.^(-PL_exp);

z1=(loc_e(1)-loc_r(:,1)).^2;
z2=(loc_e(2)-loc_r(:,2)).^2;
z=sqrt(z1+z2);
var_re=z.^(-PL_exp);

%===========================
%simulation start
%===========================
for ii=1:length(P_dB)
    Ps=10^(P_dB(ii)/10);
    Pr=Ps;
    Pj=Ps/100;    %jammer power less than relay node power

    outage_cs=0;
    outage_os=0;
    outage_ss=0;
    outage_osj=0;
    outage_ssj=0;
    outage_oscj=0;
    outage_ow=0;
    outage_sw=0;

    for jj=1:Num_Trials

%=================================================================
        %Channel coefficients
%=================================================================
        %broadcasting phase
        hsd=sqrt(var_sd/2).*(randn(1,1)+i*randn(1,1));
%S-->destination
```

```
          hsrs=sqrt(var_sr/2).*(randn(N,1)+i*randn(N,1));
%S-->relays
          hse=sqrt(var_se/2).*(randn(1,1)+i*randn(1,1));
%S-->eavesdropper

          %cooperative links
          hrsd=sqrt(var_rd/2).*(randn(N,1)+i*randn(N,1));
%R-->destination
          hrse=sqrt(var_re/2).*(randn(N,1)+i*randn(N,1));
%R-->eavesdropper

%=============================================================
          %instantaneous SNR calculations
%=============================================================
          snr_sd=Ps*abs(hsd).^2;
          snr_srs=Ps*abs(hsrs).^2;
          snr_se=Ps*abs(hse).^2;
          snr_rsd=Pr*abs(hrsd).^2;
          snr_jsd=Pj*abs(hrsd).^2;
          snr_rse=Pr*abs(hrse).^2;
          snr_jse=Pj*abs(hrse).^2;
%=============================================================
          %relay selection process
%=============================================================
          Csrs=0.5*log2(1+snr_srs);
          y=Csrs>R0;    %all valid relays
%=============================================================
          %selection techniques without jamming
%=============================================================
          %case1:Conventional Selection(CS)

          [q1 q2]=max(1+snr_sd+ y.* snr_rsd);

          Q1=(q1==0);          %Q1=1 ==> all relays fails to
decode correctly

          Cs_cs(1,jj)=0.5*max(0,(~Q1)*log2(1+snr_sd+snr_
rsd(q2))-(~Q1)*log2(1+snr_se+snr_rse(q2)))  ;

          outage_cs=outage_cs+(Rs>Cs_cs(1,jj));

          %case2:Optimal Selection(OS)

          [q1 q2]=max((1+snr_sd+y.* snr_rsd)./(1+snr_se+y.*
snr_rse));

          Q1=(q1==0);          %Q1=1 ==> all relays fails to
decode correctly

          Cs_os(1,jj)=0.5*max(0,(~Q1)*log2(1+snr_sd+snr_rsd
(q2))-(~Q1)*log2(1+snr_se+snr_rse(q2)));
```

```
        outage_os=outage_os+(Rs>Cs_os(1,jj));

        %case3:Suboptimal Selection(SS)

        [q1 q2]=max((1+snr_sd+y.* snr_rsd)./((1+mean(snr_
se))+y.*(mean (snr_rse))));

        Q1=(q1==0);            %Q1=1 ==> all relays fails to
decode correctly

Cs_ss(1,jj)=0.5*max(0,(~Q1)*log2(1+snr_sd+snr_rsd(q2))-
(~Q1)*log2(1+mean(snr_se)+mean(snr_rse(q2)))) ;

        outage_ss=outage_ss+(Rs>Cs_ss(1,jj));

%===============================================================
        %selection techniques for jamming
%===============================================================

    %case1:Optimal selection with jamming(OSJ)-individual
selection

        [q3 q4]=max(snr_jse./snr_jsd);              %selecting
jammer in first phase
        [q1 q2]=max(y.*(snr_rsd./snr_rse));         %selecting
relay in second phase
        Q1=(q1==0);                                 %Q1=1--->all
relays fail to decode correctly
        temp=ones(N,1);
        temp(q2,1)=0;
        [q5 q6]=max(temp.*snr_jse./snr_jsd);        %selecting
jammer in second phase

Cs_osj(1,jj)=0.5*max(0,(~Q1)*log2(1+(snr_sd./(1+snr_
jsd(q4)))+(snr_rsd(q2)./(1+snr_jsd(q6))))-(~Q1)*log2(1+(snr_
se./(1+snr_jse(q4)))+(snr_rse(q2)./(1+snr_jse(q6)))));

        outage_osj=outage_osj+(Rs>Cs_osj(1,jj));

        %case2:Suboptimal selection with jamming(SSJ)

        [q3 q4]=max(mean(snr_jse)./snr_jsd); %selecting
jammer in first phase
        [q1 q2]=max(y.*(snr_rsd./mean(snr_rse))); %selecting
relay in second phase
        Q1=(q1==0);                                 %Q1=1---
>all relays fail to decode correctly
        temp=ones(N,1);
        temp(q2,1)=0;
        [q5 q6]=max(temp.*(mean(snr_jse))./snr_jsd); %selecting
jammer in second phase
```

```
Cs_ssj(1,jj)=0.5*max(0,(~Q1)*log2(1+(snr_sd./(1+snr_
jsd(q4)))+(snr_rsd(q2)./(1+snr_jsd(q6))))-
(~Q1)*log2(1+(mean(snr_se)./
(1+mean(snr_jse(q4))))+(mean(snr_rse(q2))./
(1+mean(snr_jse(q6)))))));

        outage_ssj=outage_ssj+(Rs>Cs_ssj(1,jj));

%=================================================================
        %Optimal selection with controlled jamming(OSCJ)
%=================================================================
        [q3 q4]=max((snr_jse));                    %selecting
jammer in first phase
        [q1 q2]=max(y.*(snr_rsd./snr_rse));        %selecting
relay in second phase
        Q1=(q1==0);                                    %Q1=1---
>all relays fail to decode correctly
        temp=ones(N,1);
        temp(q2,1)=0;
        [q5 q6]=max(temp.*snr_jse); %selecting
jammer in second phase

Cs_oscj(1,jj)=0.5*max(0,(~Q1)*log2(1+snr_sd+snr_rsd(q2))-
(~Q1)*log2(1+(snr_se./(1+snr_jse(q4)))+(snr_rse(q2)./
(1+snr_jse(q6)))));

        outage_oscj=outage_oscj+(Rs>Cs_oscj(1,jj));

%=================================================================
        %Hybrid schemes(switching between jamming & non
jamming relay
        %selection techniques)
%=================================================================

        %case1:Optimal switching(OW)

        [q3 q4]=max(snr_jse./snr_jsd); %selecting
jammer in first phase
        [q1 q2]=max(y.*(snr_rsd./snr_rse)); %selecting
relay in second phase

        Q1=(q1==0);                                    %Q1=1---
>all relays fail to decode correctly
        temp=ones(N,1);
        temp(q2,1)=0;
        [q5 q6]=max(temp.*snr_jse./snr_jsd);       %selecting
jammer in second phase

        if((snr_jse(q4)>snr_jsd(q4))&(snr_jse(q6)>snr_jsd(q6)))
```

```
Cs_ow(1,jj)=0.5*max(0,(~Q1)*log2(1+(snr_sd./(1+snr_
jsd(q4)))+(snr_rsd(q2)./(1+snr_jsd(q6))))-(~Q1)*log2
(1+(snr_se./(1+snr_jse(q4)))+(snr_rse(q2)./(1+snr_jse(q6))))));

        else

            [q1 q2]=max((1+snr_sd+y.* snr_rsd)./(1+snr_se+y.*
snr_rse));

            Cs_ow(1,jj)=0.5*max(0,(~Q1)*log2(1+snr_sd+snr_rsd
(q2))-(~Q1)*log2(1+snr_se+snr_rse(q2)));

        end

        outage_ow=outage_ow+(Rs>Cs_ow(1,jj));

        %case2:Suboptimal switching(SW)

        [q3 q4]=max(mean(snr_jse)./snr_jsd);
%selecting jammer in first phase
        [q1 q2]=max(y.*(snr_rsd./mean(snr_rse)));
%selecting relay in second phase

        Q1=(q1==0);                                    %Q1=1-
-->all relays fail to decode correctly
        temp=ones(N,1);
        temp(q2,1)=0;
        [q5 q6]=max(temp.*(mean(snr_jse))./snr_jsd);
%selecting jammer in second phase
if((mean(snr_jse(q4))>snr_jsd(q4))&(mean(snr_jse(q6))>snr_
jsd(q6)))

Cs_sw(1,jj)=0.5*max(0,(~Q1)*log2(1+(snr_sd./(1+snr_
jsd(q4)))+(snr_rsd(q2)./(1+snr_jsd(q6))))-
(~Q1)*log2(1+(mean(snr_se)./(1+mean(snr_jse(q4))))+
(mean(snr_rse(q2))./(1+mean(snr_jse(q6))))));

        else

            [q1 q2]=max((1+snr_sd+y.*
snr_rsd)./((1+mean(snr_se))+y.*(mean (snr_rse))));
            Cs_sw(1,jj)=
0.5*max(0,(~Q1)*log2(1+snr_sd+snr_rsd(q2))-
(~Q1)*log2(1+mean(snr_se)+mean(snr_rse(q2))));

        end

        outage_sw=outage_sw+(Rs>Cs_sw(1,jj));

        end
```

```
      out_cs(ii)=outage_cs/Num_Trials
      c_cs(ii)=mean(Cs_cs)

      out_os(ii)=outage_os/Num_Trials
      c_os(ii)=mean(Cs_os)

      out_ss(ii)=outage_ss/Num_Trials
      c_ss(ii)=mean(Cs_ss)

      out_osj(ii)=outage_osj/Num_Trials
      c_osj(ii)=mean(Cs_osj)

      out_ssj(ii)=outage_ssj/Num_Trials
      c_ssj(ii)=mean(Cs_ssj)

      out_oscj(ii)=outage_oscj/Num_Trials
      c_oscj(ii)=mean(Cs_oscj)

      out_ow(ii)=outage_ow/Num_Trials
      c_ow(ii)=mean(Cs_ow)

      out_sw(ii)=outage_sw/Num_Trials
      c_sw(ii)=mean(Cs_sw)

 end
 plot( P_dB, c_cs,'k-o',P_dB, c_os,'b-x',P_dB, c_ss,'r-s',P_dB,
c_osj,'g:<',...
      P_dB, c_ow,'b-.+',P_dB, c_ssj,'c:>', P_dB, c_sw,'r-.d',P_
dB, c_oscj,'m-p')

 xlabel('P[dB]');
 ylabel('secrecy ergodic capacity');
 grid on; box on;
 axis([0 40 0 3]);
 legend ('CS','OS','SS','OSJ','OW','SSJ','SW','OSCJ',2);
 figure

 semilogy(P_dB,out_cs,'k-o',P_dB, out_os,'b-x',P_dB, out_ss,
'r-s',P_dB, out_osj,'g:<',...
      P_dB, out_ow,'b-.+',P_dB, out_ssj,'c:>', P_dB,out_sw,
'r-.d',P_dB, out_oscj,'m-p')

 xlabel('P[dB]');
 ylabel('secrecy outage probability');
 grid on; box on;
 axis([0 40 10^-3 1]);
  legend ('CS','OS','SS','OSJ','OW','SSJ','SW','OSCJ',3);
 figure
```

```
%================================================================
                % Second case "Eavesdropper near source"
%================================================================


 %relay selection techniques for implementing security at
physical(PHY)layer
 %direct links are available&unsecure broadcast phase is
assumed
 %secrecy outage probability and secrecy ergodic capacity
versus (P)

 clear all
 clc
.
%================================================================
 %system parameters
.
%================================================================
 N=4;                   %number of relay nodes
 M=1;                   %number of eavesdroppers
 Rs=.1;                 %target secrecy rate (BPCU)-for outage
probability calculation
 R0=2;                  %Transmission Rate-to check which relay
can decode correctly
 P_dB=0:5:40;           %Transmitted power (dB)
 Num_Trials=3000;       %number of trials
 PL_exp=3;              %path loss exponent
 N0=1;                  %noise variance

 loc_s=[0 0];
 loc_e=[.2 .2];
 loc_d=[1 0];
 loc_r=[.41 .58; .58 .44; .7 .35; .66 .7];

 var_sd=(sqrt((loc_s(1)-loc_d(1)).^2+(loc_s(2)-
loc_d(2)).^2)).^(-PL_exp);
 var_se=(sqrt((loc_s(1)-loc_e(1)).^2+(loc_s(2)-
loc_e(2)).^2)).^(-PL_exp);

 z1=(loc_s(1)-loc_r(:,1)).^2;
 z2=(loc_s(2)-loc_r(:,2)).^2;
 z=sqrt(z1+z2);
 var_sr=z.^(-PL_exp);

 z1=(loc_d(1)-loc_r(:,1)).^2;
 z2=(loc_d(2)-loc_r(:,2)).^2;
 z=sqrt(z1+z2);
 var_rd=z.^(-PL_exp);
```

```
z1=(loc_e(1)-loc_r(:,1)).^2;
z2=(loc_e(2)-loc_r(:,2)).^2;
z=sqrt(z1+z2);
var_re=z.^(-PL_exp);

%=========================
%simulation start
%=========================
for ii=1:length(P_dB)
    Ps=10^(P_dB(ii)/10);
    Pr=Ps;
    Pj=Ps/100;    %jammer power less than relay node power

    outage_cs=0;
    outage_os=0;
    outage_ss=0;
    outage_osj=0;
    outage_ssj=0;
    outage_oscj=0;
    outage_ow=0;
    outage_sw=0;

    for jj=1:Num_Trials
%===============================================================
        %Channel coefficients
%===============================================================
        %broadcasting phase
        hsd=sqrt(var_sd/2).*(randn(1,1)+i*randn(1,1));
%S-->destination
        hsrs=sqrt(var_sr/2).*(randn(N,1)+i*randn(N,1));
%S-->relays
        hse=sqrt(var_se/2).*(randn(1,1)+i*randn(1,1));
%S-->eavesdropper

        %cooperative links
        hrsd=sqrt(var_rd/2).*(randn(N,1)+i*randn(N,1));
%R-->destination
        hrse=sqrt(var_re/2).*(randn(N,1)+i*randn(N,1));
%R-->eavesdropper

%===============================================================
        %instantaneous SNR calculations
%===============================================================
        snr_sd=Ps*abs(hsd).^2;
        snr_srs=Ps*abs(hsrs).^2;
        snr_se=Ps*abs(hse).^2;
        snr_rsd=Pr*abs(hrsd).^2;
        snr_jsd=Pj*abs(hrsd).^2;
        snr_rse=Pr*abs(hrse).^2;
        snr_jse=Pj*abs(hrse).^2;
```

```
%===============================================================
          %relay selection process
%===============================================================
          Csrs=0.5*log2(1+snr_srs);
          y=Csrs>R0;    %all valid relays
%===============================================================
          %selection techniques without jamming
%===============================================================
          %case1:Conventional Selection(CS)

          [q1 q2]=max(1+snr_sd+ y.* snr_rsd);

          Q1=(q1==0);            %Q1=1 ==> all relays fails to
decode correctly

          Cs_cs(1,jj)=0.5*max(0,(~Q1)*log2(1+snr_sd+snr_rsd(q2))-
(~Q1)*log2(1+snr_se+snr_rse(q2)))  ;


          outage_cs=outage_cs+(Rs>Cs_cs(1,jj));


          %case2:Optimal Selection(OS)


          [q1 q2]=max((1+snr_sd+y.* snr_rsd)./(1+snr_se+y.*
snr_rse));

          Q1=(q1==0);            %Q1=1 ==> all relays fails to
decode correctly

        Cs_os(1,jj)=0.5*max(0,(~Q1)*log2(1+snr_sd+snr_rsd
(q2))-(~Q1)*log2(1+snr_se+snr_rse(q2)));

          outage_os=outage_os+(Rs>Cs_os(1,jj));

           %case3:Suboptimal Selection(SS)

          [q1 q2]=max((1+snr_sd+y.*
snr_rsd)./((1+mean(snr_se))+y.*(mean (snr_rse))));

          Q1=(q1==0);            %Q1=1 ==> all relays fails to
decode correctly

Cs_ss(1,jj)=0.5*max(0,(~Q1)*log2(1+snr_sd+snr_rsd(q2))-
(~Q1)*log2(1+mean(snr_se)+mean(snr_rse(q2))))  ;

          outage_ss=outage_ss+(Rs>Cs_ss(1,jj));
```

```
%==============================================================
          %selection techniques for jamming
%==============================================================

     %case1:Optimal selection with jamming(OSJ)-individual
selection

          [q3 q4]=max(snr_jse./snr_jsd);              %selecting
jammer in first phase
          [q1 q2]=max(y.*(snr_rsd./snr_rse));          %selecting
relay in second phase
          Q1=(q1==0);                                %Q1=1--->all
relays fail to decode correctly
          temp=ones(N,1);
          temp(q2,1)=0;
          [q5 q6]=max(temp.*snr_jse./snr_jsd);        %selecting
jammer in second phase

Cs_osj(1,jj)=0.5*max(0,(~Q1)*log2(1+(snr_sd./(1+snr_
jsd(q4)))+(snr_rsd(q2)./(1+snr_jsd(q6))))-
(~Q1)*log2(1+(snr_se./(1+snr_jse(q4)))+(snr_rse(q2)./
(1+snr_jse(q6)))));

          outage_osj=outage_osj+(Rs>Cs_osj(1,jj));

          %case2:Suboptimal selection with jamming(SSJ)

          [q3 q4]=max(mean(snr_jse)./snr_jsd); %selecting
jammer in first phase
          [q1 q2]=max(y.*(snr_rsd./mean(snr_rse))); %selecting
relay in second phase
          Q1=(q1==0);                                %Q1=1---
>all relays fail to decode correctly
          temp=ones(N,1);
          temp(q2,1)=0;
          [q5 q6]=max(temp.*(mean(snr_jse))./snr_jsd); %selecting
jammer in second phase

Cs_ssj(1,jj)=0.5*max(0,(~Q1)*log2(1+(snr_sd./(1+snr_jsd(q4)))+
(snr_rsd(q2)./(1+snr_jsd(q6))))-(~Q1)*log2(1+(mean(snr_se)./
(1+mean(snr_jse(q4))))+(mean(snr_rse(q2))./
(1+mean(snr_jse(q6))))));

          outage_ssj=outage_ssj+(Rs>Cs_ssj(1,jj));

%==============================================================
          %Optimal selection with controlled jamming(OSCJ)
%==============================================================
          [q3 q4]=max((snr_jse));                     %selecting
jammer in first phase
```

```
        [q1 q2]=max(y.*(snr_rsd./snr_rse));           %selecting
relay in second phase
        Q1=(q1==0);                                    %Q1=1--->all
relays fail to decode correctly
        temp=ones(N,1);
        temp(q2,1)=0;
        [q5 q6]=max(temp.*snr_jse); %selecting
jammer in second phase

Cs_oscj(1,jj)=0.5*max(0,(~Q1)*log2(1+snr_sd+snr_rsd(q2))-
(~Q1)*log2(1+(snr_se./(1+snr_jse(q4)))+(snr_rse(q2)./
(1+snr_jse(q6)))));

        outage_oscj=outage_oscj+(Rs>Cs_oscj(1,jj));

%================================================================
        %Hybrid schemes(switching between jamming & non jamming
relay
        %selection techniques)
%================================================================

        %case1:Optimal switching(OW)

        [q3 q4]=max(snr_jse./snr_jsd); %selecting
jammer in first phase
        [q1 q2]=max(y.*(snr_rsd./snr_rse)); %selecting
relay in second phase

        Q1=(q1==0);                                    %Q1=1---
>all relays fail to decode correctly
        temp=ones(N,1);
        temp(q2,1)=0;
        [q5 q6]=max(temp.*snr_jse./snr_jsd);          %selecting
jammer in second phase

    if((snr_jse(q4)>snr_jsd(q4))&(snr_jse(q6)>snr_jsd(q6)))

Cs_ow(1,jj)=0.5*max(0,(~Q1)*log2(1+(snr_sd./(1+snr_
jsd(q4)))+(snr_rsd(q2)./(1+snr_jsd(q6))))-
(~Q1)*log2(1+(snr_se./(1+snr_jse(q4)))+(snr_rse(q2)./
(1+snr_jse(q6)))));

        else

        [q1 q2]=max((1+snr_sd+y.* snr_rsd)./(1+snr_se+y.*
snr_rse));

        Cs_ow(1,jj)=0.5*max(0,(~Q1)*log2(1+snr_sd+snr_
rsd(q2))-(~Q1)*log2(1+snr_se+snr_rse(q2)));

        end
```

```
        outage_ow=outage_ow+(Rs>Cs_ow(1,jj));

    %case2:Suboptimal switching(SW)

    [q3 q4]=max(mean(snr_jse)./snr_jsd);
%selecting jammer in first phase
    [q1 q2]=max(y.*(snr_rsd./mean(snr_rse)));
%selecting relay in second phase

    Q1=(q1==0);                                %Q1=1---
>all relays fail to decode correctly
    temp=ones(N,1);
    temp(q2,1)=0;
    [q5 q6]=max(temp.*(mean(snr_jse))./snr_jsd);
%selecting jammer in second phase

if((mean(snr_jse(q4))>snr_jsd(q4))&(mean(snr_jse(q6))>snr_
jsd(q6)))

Cs_sw(1,jj)=0.5*max(0,(~Q1)*log2(1+(snr_sd./(1+snr_
jsd(q4)))+(snr_rsd(q2)./(1+snr_jsd(q6))))-(~Q1)*log2
(1+(mean(snr_se)./(1+mean(snr_jse(q4))))+(mean(snr_rse(q2))./
(1+mean(snr_jse(q6)))))));

    else

    [q1 q2]=max((1+snr_sd+y.*
snr_rsd)./((1+mean(snr_se))+y.*(mean (snr_rse))));
    Cs_sw(1,jj)=
    0.5*max(0,(~Q1)*log2(1+snr_sd+snr_rsd(q2))-
(~Q1)*log2(1+mean(snr_se)+mean(snr_rse(q2)))) ;

    end

    outage_sw=outage_sw+(Rs>Cs_sw(1,jj));

    end

out_cs(ii)=outage_cs/Num_Trials
c_cs(ii)=mean(Cs_cs)

 out_os(ii)=outage_os/Num_Trials
c_os(ii)=mean(Cs_os)

out_ss(ii)=outage_ss/Num_Trials
c_ss(ii)=mean(Cs_ss)

out_osj(ii)=outage_osj/Num_Trials
c_osj(ii)=mean(Cs_osj)

out_ssj(ii)=outage_ssj/Num_Trials
```

```
    c_ssj(ii)=mean(Cs_ssj)

    out_oscj(ii)=outage_oscj/Num_Trials
    c_oscj(ii)=mean(Cs_oscj)

    out_ow(ii)=outage_ow/Num_Trials
    c_ow(ii)=mean(Cs_ow)

    out_sw(ii)=outage_sw/Num_Trials
    c_sw(ii)=mean(Cs_sw)

end
plot( P_dB, c_cs,'k-o',P_dB, c_os,'b-x',P_dB, c_ss,'r-s',P_dB,
c_osj,'g:<',...
    P_dB, c_ow,'b-.+',P_dB, c_ssj,'c:>', P_dB, c_sw,'r-
.d',P_dB,c_oscj,'m-p')

xlabel('P[dB]');
ylabel('secrecy ergodic capacity');
grid on; box on;
axis([0 40 0 3]);
legend ('CS','OS','SS','OSJ','OW','SSJ','SW','OSCJ',2);
figure

semilogy(P_dB,out_cs,'k-o',P_dB, out_os,'b-x',P_dB, out_
ss,'r-s',P_dB, out_osj,'g:<',...
    P_dB, out_ow,'b-.+',P_dB, out_ssj,'c:>', P_dB,out_sw,'r-
.d',P_dB, out_oscj,'m-p')

xlabel('P[dB]');
ylabel('secrecy outage probability');
grid on; box on;
axis([0 40 10^-3 1]);
  legend ('CS','OS','SS','OSJ','OW','SSJ','SW','OSCJ',3);
figure


%==================================================================
%              % Number of Eavesdropper = 2"
%==================================================================

clear all
clc

%==================================================================
%system parameters
%==================================================================
N=4;                    %number of relay nodes
M=2;                    %number of eavesdroppers
```

```
 Rs=.1;                   %target secrecy rate (BPCU)-for outage
probability calculation
 R0=2;                    %Transmission Rate-to check which relay can
decode correctly
 P_dB=0:5:40;             %Transmitted power (dB)
 Num_Trials=3000;  %number of trials
 PL_exp=3;                %path loss exponent
 N0=1;                    %noise variance

 loc_s=[0 0];
 loc_e=[0 1;1 1];
 loc_d=[1 0];
 loc_r=[.41 .58; .58 .44; .7 .35; .66 .7];

 var_sd=(sqrt((loc_s(1)-loc_d(1)).^2+(loc_s(2)-
 loc_d(2)).^2)).^(-PL_exp);
 z1=(loc_s(1)-loc_e(:,1)).^2;
 z2=(loc_s(2)-loc_e(:,2)).^2;
 z=sqrt(z1+z2);
 var_ses=z.^(-PL_exp);

 z1=(loc_s(1)-loc_r(:,1)).^2;
 z2=(loc_s(2)-loc_r(:,2)).^2;
 z=sqrt(z1+z2);
 var_srs=z.^(-PL_exp);

 z1=(loc_d(1)-loc_r(:,1)).^2;
 z2=(loc_d(2)-loc_r(:,2)).^2;
 z=sqrt(z1+z2);
 var_rsd=z.^(-PL_exp);

 z1=(loc_r(:,1)-loc_e(1,1)).^2;
 z2=(loc_r(:,2)-loc_e(1,2)).^2;
 z11=sqrt(z1+z2);
 z3=(loc_r(:,1)-loc_e(2,1)).^2;
 z4=(loc_r(:,2)-loc_e(2,2)).^2;
 z22=sqrt(z3+z4);
 z=[z11 z22];
 var_rses=z.^(-PL_exp);

 %===========================
 %simulation start
 %===========================
 for ii=1:length(P_dB)
     Ps=10^(P_dB(ii)/10);
     Pr=Ps;
     Pj=Ps/100;    %jammer power less than relay node power

     outage_os=0;
     outage_osj=0;
```

```
        outage_oscj=0;

        for jj=1:Num_Trials

%================================================================
        %Channel coefficients
%================================================================
        %broadcasting phase
        hsd=sqrt(var_sd/2).*(randn(1,1)+i*randn(1,1));
%S-->destination
        hsrs=sqrt(var_srs/2).*(randn(N,1)+i*randn(N,1));
%S-->relays
        hses=sqrt(var_ses/2).*(randn(M,1)+i*randn(M,1));
        %cooperative links
        hrsd=sqrt(var_rsd/2).*(randn(N,1)+i*randn(N,1));
%R-->destination
        hrses=sqrt(var_rses/2).*(randn(N,M)+i*randn(N,M));
% R-->eavesdropper

%================================================================
        %instantaneous SNR calculations
%================================================================
        snr_sd=Ps*abs(hsd).^2;
        snr_srs=Ps*abs(hsrs).^2;
        snr_ses=Ps*abs(hses).^2;
        snr_rsd=Pr*abs(hrsd).^2;
        snr_jsd=Pj*abs(hrsd).^2;
        snr_rses=Pr*abs(hrses).^2;
        snr_jses=Pj*abs(hrses).^2;

%================================================================
        %relay selection process
%================================================================
        Csrs=0.5*log2(1+snr_srs);
        y=Csrs>R0;    %all valid relays

%================================================================
        %Optimal selection technique without jamming(OS)
%================================================================

        F=[];
        for k1=1:N

        C=0.5*max(0,log2(1+snr_sd+y(k1)*snr_rsd(k1))-log2
(1+sum(snr_ses+y(k1)*snr_rses(k1,:)')));
        F=[F C];
        end
        Cs_os(1,jj)=max(F);
        outage_os=outage_os+(Rs>Cs_os(1,jj));
```

```
%================================================================
        % Selection Techniques with jamming
%================================================================

            %case1:Conventional jamming(OSJ)

            [q3 q4]=max(sum(snr_jses,2)./snr_jsd);
%selecting jammer in first phase
            [q1 q2]=max(y.*(snr_rsd./sum(snr_rses,2)));
%selecting relay in second phase
            Q1=(q1==0);                              %Q1=1---
>all relays fail to decode correctly
            temp=ones(N,1);
            temp(q2,1)=0;
            [q5 q6]=max(temp.*(sum(snr_jses,2)./snr_jsd));
%selecting jammer in second phase

Cs_osj(1,jj)=0.5*max(0,log2(1+(snr_sd./(1+snr_jsd
(q4)))+(y(q2)*snr_rsd(q2)./(1+snr_jsd(q6))))-log2(1+sum((snr_
ses./(1+snr_jses(q4,:)'))+((y(q2)*(snr_rses(q2,:)'))./
(1+snr_jses(q6,:)')))));

            outage_osj=outage_osj+(Rs>Cs_osj(1,jj));

            %case2: controlled jamming(OSCJ)

            [q3 q4]=max(sum(snr_jses,2));
%selecting jammer in first phase
            [q1 q2]=max(y.*(snr_rsd./sum(snr_rses,2)));
%selecting relay in second phase
            Q1=(q1==0);                              %Q1=1---
>all relays fail to decode correctly
            temp=ones(N,1);
            temp(q2,1)=0;
            [q5 q6]=max(temp.*(sum(snr_jses,2)));
%selecting jammer in second phase

Cs_oscj(1,jj)=0.5*max(0,log2(1+snr_sd+(y(q2)*snr_rsd(q2)))-
log2(1+sum((snr_ses./(1+snr_jses(q4,:)'))+((y(q2)*snr_
rses(q2,:)')./(1+snr_jses(q6,:)')))));

            outage_oscj=outage_oscj+(Rs>Cs_oscj(1,jj));

    end

    out_os(ii)=outage_os/Num_Trials
    c_os(ii)=mean(Cs_os)

    out_osj(ii)=outage_osj/Num_Trials
    c_osj(ii)=mean(Cs_osj)
```

```
      out_oscj(ii)=outage_oscj/Num_Trials
      c_oscj(ii)=mean(Cs_oscj)

end
  plot( P_dB, c_os,'b-x',P_dB, c_osj,'g:<',P_dB, c_oscj,'m-p')

xlabel('P[dB]');
ylabel('secrecy ergodic capacity');
grid on; box on;
axis([0 40 0 3]);
legend ('OS','OSJ','OSCJ',2);
figure

semilogy(P_dB, out_os,'b-x',P_dB, out_osj,'g:<',P_dB,
out_oscj,'m-p')

xlabel('P[dB]');
ylabel('secrecy outage probability');
grid on; box on;
axis([0 40 10^-3 1]);
  legend ('OS','OSJ','OSCJ',3);
figure


...


%=================================================================
                % Number of Eavesdropper = 3"
%=================================================================

        clear all
  clc
  .
%=================================================================
%system parameters
%=================================================================
 N=4;                %number of relay nodes
 M=3;                %number of eavesdroppers
 Rs=.1;              %target secrecy rate (BPCU)-for outage
probability calculation
 R0=2;               %Transmission Rate-to check which relay
can decode correctly
 P_dB=0:5:40;        %Transmitted power (dB)
 Num_Trials=3000;   %number of trials
 PL_exp=3;           %path loss exponent
 N0=1;               %noise variance

 loc_s=[0 0];
 loc_e=[0 1;1 1;.5 .5];
```

```
loc_d=[1 0];
loc_r=[.41 .58; .58 .44; .7 .35; .66 .7];

var_sd=(sqrt((loc_s(1)-loc_d(1)).^2+(loc_s(2)-
loc_d(2)).^2)).^(-PL_exp);
z1=(loc_s(1)-loc_e(:,1)).^2;
z2=(loc_s(2)-loc_e(:,2)).^2;
z=sqrt(z1+z2);
var_ses=z.^(-PL_exp);

z1=(loc_s(1)-loc_r(:,1)).^2;
z2=(loc_s(2)-loc_r(:,2)).^2;
z=sqrt(z1+z2);
var_sr=z.^(-PL_exp);

z1=(loc_d(1)-loc_r(:,1)).^2;
z2=(loc_d(2)-loc_r(:,2)).^2;
z=sqrt(z1+z2);
var_rd=z.^(-PL_exp);

z1=(loc_r(:,1)-loc_e(1,1)).^2;
z2=(loc_r(:,2)-loc_e(1,2)).^2;
z11=sqrt(z1+z2);
z3=(loc_r(:,1)-loc_e(2,1)).^2;
z4=(loc_r(:,2)-loc_e(2,2)).^2;
z22=sqrt(z3+z4);
z5=(loc_r(:,1)-loc_e(3,1)).^2;
z6=(loc_r(:,2)-loc_e(3,2)).^2;
z33=sqrt(z5+z6);
z=[z11 z22 z33];
var_rses=z.^(-PL_exp);

%===========================
%simulation start
%===========================
for ii=1:length(P_dB)
    Ps=10^(P_dB(ii)/10);
    Pr=Ps;
    Pj=Ps/100;    %jammer power less than relay node power

    outage_os=0;
    outage_osj=0;
    outage_oscj=0;

    for jj=1:Num_Trials

%================================================================
        %Channel coefficients
%================================================================
        %broadcasting phase
```

```
        hsd=sqrt(var_sd/2).*(randn(1,1)+i*randn(1,1));
%S-->destination
        hsrs=sqrt(var_sr/2).*(randn(N,1)+i*randn(N,1));
%S-->relays
        hses=sqrt(var_ses/2).*(randn(M,1)+i*randn(M,1));
        %cooperative links
        hrsd=sqrt(var_rd/2).*(randn(N,1)+i*randn(N,1));
%R-->destination
        hrses=sqrt(var_rses/2).*(randn(N,M)+i*randn(N,M));
% R-->eavesdropper


%=============================================================
        %instantaneous SNR calculations
%=============================================================
        snr_sd=Ps*abs(hsd).^2;
        snr_srs=Ps*abs(hsrs).^2;
        snr_ses=Ps*abs(hses).^2;
        snr_rsd=Pr*abs(hrsd).^2;
        snr_jsd=Pj*abs(hrsd).^2;
        snr_rses=Pr*abs(hrses).^2;
        snr_jses=Pj*abs(hrses).^2;


%=============================================================
        %relay selection process
%=============================================================
        Csrs=0.5*log2(1+snr_srs);
        y=Csrs>R0;    %all valid relays


%=============================================================
        %Optimal selection technique without jamming(OS)
%=============================================================

        F=[];
        for k1=1:N

        C=0.5*max(0,log2(1+snr_sd+y(k1)*snr_rsd(k1))-
log2(1+sum(snr_ses+y(k1)*snr_rses(k1,:)')));
        F=[F C];
        end
        Cs_os(1,jj)=max(F);
        outage_os=outage_os+(Rs>Cs_os(1,jj));


%=============================================================
        % Selection Techniques with jamming
%=============================================================

        %case1:Conventional jamming(OSJ)

        [q3 q4]=max(sum(snr_jses,2)./snr_jsd);
%selecting jammer in first phase
```

```
          [q1 q2]=max(y.*(snr_rsd./sum(snr_rses,2)));
%selecting relay in second phase
          Q1=(q1==0);                              %Q1=1---
>all relays fail to decode correctly
          temp=ones(N,1);
          temp(q2,1)=0;
          [q5 q6]=max(temp.*(sum(snr_jses,2)./snr_jsd));
%selecting jammer in second phase

Cs_osj(1,jj)=0.5*max(0,log2(1+(snr_sd./(1+snr_
jsd(q4)))+(y(q2)*snr_rsd(q2)./(1+snr_jsd(q6))))-
log2(1+sum((snr_ses./(1+snr_jses(q4,:)'))+
((y(q2)*(snr_rses(q2,:)'))./(1+snr_jses(q6,:)')))));

          outage_osj=outage_osj+(Rs>Cs_osj(1,jj));

          %case2: controlled jamming(OSCJ)

          [q3 q4]=max(sum(snr_jses,2));
%selecting jammer in first phase
          [q1 q2]=max(y.*(snr_rsd)./sum(snr_rses,2));
%selecting relay in second phase
          Q1=(q1==0);
%Q1=1--->all relays fail to decode correctly
          temp=ones(N,1);
          temp(q2,1)=0;
          [q5 q6]=max(temp.*(sum(snr_jses,2)));
%selecting jammer in second phase

Cs_oscj(1,jj)=0.5*max(0,log2(1+snr_sd+(y(q2)*snr_rsd(q2)))-
log2(1+sum((snr_ses./(1+snr_jses(q4,:)'))+((y(q2)*snr_
rses(q2,:)')./(1+snr_jses(q6,:)')))));

          outage_oscj=outage_oscj+(Rs>Cs_oscj(1,jj));

     end

     out_os(ii)=outage_os/Num_Trials
     c_os(ii)=mean(Cs_os)

     out_osj(ii)=outage_osj/Num_Trials
     c_osj(ii)=mean(Cs_osj)

     out_oscj(ii)=outage_oscj/Num_Trials
     c_oscj(ii)=mean(Cs_oscj)

 end
   plot( P_dB, c_os,'b-x',P_dB, c_osj,'g:<',P_dB, c_oscj,'m-p')
```

```
xlabel('P[dB]');
ylabel('secrecy ergodic capacity');
grid on; box on;
axis([0 40 0 3]);
legend ('OS','OSJ','OSCJ',2);
figure

semilogy(P_dB, out_os,'b-x',P_dB, out_osj,'g:<',P_dB,
out_oscj,'m-p')

xlabel('P[dB]');
ylabel('secrecy outage probability');
grid on; box on;
axis([0 40 10^-3 1]);
  legend ('OS','OSJ','OSCJ',3);
figure

%==============================================================
%                  % Relays close to Destination
%==============================================================
    %relay selection techniques for implementing security at
physical(PHY)layer
    %direct links are available&unsecure broadcast phase is assumed
    %secrecy outage probability and secrecy ergodic capacity
versus (P)

clear all
clc
.
%==============================================================
%system parameters
.
%==============================================================
N=4;                    %number of relay nodes
M=1;                    %number of eavesdroppers
Rs=.1;                  %target secrecy rate (BPCU)-for outage
probability calculation
R0=2;                   %Transmission Rate-to check which relay can
decode correctly
P_dB=0:5:40;            %Transmitted power (dB)
Num_Trials=3000;        %number of trials
PL_exp=3;               %path loss exponent
N0=1;                   %noise variance

loc_s=[0 0];
loc_e=[0 1];
loc_d=[1 0];
loc_r=[.6 .2;.68 .48;.7 .25;.75 .5];
```

```
 var_sd=(sqrt((loc_s(1)-loc_d(1)).^2+(loc_s(2)-
loc_d(2)).^2)).^(-PL_exp);

 var_se=(sqrt((loc_s(1)-loc_e(1)).^2+(loc_s(2)-
loc_e(2)).^2)).^(-PL_exp);

 z1=(loc_s(1)-loc_r(:,1)).^2;
 z2=(loc_s(2)-loc_r(:,2)).^2;
 z=sqrt(z1+z2);
 var_sr=z.^(-PL_exp);

 z1=(loc_d(1)-loc_r(:,1)).^2;
 z2=(loc_d(2)-loc_r(:,2)).^2;
 z=sqrt(z1+z2);
 var_rd=z.^(-PL_exp);

 z1=(loc_e(1)-loc_r(:,1)).^2;
 z2=(loc_e(2)-loc_r(:,2)).^2;
 z=sqrt(z1+z2);
 var_re=z.^(-PL_exp);

 %==========================
 %simulation start
 %==========================
 for ii=1:length(P_dB)
     Ps=10^(P_dB(ii)/10);
     Pr=Ps;
     Pj=Ps/100;    %jammer power less than relay node power

     outage_cs=0;
     outage_os=0;
     outage_ss=0;
     outage_osj=0;
     outage_ssj=0;
     outage_oscj=0;
     outage_ow=0;
     outage_sw=0;

     for jj=1:Num_Trials

 %=================================================================
           %Channel coefficients
 %=================================================================
           %broadcasting phase
           hsd=sqrt(var_sd/2).*(randn(1,1)+i*randn(1,1));
 %S-->destination
           hsrs=sqrt(var_sr/2).*(randn(N,1)+i*randn(N,1));
 %S-->relays
           hse=sqrt(var_se/2).*(randn(1,1)+i*randn(1,1));
 %S-->eavesdropper
```

```
        %cooperative links
        hrsd=sqrt(var_rd/2).*(randn(N,1)+i*randn(N,1));
%R-->destination
        hrse=sqrt(var_re/2).*(randn(N,1)+i*randn(N,1));
%R-->eavesdropper

%================================================================
        %instantaneous SNR calculations
%================================================================
        snr_sd=Ps*abs(hsd).^2;
        snr_srs=Ps*abs(hsrs).^2;
        snr_se=Ps*abs(hse).^2;
        snr_rsd=Pr*abs(hrsd).^2;
        snr_jsd=Pj*abs(hrsd).^2;
        snr_rse=Pr*abs(hrse).^2;
        snr_jse=Pj*abs(hrse).^2;

%================================================================
        %relay selection process
%================================================================
        Csrs=0.5*log2(1+snr_srs);
        y=Csrs>R0;    %all valid relays

%================================================================
        %selection techniques without jamming
%================================================================
        %case1:Conventional Selection(CS)

        [q1 q2]=max(1+snr_sd+ y.* snr_rsd);

        Q1=(q1==0);          %Q1=1 ==> all relays fails to
decode correctly

        Cs_cs(1,jj)=0.5*max(0,(~Q1)*log2(1+snr_sd+snr_
rsd(q2))-(~Q1)*log2(1+snr_se+snr_rse(q2))) ;

        outage_cs=outage_cs+(Rs>Cs_cs(1,jj));

        %case2:Optimal Selection(OS)

        [q1 q2]=max((1+snr_sd+y.* snr_rsd)./(1+snr_se+y.*
snr_rse));

        Q1=(q1==0);          %Q1=1 ==> all relays fails to
decode correctly

        Cs_os(1,jj)=0.5*max(0,(~Q1)*log2(1+snr_sd+snr_rsd
(q2))-(~Q1)*log2(1+snr_se+snr_rse(q2)));

        outage_os=outage_os+(Rs>Cs_os(1,jj));
```

```
%case3:Suboptimal Selection(SS)

      [q1 q2]=max((1+snr_sd+y.*
snr_rsd)./((1+mean(snr_se))+y.*(mean (snr_rse))));

      Q1=(q1==0);           %Q1=1 ==> all relays fails to
decode correctly

Cs_ss(1,jj)=0.5*max(0,(~Q1)*log2(1+snr_sd+snr_rsd(q2))-
(~Q1)*log2(1+mean(snr_se)+mean(snr_rse(q2)))) ;

      outage_ss=outage_ss+(Rs>Cs_ss(1,jj));

%================================================================
      %selection techniques for jamming
%================================================================

      %case1:Optimal selection with jamming(OSJ)-individual
selection

      [q3 q4]=max(snr_jse./snr_jsd);           %selecting
jammer in first phase
      [q1 q2]=max(y.*(snr_rsd./snr_rse));       %selecting
relay in second phase
      Q1=(q1==0);                             %Q1=1---
>all relays fail to decode correctly
      temp=ones(N,1);
      temp(q2,1)=0;
      [q5 q6]=max(temp.*snr_jse./snr_jsd);      %selecting
jammer in second phase

Cs_osj(1,jj)=0.5*max(0,(~Q1)*log2(1+(snr_sd./(1+snr_
jsd(q4)))+(snr_rsd(q2)./(1+snr_jsd(q6))))-(~Q1)*log2(1+(snr_
se./(1+snr_jse(q4)))+(snr_rse(q2)./(1+snr_jse(q6)))))) ;

      outage_osj=outage_osj+(Rs>Cs_osj(1,jj));

      %case2:Suboptimal selection with jamming(SSJ)

      [q3 q4]=max(mean(snr_jse)./snr_jsd); %selecting
jammer in first phase
      [q1 q2]=max(y.*(snr_rsd./mean(snr_rse))); %selecting
relay in second phase
      Q1=(q1==0);                             %Q1=1---
>all relays fail to decode correctly
      temp=ones(N,1);
      temp(q2,1)=0;
      [q5 q6]=max(temp.*(mean(snr_jse))./snr_jsd);
%selecting jammer in second phase
```

```
Cs_ssj(1,jj)=0.5*max(0,(~Q1)*log2(1+(snr_sd./(1+snr_
jsd(q4)))+(snr_rsd(q2)./(1+snr_jsd(q6))))-
(~Q1)*log2(1+(mean(snr_se)./(1+mean(snr_jse(q4))))+
(mean(snr_rse(q2))./(1+mean(snr_jse(q6)))))));

        outage_ssj=outage_ssj+(Rs>Cs_ssj(1,jj));

%===============================================================
        %Optimal selection with controlled jamming(OSCJ)
%===============================================================
        [q3 q4]=max((snr_jse));                 %selecting
jammer in first phase
        [q1 q2]=max(y.*(snr_rsd./snr_rse));     %selecting
relay in second phase
        Q1=(q1==0);                             %Q1=1---
>all relays fail to decode correctly
        temp=ones(N,1);
        temp(q2,1)=0;
        [q5 q6]=max(temp.*snr_jse); %selecting
jammer in second phase

Cs_oscj(1,jj)=0.5*max(0,(~Q1)*log2(1+snr_sd+snr_rsd(q2))-
(~Q1)*log2(1+(snr_se./(1+snr_jse(q4)))+(snr_rse(q2)./
(1+snr_jse(q6)))));

        outage_oscj=outage_oscj+(Rs>Cs_oscj(1,jj));

%===============================================================
        %Hybrid schemes(switching between jamming & non
jamming relay
        %selection techniques)
%===============================================================

         %case1:Optimal switching(OW)

        [q3 q4]=max(snr_jse./snr_jsd); %selecting
jammer in first phase
        [q1 q2]=max(y.*(snr_rsd./snr_rse)); %selecting
relay in second phase

        Q1=(q1==0);                             %Q1=1---
>all relays fail to decode correctly
        temp=ones(N,1);
        temp(q2,1)=0;
        [q5 q6]=max(temp.*snr_jse./snr_jsd);    %selecting
jammer in second phase

         if((snr_jse(q4)>snr_jsd(q4))&(snr_jse(q6)>snr_jsd(q6)))
```

```
Cs_ow(1,jj)=0.5*max(0,(~Q1)*log2(1+(snr_sd./(1+snr_
jsd(q4)))+(snr_rsd(q2)./(1+snr_jsd(q6))))-(~Q1)*log2(1+(snr_
se./(1+snr_jse(q4)))+(snr_rse(q2)./(1+snr_jse(q6)))));

        else

            [q1 q2]=max((1+snr_sd+y.*
snr_rsd)./(1+snr_se+y.* snr_rse));

            Cs_ow(1,jj)=
0.5*max(0,(~Q1)*log2(1+snr_sd+snr_rsd(q2))-(~Q1)*log2(1+snr_
se+snr_rse(q2)));

        end

        outage_ow=outage_ow+(Rs>Cs_ow(1,jj));

        %case2:Suboptimal switching(SW)

        [q3 q4]=max(mean(snr_jse)./snr_jsd);
%selecting jammer in first phase
        [q1 q2]=max(y.*(snr_rsd./mean(snr_rse)));
%selecting relay in second phase

            Q1=(q1==0);                              %Q1=1-
-->all relays fail to decode correctly
            temp=ones(N,1);
            temp(q2,1)=0;
            [q5 q6]=max(temp.*(mean(snr_jse))./snr_jsd);
%selecting jammer in second phase

if((mean(snr_jse(q4))>snr_jsd(q4))&(mean(snr_jse(q6))>snr_
jsd(q6)))

Cs_sw(1,jj)=0.5*max(0,(~Q1)*log2(1+(snr_sd./(1+snr_jsd(q4)))+
(snr_rsd(q2)./(1+snr_jsd(q6))))-(~Q1)*log2(1+(mean(snr_se)./
(1+mean(snr_jse(q4))))+(mean(snr_rse(q2))./
(1+mean(snr_jse(q6))))));

        else

            [q1 q2]=max((1+snr_sd+y.*
snr_rsd)./((1+mean(snr_se))+y.*(mean (snr_rse))));
            Cs_sw(1,jj)=
0.5*max(0,(~Q1)*log2(1+snr_sd+snr_rsd(q2))-
(~Q1)*log2(1+mean(snr_se)+mean(snr_rse(q2)))) ;

        end

        outage_sw=outage_sw+(Rs>Cs_sw(1,jj));

        end
```

```
        out_cs(ii)=outage_cs/Num_Trials
        c_cs(ii)=mean(Cs_cs)

         out_os(ii)=outage_os/Num_Trials
        c_os(ii)=mean(Cs_os)

        out_ss(ii)=outage_ss/Num_Trials
        c_ss(ii)=mean(Cs_ss)

        out_osj(ii)=outage_osj/Num_Trials
        c_osj(ii)=mean(Cs_osj)

        out_ssj(ii)=outage_ssj/Num_Trials
        c_ssj(ii)=mean(Cs_ssj)

        out_oscj(ii)=outage_oscj/Num_Trials
        c_oscj(ii)=mean(Cs_oscj)

        out_ow(ii)=outage_ow/Num_Trials
        c_ow(ii)=mean(Cs_ow)

        out_sw(ii)=outage_sw/Num_Trials
        c_sw(ii)=mean(Cs_sw)

 end
 plot( P_dB, c_cs,'k-o',P_dB, c_os,'b-x',P_dB, c_ss,'r-s',P_
dB, c_osj,'g:<',...
      P_dB, c_ow,'b-.+',P_dB, c_ssj,'c:>', P_dB, c_sw,'r-.d',P_
dB, c_oscj,'m-p')

 xlabel('P[dB]');
 ylabel('secrecy ergodic capacity');
 grid on; box on;
 axis([0 40 0 3]);
 legend ('CS','OS','SS','OSJ','OW','SSJ','SW','OSCJ',2);
 figure

 semilogy(P_dB,out_cs,'k-o',P_dB, out_os,'b-x',P_dB, out_
ss,'r-s',P_dB, out_osj,'g:<',...
      P_dB, out_ow,'b-.+',P_dB, out_ssj,'c:>', P_dB,out_sw,'r-
.d',P_dB, out_oscj,'m-p')

 xlabel('P[dB]');
 ylabel('secrecy outage probability');
 grid on; box on;
 axis([0 40 10^-3 1]);
  legend ('CS','OS','SS','OSJ','OW','SSJ','SW','OSCJ',3);
 figure
```

```
%================================================================
                % Relays close to Eavesdropper
%================================================================
 %relay selection techniques for implementing security at
physical(PHY)layer
 %direct links are available&unsecure broadcast phase is
assumed
 %secrecy outage probability and secrecy ergodic capacity
versus (P)

 clear all
 clc

%================================================================
 %system parameters
%================================================================
 N=4;                   %number of relay nodes
 M=1;                   %number of eavesdroppers
 Rs=.1;                 %target secrecy rate (BPCU)-for outage
probability calculation
 R0=2;                  %Transmission Rate-to check which relay
can decode correctly
 P_dB-0:5:40;           %Transmitted power (dB)
 Num_Trials=3000;       %number of trials
 PL_exp=3;              %path loss exponent
 N0=1;                  %noise variance

 loc_s=[0 0];
 loc_e=[0 1];
 loc_d=[1 0];
 loc_r=[.2 .75;.2 .8;.3 .95;.4 .85];

 var_sd=(sqrt((loc_s(1)-loc_d(1)).^2+(loc_s(2)-
loc_d(2)).^2)).^(-PL_exp);

 var_se=(sqrt((loc_s(1)-loc_e(1)).^2+(loc_s(2)-
loc_e(2)).^2)).^(-PL_exp);

 z1=(loc_s(1)-loc_r(:,1)).^2;
 z2=(loc_s(2)-loc_r(:,2)).^2;
 z=sqrt(z1+z2);
 var_sr=z.^(-PL_exp);

 z1=(loc_d(1)-loc_r(:,1)).^2;
 z2=(loc_d(2)-loc_r(:,2)).^2;
 z=sqrt(z1+z2);
 var_rd=z.^(-PL_exp);

 z1=(loc_e(1)-loc_r(:,1)).^2;
 z2=(loc_e(2)-loc_r(:,2)).^2;
```

```
z=sqrt(z1+z2);
var_re=z.^(-PL_exp);

%==========================
%simulation start
%==========================
for ii=1:length(P_dB)
    Ps=10^(P_dB(ii)/10);
    Pr=Ps;
    Pj=Ps/100;    %jammer power less than relay node power

    outage_cs=0;
    outage_os=0;
    outage_ss=0;
    outage_osj=0;
    outage_ssj=0;
    outage_oscj=0;
    outage_ow=0;
    outage_sw=0;

    for jj=1:Num_Trials

%================================================================
        %Channel coefficients
%================================================================
        %broadcasting phase
        hsd=sqrt(var_sd/2).*(randn(1,1)+i*randn(1,1));
%S-->destination
        hsrs=sqrt(var_sr/2).*(randn(N,1)+i*randn(N,1));
%S-->relays
        hse=sqrt(var_se/2).*(randn(1,1)+i*randn(1,1));
%S-->eavesdropper

        %cooperative links
        hrsd=sqrt(var_rd/2).*(randn(N,1)+i*randn(N,1));
%R-->destination
        hrse=sqrt(var_re/2).*(randn(N,1)+i*randn(N,1));
%R-->eavesdropper

%================================================================
        %instantaneous SNR calculations
%================================================================
        snr_sd=Ps*abs(hsd).^2;
        snr_srs=Ps*abs(hsrs).^2;
        snr_se=Ps*abs(hse).^2;
        snr_rsd=Pr*abs(hrsd).^2;
        snr_jsd=Pj*abs(hrsd).^2;
        snr_rse=Pr*abs(hrse).^2;
        snr_jse=Pj*abs(hrse).^2;
```

```
%===============================================================
          %relay selection process

%===============================================================
          Csrs=0.5*log2(1+snr_srs);
          y=Csrs>R0;    %all valid relays

%===============================================================
          %selection techniques without jamming

%---------------------------------------------------------------
          %case1:Conventional Selection(CS)

          [q1 q2]=max(1+snr_sd+ y.* snr_rsd);

          Q1=(q1==0);        %Q1=1 ==> all relays fails to
decode correctly

          Cs_cs(1,jj)=0.5*max(0,(~Q1)*log2(1+snr_sd+snr_
rsd(q2))-(~Q1)*log2(1+snr_se+snr_rse(q2)))  ;

          outage_cs=outage_cs+(Rs>Cs_cs(1,jj));

          %case2:Optimal Selection(OS)

          [q1 q2]=max((1+snr_sd+y.* snr_rsd)./(1+snr_se+y.*
snr_rse));

          Q1=(q1==0);        %Q1=1 ==> all relays fails to
decode correctly

          Cs_os(1,jj)=0.5*max(0,(~Q1)*log2(1+snr_sd+snr_rsd
(q2))-(~Q1)*log2(1+snr_se+snr_rse(q2)));

          outage_os=outage_os+(Rs>Cs_os(1,jj));

          %case3:Suboptimal Selection(SS)

          [q1 q2]=max((1+snr_sd+y.* snr_rsd)./((1+mean(snr_
se))+y.*(mean (snr_rse))));

          Q1=(q1==0);        %Q1=1 ==> all relays fails to
decode correctly

Cs_ss(1,jj)=0.5*max(0,(~Q1)*log2(1+snr_sd+snr_rsd(q2))-
(~Q1)*log2(1+mean(snr_se)+mean(snr_rse(q2))))  ;

          outage_ss=outage_ss+(Rs>Cs_ss(1,jj));
```

```
%===============================================================
            %selection techniques for jamming
%===============================================================

        %case1:Optimal selection with jamming(OSJ)-individual
selection

            [q3 q4]=max(snr_jse./snr_jsd);              %selecting
jammer in first phase
            [q1 q2]=max(y.*(snr_rsd./snr_rse));          %selecting
relay in second phase
            Q1=(q1==0);                                  %Q1=1---
>all relays fail to decode correctly
            temp=ones(N,1);
            temp(q2,1)=0;
            [q5 q6]=max(temp.*snr_jse./snr_jsd);         %selecting
jammer in second phase

Cs_osj(1,jj)=0.5*max(0,(~Q1)*log2(1+(snr_sd./(1+snr_
jsd(q4)))+(snr_rsd(q2)./(1+snr_jsd(q6))))-
(~Q1)*log2(1+(snr_se./(1+snr_jse(q4)))+(snr_rse(q2)./
(1+snr_jse(q6)))));

            outage_osj=outage_osj+(Rs>Cs_osj(1,jj));

            %case2:Suboptimal selection with jamming(SSJ)

            [q3 q4]=max(mean(snr_jse)./snr_jsd);
%selecting jammer in first phase
            [q1 q2]=max(y.*(snr_rsd./mean(snr_rse)));
%selecting relay in second phase
            Q1=(q1==0);                                  %Q1=1---
>all relays fail to decode correctly
            temp=ones(N,1);
            temp(q2,1)=0;
            [q5 q6]=max(temp.*(mean(snr_jse))./snr_jsd);
%selecting jammer in second phase

Cs_ssj(1,jj)=0.5*max(0,(~Q1)*log2(1+(snr_sd./(1+snr_
jsd(q4)))+(snr_rsd(q2)./(1+snr_jsd(q6))))-(~Q1)*log2
(1+(mean(snr_se)./(1+mean(snr_jse(q4))))+(mean(snr_rse(q2))./
(1+mean(snr_jse(q6))))));

            outage_ssj=outage_ssj+(Rs>Cs_ssj(1,jj));

%===============================================================
            %Optimal selection with controlled jamming(OSCJ)
%===============================================================
            [q3 q4]=max((snr_jse));                       %selecting
jammer in first phase
```

```
          [q1 q2]=max(y.*(snr_rsd./snr_rse));          %selecting
relay in second phase
          Q1=(q1==0);                                  %Q1=1---
>all relays fail to decode correctly
          temp=ones(N,1);
          temp(q2,1)=0;
          [q5 q6]=max(temp.*snr_jse);
%selecting jammer in second phase

Cs_oscj(1,jj)=0.5*max(0,(~Q1)*log2(1+snr_sd+snr_rsd(q2))-
(~Q1)*log2(1+(snr_se./(1+snr_jse(q4)))+(snr_rse(q2)./
(1+snr_jse(q6))))));

          outage_oscj=outage_oscj+(Rs>Cs_oscj(1,jj));

%================================================================
          %Hybrid schemes(switching between jamming & non
jamming relay
          %selection techniques)

%================================================================

          %case1:Optimal switching(OW)

          [q3 q4]=max(snr_jse./snr_jsd); %selecting
jammer in first phase
          [q1 q2]=max(y.*(snr_rsd./snr_rse)); %selecting
relay in second phase

          Q1=(q1==0);                                  %Q1=1---
>all relays fail to decode correctly
          temp=ones(N,1);
          temp(q2,1)=0;
          [q5 q6]=max(temp.*snr_jse./snr_jsd);          %selecting
jammer in second phase

          if((snr_jse(q4)>snr_jsd(q4))&(snr_jse(q6)>snr_jsd(q6)))

Cs_ow(1,jj)=0.5*max(0,(~Q1)*log2(1+(snr_sd./(1+snr_
jsd(q4)))+(snr_rsd(q2)./(1+snr_jsd(q6))))-
(~Q1)*log2(1+(snr_se./(1+snr_jse(q4)))+(snr_rse(q2)./
(1+snr_jse(q6))))));

        else

          [q1 q2]=max((1+snr_sd+y.* snr_rsd)./(1+snr_se+y.*
snr_rse));

          Cs_ow(1,jj)=0.5*max(0,(~Q1)*log2(1+snr_sd+snr_
rsd(q2))-(~Q1)*log2(1+snr_se+snr_rse(q2)));

        end
```

```
      outage_ow=outage_ow+(Rs>Cs_ow(1,jj));

    %case2:Suboptimal switching(SW)

  [q3 q4]=max(mean(snr_jse)./snr_jsd);
%selecting jammer in first phase
  [q1 q2]=max(y.*(snr_rsd./mean(snr_rse)));
%selecting relay in second phase

    Q1=(q1==0);                                    %Q1=1-
-->all relays fail to decode correctly
    temp=ones(N,1);
    temp(q2,1)=0;
    [q5 q6]=max(temp.*(mean(snr_jse))./snr_jsd);
%selecting jammer in second phase

if((mean(snr_jse(q4))>snr_jsd(q4))&(mean(snr_jse(q6))>snr_
jsd(q6)))

Cs_sw(1,jj)=0.5*max(0,(~Q1)*log2(1+(snr_sd./(1+snr_
jsd(q4)))+(snr_rsd(q2)./(1+snr_jsd(q6))))-
(~Q1)*log2(1+(mean(snr_se)./(1+mean(snr_jse(q4))))+
(mean(snr_rse(q2))./(1+mean(snr_jse(q6))))));

    else

    [q1 q2]=max((1+snr_sd+y.* snr_rsd)./((1+mean(snr_
se))+y.*(mean (snr_rse))));
    Cs_sw(1,jj)=
0.5*max(0,(~Q1)*log2(1+snr_sd+snr_rsd(q2))-
(~Q1)*log2(1+mean(snr_se)+mean(snr_rse(q2)))) ;

    end

    outage_sw=outage_sw+(Rs>Cs_sw(1,jj));

    end

  out_cs(ii)=outage_cs/Num_Trials
  c_cs(ii)=mean(Cs_cs)

   out_os(ii)=outage_os/Num_Trials
  c_os(ii)=mean(Cs_os)

  out_ss(ii)=outage_ss/Num_Trials
  c_ss(ii)=mean(Cs_ss)

  out_osj(ii)=outage_osj/Num_Trials
  c_osj(ii)=mean(Cs_osj)
```

```
    out_ssj(ii)=outage_ssj/Num_Trials
    c_ssj(ii)=mean(Cs_ssj)

    out_oscj(ii)=outage_oscj/Num_Trials
    c_oscj(ii)=mean(Cs_oscj)

    out_ow(ii)=outage_ow/Num_Trials
    c_ow(ii)=mean(Cs_ow)

    out_sw(ii)=outage_sw/Num_Trials
    c_sw(ii)=mean(Cs_sw)

end
plot( P_dB, c_cs,'k-o',P_dB, c_os,'b-x',P_dB, c_ss,'r-s',P_
dB, c_osj,'g:<',...
    P_dB, c_ow,'b-.+',P_dB, c_ssj,'c:>', P_dB, c_sw,'r-.d',P_
dB, c_oscj,'m-p')

xlabel('P[dB]');
ylabel('secrecy ergodic capacity');
grid on; box on;
axis([0 40 0 3]);
legend ('CS','OS','SS','OSJ','OW','SSJ','SW','OSCJ',2);
figure

semilogy(P_dB,out_cs,'k-o',P_dB, out_os,'b-x',P_dB, out_
ss,'r-s',P_dB, out_osj,'g:<',...
    P_dB, out_ow,'b-.+',P_dB, out_ssj,'c:>', P_dB,out_sw,'r-
.d',P_dB, out_oscj,'m-p')

xlabel('P[dB]');
ylabel('secrecy outage probability');
grid on; box on;
axis([0 40 10^-3 1]);
  legend ('CS','OS','SS','OSJ','OW','SSJ','SW','OSCJ',3);
figure

%==================================================================
% Relays in the middle between Destination & Eavesdropper
%==================================================================

        %relay selection techniques for implementing security
at physical(PHY)layer
%direct links are available&unsecure broadcast phase is
assumed
%secrecy outage probability and secrecy ergodic capacity
versus (P)

clear all
clc
.
```

```
%============================================================
%system parameters
%============================================================
 N=4;                    %number of relay nodes
 M=1;                    %number of eavesdroppers
 Rs=.1;                  %target secrecy rate (BPCU)-for outage
probability calculation
 R0=2;                   %Transmission Rate-to check which relay
can decode correctly
 P_dB=0:5:40;            %Transmitted power (dB)
 Num_Trials=3000;    %number of trials
 PL_exp=3;               %path loss exponent
 N0=1;                   %noise variance

 loc_s=[0 0];
 loc_e=[0 1];
 loc_d=[1 0];
 loc_r=[.41 .58; .58 .44; .7 .35; .66 .7];

 var_sd=(sqrt((loc_s(1)-loc_d(1)).^2+(loc_s(2)-
loc_d(2)).^2)).^(-PL_exp);

 var_se=(sqrt((loc_s(1)-loc_e(1)).^2+(loc_s(2)-
loc_e(2)).^2)).^(-PL_exp);

 z1=(loc_s(1)-loc_r(:,1)).^2;
 z2=(loc_s(2)-loc_r(:,2)).^2;
 z=sqrt(z1+z2);
 var_sr=z.^(-PL_exp);

 z1=(loc_d(1)-loc_r(:,1)).^2;
 z2=(loc_d(2)-loc_r(:,2)).^2;
 z=sqrt(z1+z2);
 var_rd=z.^(-PL_exp);

 z1=(loc_e(1)-loc_r(:,1)).^2;
 z2=(loc_e(2)-loc_r(:,2)).^2;
 z=sqrt(z1+z2);
 var_re=z.^(-PL_exp);

 %===========================
 %simulation start
 %===========================
 for ii=1:length(P_dB)
     Ps=10^(P_dB(ii)/10);
     Pr=Ps;
     Pj=Ps/100;    %jammer power less than relay node power

     outage_cs=0;
     outage_os=0;
```

```
        outage_ss=0;
        outage_osj=0;
        outage_ssj=0;
        outage_oscj=0;
        outage_ow=0;
        outage_sw=0;

        for jj=1:Num_Trials

%===============================================================
        %Channel coefficients
%===============================================================
        %broadcasting phase
        hsd=sqrt(var_sd/2).*(randn(1,1)+i*randn(1,1));
%S-->destination
        hsrs=sqrt(var_sr/2).*(randn(N,1)+i*randn(N,1));
%S-->relays
        hse=sqrt(var_se/2).*(randn(1,1)+i*randn(1,1));
%S-->eavesdropper

        %cooperative links
        hrsd=sqrt(var_rd/2).*(randn(N,1)+i*randn(N,1));
%R-->destination
        hrse=sqrt(var_re/2).*(randn(N,1)+i*randn(N,1));
%R-->eavesdropper

%===============================================================
        %instantaneous SNR calculations

%===============================================================
        snr_sd=Ps*abs(hsd).^2;
        snr_srs=Ps*abs(hsrs).^2;
        snr_se=Ps*abs(hse).^2;
        snr_rsd=Pr*abs(hrsd).^2;
        snr_jsd=Pj*abs(hrsd).^2;
        snr_rse=Pr*abs(hrse).^2;
        snr_jse=Pj*abs(hrse).^2;

%===============================================================
        %relay selection process

%===============================================================
        Csrs=0.5*log2(1+snr_srs);
        y=Csrs>R0;   %all valid relays

%===============================================================
        %selection techniques without jamming
```

```
%================================================================
         %case1:Conventional Selection(CS)

         [q1 q2]=max(1+snr_sd+ y.* snr_rsd);

         Q1=(q1==0);          %Q1=1 ==> all relays fails to
decode correctly

         Cs_cs(1,jj)=0.5*max(0,(~Q1)*log2(1+snr_sd+snr_
rsd(q2))-(~Q1)*log2(1+snr_se+snr_rse(q2))) ;

         outage_cs=outage_cs+(Rs>Cs_cs(1,jj));

         %case2:Optimal Selection(OS)

         [q1 q2]=max((1+snr_sd+y.* snr_rsd)./(1+snr_se+y.*
snr_rse));

         Q1=(q1==0);          %Q1=1 ==> all relays fails to
decode correctly

         Cs_os(1,jj)=0.5*max(0,(~Q1)*log2(1+snr_sd+snr_
rsd(q2))-(~Q1)*log2(1+snr_se+snr_rse(q2)));

         outage_os=outage_os+(Rs>Cs_os(1,jj));

          %case3:Suboptimal Selection(SS)

         [q1 q2]=max((1+snr_sd+y.* snr_rsd)./((1+mean(snr_
se))+y.*(mean (snr_rse))));

         Q1=(q1==0);          %Q1=1 ==> all relays fails to
decode correctly

Cs_ss(1,jj)=0.5*max(0,(~Q1)*log2(1+snr_sd+snr_rsd(q2))-
(~Q1)*log2(1+mean(snr_se)+mean(snr_rse(q2)))) ;

         outage_ss=outage_ss+(Rs>Cs_ss(1,jj));

%================================================================
         %selection techniques for jamming

%================================================================

     %case1:Optimal selection with jamming(OSJ)-individual
selection

         [q3 q4]=max(snr_jse./snr_jsd);                %selecting
jammer in first phase
         [q1 q2]=max(y.*(snr_rsd./snr_rse));           %selecting
relay in second phase
```

```
        Q1=(q1==0);                                        %Q1=1---
>all relays fail to decode correctly
            temp=ones(N,1);
            temp(q2,1)=0;
            [q5 q6]=max(temp.*snr_jse./snr_jsd);        %selecting
jammer in second phase

Cs_osj(1,jj)=0.5*max(0,(~Q1)*log2(1+(snr_sd./(1+snr_
jsd(q4)))+(snr_rsd(q2)./(1+snr_jsd(q6))))-
(~Q1)*log2(1+(snr_se./(1+snr_jse(q4)))+(snr_rse(q2)./
(1+snr_jse(q6))))));

        outage_osj=outage_osj+(Rs>Cs_osj(1,jj));

        %case2:Suboptimal selection with jamming(SSJ)

        [q3 q4]=max(mean(snr_jse)./snr_jsd);
%selecting jammer in first phase
        [q1 q2]=max(y.*(snr_rsd./mean(snr_rse)));
%selecting relay in second phase
        Q1=(q1==0);                                        %Q1=1---
>all relays fail to decode correctly
            temp=ones(N,1);
            temp(q2,1)=0;
            [q5 q6]=max(temp.*(mean(snr_jse))./snr_jsd);
%selecting jammer in second phase

Cs_ssj(1,jj)=0.5*max(0,(~Q1)*log2(1+(snr_sd./(1+snr_jsd(q4)))+
(snr_rsd(q2)./(1+snr_jsd(q6))))-(~Q1)*log2(1+(mean(snr_se)./
(1+mean(snr_jse(q4))))+(mean(snr_rse(q2))./(1+mean(snr_
jse(q6))))));

        outage_ssj=outage_ssj+(Rs>Cs_ssj(1,jj));

%===============================================================
        %Optimal selection with controlled jamming(OSCJ)
%===============================================================
        [q3 q4]=max((snr_jse));                         %selecting
jammer in first phase
        [q1 q2]=max(y.*(snr_rsd./snr_rse));             %selecting
relay in second phase
        Q1=(q1==0);                                        %Q1=1---
>all relays fail to decode correctly
            temp=ones(N,1);
            temp(q2,1)=0;
            [q5 q6]=max(temp.*snr_jse);
%selecting jammer in second phase

Cs_oscj(1,jj)=0.5*max(0,(~Q1)*log2(1+snr_sd+snr_rsd(q2))-
(~Q1)*log2(1+(snr_se./(1+snr_jse(q4)))+(snr_rse(q2)./
(1+snr_jse(q6)))));
```

```
            outage_oscj=outage_oscj+(Rs>Cs_oscj(1,jj));

%================================================================
            %Hybrid schemes(switching between jamming & non
jamming relay
            %selection techniques)
%================================================================

            %case1:Optimal switching(OW)

            [q3 q4]=max(snr_jse./snr_jsd); %selecting
jammer in first phase
            [q1 q2]=max(y.*(snr_rsd./snr_rse)); %selecting
relay in second phase

            Q1=(q1==0);                                %Q1=1---
>all relays fail to decode correctly
            temp=ones(N,1);
            temp(q2,1)=0;
            [q5 q6]=max(temp.*snr_jse./snr_jsd);        %selecting
jammer in second phase

            if((snr_jse(q4)>snr_jsd(q4))&(snr_jse(q6)>snr_jsd(q6)))

Cs_ow(1,jj)=0.5*max(0,(~Q1)*log2(1+(snr_sd./(1+snr_
jsd(q4)))+(snr_rsd(q2)./(1+snr_jsd(q6))))-(~Q1)*log2(1+(snr_
se./(1+snr_jse(q4)))+(snr_rse(q2)./(1+snr_jse(q6)))));

        else

            [q1 q2]=max((1+snr_sd+y.* snr_rsd)./(1+snr_se+y.*
snr_rse));

            Cs_ow(1,jj)=0.5*max(0,(~Q1)*log2(1+snr_sd+snr_rsd
(q2))-(~Q1)*log2(1+snr_se+snr_rse(q2)));

        end

            outage_ow=outage_ow+(Rs>Cs_ow(1,jj));

            %case2:Suboptimal switching(SW)

            [q3 q4]=max(mean(snr_jse)./snr_jsd);
%selecting jammer in first phase
            [q1 q2]=max(y.*(snr_rsd./mean(snr_rse)));
%selecting relay in second phase

            Q1=(q1==0);                                %Q1=1-
-->all relays fail to decode correctly
            temp=ones(N,1);
            temp(q2,1)=0;
```

```
        [q5 q6]=max(temp.*(mean(snr_jse))./snr_jsd);
%selecting jammer in second phase

if((mean(snr_jse(q4))>snr_jsd(q4))&(mean(snr_jse(q6))>snr_
jsd(q6)))

Cs_sw(1,jj)=0.5*max(0,(~Q1)*log2(1+(snr_sd./(1+snr_jsd(q4)))+
(snr_rsd(q2)./(1+snr_jsd(q6))))-(~Q1)*log2(1+(mean(snr_se)./
(1+mean(snr_jse(q4))))+(mean(snr_rse(q2))./
(1+mean(snr_jse(q6))))));

        else

        [q1 q2]=max((1+snr_sd+y.* snr_rsd)./((1+mean(snr_
se))+y.*(mean (snr_rse)))));
        Cs_sw(1,jj)=
0.5*max(0,(~Q1)*log2(1+snr_sd+snr_rsd(q2))-
(~Q1)*log2(1+mean(snr_se)+mean(snr_rse(q2)))) ;

        end

        outage_sw=outage_sw+(Rs>Cs_sw(1,jj));

        end

    out_cs(ii)=outage_cs/Num_Trials
    c_cs(ii)=mean(Cs_cs)

     out_os(ii)=outage_os/Num_Trials
    c_os(ii)=mean(Cs_os)

    out_ss(ii)=outage_ss/Num_Trials
    c_ss(ii)=mean(Cs_ss)

    out_osj(ii)=outage_osj/Num_Trials
    c_osj(ii)=mean(Cs_osj)

    out_ssj(ii)=outage_ssj/Num_Trials
    c_ssj(ii)=mean(Cs_ssj)

    out_oscj(ii)=outage_oscj/Num_Trials
    c_oscj(ii)=mean(Cs_oscj)

    out_ow(ii)=outage_ow/Num_Trials
    c_ow(ii)=mean(Cs_ow)

    out_sw(ii)=outage_sw/Num_Trials
    c_sw(ii)=mean(Cs_sw)

end
```

```
plot( P_dB, c_cs,'k-o',P_dB, c_os,'b-x',P_dB, c_ss,'r-s',P_
dB, c_osj,'g:<',...
     P_dB, c_ow,'b-.+',P_dB, c_ssj,'c:>', P_dB, c_sw,'r-.d',P_
dB, c_oscj,'m-p')

xlabel('P[dB]');
ylabel('secrecy ergodic capacity');
grid on; box on;
axis([0 40 0 3]);
legend ('CS','OS','SS','OSJ','OW','SSJ','SW','OSCJ',2);
figure

semilogy(P_dB,out_cs,'k-o',P_dB, out_os,'b-x',P_dB, out_
ss,'r-s',P_dB, out_osj,'g:<',...
     P_dB, out_ow,'b-.+',P_dB, out_ssj,'c:>', P_dB,out_sw,'r-
.d',P_dB, out_oscj,'m-p')

xlabel('P[dB]');
ylabel('secrecy outage probability');
grid on; box on;
axis([0 40 10^-3 1]);
  legend ('CS','OS','SS','OSJ','OW','SSJ','SW','OSCJ',3);
figure
```

Appendix B: MATLAB®
Simulation Codes
for Chapter 5

```
%****************************************************************
% ------- Eavesdropper located close to both sources-----------
%****************************************************************

%Two-way relay networks
 clear all
 clc
 %==================
 %system parameters
 %==================
 N=4;                    %number of relay nodes
 M=1;                    %number of eavesdroppers
 Rs=0.1;                 %target secrecy rate (BPCU)-for outage
probability calculation
 R0=2;                   %Transmission Rate-to check which relay
can decode correctly
 P_dB=0:5:40;            %Transmitted power (dB)
 Num_Trials=8000;        %number of trials
 PL_exp=3;               %path loss exponent
 N0=1;                   %noise variance

 loc_s1=[0 0];
 loc_e=[0.5 0.2];
 loc_s2=[1 0];
 loc_r=[0.35 0.3;0.4 0.65;0.6 0.4;0.65 0.7];

 var_s1e=(sqrt((loc_s1(1)-loc_e(1)).^2+(loc_s1(2)-
loc_e(2)).^2)).^(-PL_exp);
 var_s2e=(sqrt((loc_s2(1)-loc_e(1)).^2+(loc_s2(2)-
loc_e(2)).^2)).^(-PL_exp);
```

```
z1=(loc_s1(1)-loc_r(:,1)).^2;
z2=(loc_s1(2)-loc_r(:,2)).^2;
z=sqrt(z1+z2);
var_s1r=z.^(-PL_exp);

z1=(loc_s2(1)-loc_r(:,1)).^2;
z2=(loc_s2(2)-loc_r(:,2)).^2;
z=sqrt(z1+z2);
var_s2r=z.^(-PL_exp);

z1=(loc_e(1)-loc_r(:,1)).^2;
z2=(loc_e(2)-loc_r(:,2)).^2;
z=sqrt(z1+z2);
var_re=z.^(-PL_exp);

%==========================
%simulation start
%==========================
for ii=1:length(P_dB)
    Ps=10^(P_dB(ii)/10);
    Pr=Ps;
    Pj=Ps/100;   %jammer power less than relay node power

    outage_cs=0;
    outage_os=0;
    outage_osmmisr=0;
    outage_ss=0;
    outage_ssmmisr=0;
    outage_osj=0;
    outage_osjmmisr=0;
    outage_ssj=0;
    outage_ssjmmisr=0;
    outage_oscj=0;
    outage_ow=0;
    outage_sw=0;

    for jj=1:Num_Trials

%================================================================
        %Channel coefficients
%================================================================
        %broadcasting phase

        hs1r=sqrt(var_s1r/2).*(randn(N,1)+i*randn(N,1));
% S1-->relays
        hs2r=sqrt(var_s2r/2).*(randn(N,1)+i*randn(N,1));
% S2-->relays
        hs1e=sqrt(var_s1e/2).*(randn(1,1)+i*randn(1,1));
% S1-->eavesdropper
        hs2e=sqrt(var_s2e/2).*(randn(1,1)+i*randn(1,1));
```

```
% S2-->eavesdropper
        %cooperative links
        %hrs1=sqrt(var_s1r/2).*(randn(N,1)+i*randn(N,1));
% R-->destination1
        %hrs2=sqrt(var_s2r/2).*(randn(N,1)+i*randn(N,1));
% R-->destination2
        hre=sqrt(var_re/2).*(randn(N,1)+i*randn(N,1));
% R-->eavesdropper

%=============================================================
        %instantaneous SNR calculations
%=============================================================

        snr_s1r=Ps*abs(hs1r).^2;
        snr_s2r=Ps*abs(hs2r).^2;
        snr_s1e=Ps*abs(hs1e).^2;
        snr_s2e=Ps*abs(hs2e).^2;
        snr_rs1=Pr*abs(hs1r).^2;
        snr_rs2=Pr*abs(hs2r).^2;
        snr_js1=Pj*abs(hs1r).^2;
        snr_js2=Pj*abs(hs2r).^2;
        snr_re=Pr*abs(hre).^2;
        snr_je=Pj*abs(hre).^2;

        cs1r=0.5*log2(1+snr_s1r);
        y1=cs1r>R0;

        cs2r=0.5*log2(1+snr_s2r);
        y2=cs2r>R0;

        y=(y1&y2);

    %Two-way relay networks
    %=========================
        %Non-jamming techniques
    %=========================

    %case:optimal selection (OS)

        sinr1_os=(y.*snr_rs1);
        sinr2_os=(y.*snr_rs2);

        sinre1_os=(((snr_s1e)/(snr_s2e+1))+(y.*snr_re));
        sinre2_os=(((snr_s2e)/(snr_s1e+1))+(y.*snr_re));

    [q1
q2]=max(((1+sinr1_os)./(1+sinre2_os)).*((1+sinr2_os)./
(1+sinre1_os)));

        sinr1_osn=(y(q2).*snr_rs1(q2));
        sinr2_osn=(y(q2).*snr_rs2(q2));
```

```
sinre1_osn=(((snr_s1e)/(snr_s2e+1))+(y(q2).*snr_re(q2)));

sinre2_osn=(((snr_s2e)/(snr_s1e+1))+(y(q2).*snr_re(q2)));

    Cs_os(1,jj)=max(0,0.5*log2(1+sinr1_osn)
-0.5*log2(1+sinre2_osn)+0.5*log2(1+sinr2_osn)
-0.5*log2(1+sinre1_osn));

    outage_os=outage_os+(Rs>Cs_os(1,jj));

%case:Suboptimal selection (SS)

    sinr1_ss=(y.*snr_rs1);
    sinr2_ss=(y.*snr_rs2);

    sinre1_ss=((mean(snr_s1e)/(mean(snr_s2e)+1))+
y.*mean(snr_re));
    sinre2_ss=((mean(snr_s2e)/(mean(snr_s1e)+1))+
y.*mean(snr_re));

    [q1
q2]=max((((1+sinr1_ss)./(1+sinre2_ss)).*((1+sinr2_ss)./
(1+sinre1_ss)));

    sinr1_ssn=y(q2)*snr_s1r(q2);
    sinr2_ssn=y(q2)*snr_s2r(q2);
    sinre1_ssn=((snr_s1e./(snr_s2e+1))+ y(q2)*snr_re(q2));
    sinre2_ssn=((snr_s2e./(snr_s1e+1))+ y(q2)*snr_re(q2));

    Cs_ss(1,jj)=max(0,0.5*log2(1+sinr1_ssn)
    -0.5*log2(1+sinre2_ssn)+0.5*log2(1+sinr2_ssn)
    -0.5*log2(1+sinre1_ssn));

     outage_ss=outage_ss+(Rs>Cs_ss(1,jj));

     % case:optimal selection with max-min instantaneous
secrecy rate(OS-MMISR)

    sinr1_os=(y.*snr_rs1);
    sinr2_os=(y.*snr_rs2);

    sinre1_os=(((snr_s1e)/(snr_s2e+1))+(y.*snr_re));
    sinre2_os=(((snr_s2e)/(snr_s1e+1))+(y.*snr_re));

    [q1
q2]=max(min(((1+sinr1_os)./(1+sinre2_os)),((1+sinr2_os)./
(1+sinre1_os))));

    sinr1_osn=(y(q2).*snr_rs1(q2));
    sinr2_osn=(y(q2).*snr_rs2(q2));
```

```
sinrel_osn=(((snr_s1e)/(snr_s2e+1))+(y(q2).*snr_re(q2)));
sinre2_osn=(((snr_s2e)/(snr_s1e+1))+(y(q2).*snr_re(q2)));

        Cs_osmmisr(1,jj)=max(0,0.5*log2(1+sinr1_osn)-0.5*log2
(1+sinre2_osn)+0.5*log2(1+sinr2_osn)-0.5*log2(1+sinre1_osn));

        outage_osmmisr=outage_osmmisr+(Rs>Cs_osmmisr(1,jj));

    % case:Suboptimal selection with max-min instantaneous
secrecy
        % rate(SS-MMISR)

            sinr1_ss=(y.*snr_rs1);
            sinr2_ss=(y.*snr_rs2);

            sinre1_ss=((mean(snr_s1e)/(mean(snr_s2e)+1))+
y.*mean(snr_re));
            sinre2_ss=((mean(snr_s2e)/(mean(snr_s1e)+1))+
y.*mean(snr_re));

            [q1
q2]=max(min(((1+sinr1_ss)./(1+sinre2_ss)),((1+sinr2_ss)./
(1+sinre1_ss)))));

            sinr1_ssn=y(q2)*snr_s1r(q2);
            sinr2_ssn=y(q2)*snr_s2r(q2);
            sinre1_ssn=((snr_s1e./(snr_s2e+1))+ y(q2)*snr_re(q2));
            sinre2_ssn=((snr_s2e./(snr_s1e+1))+ y(q2)*snr_re(q2));

            Cs_ssmmisr(1,jj)=max(0,0.5*log2(1+sinr1_ssn)-0.5*log2
(1+sinre2_ssn)+0.5*log2(1+sinr2_ssn)
-0.5*log2(1+sinre1_ssn));

            outage_ssmmisr=outage_ssmmisr+(Rs>Cs_ssmmisr(1,jj));

    %=========================
     %Jamming techniques
    %=========================
  % case:optimal selection with jamming(OSJ)

        [q3 q4]=max(snr_je.*snr_je);      % selecting jammer in
first phase.

        [q1 q2]=max(y.*((snr_rs1./snr_re).*(snr_rs2./snr_re)));
% selecting relay in second phase.

    Q1=(q1==0);        %Q1=1==> all the relays fail to decode
    temp=ones(N,1);
    temp(q2,1)=0;

        [q5 q6]=max(temp.*((snr_je./snr_js1).*(snr_je./snr_
js2)));      %selecting the jammer in the second phase
```

```
    sinr1_osj=((y(q2)*snr_rs1(q2))./(1+snr_js1(q6)));
    sinr2_osj=((y(q2)*snr_rs2(q2))./(1+snr_js2(q6)));

    sinre1_osj=(snr_s1e./(1+snr_s2e+snr_je(q4)))+((y(q2)*snr_
re(q2))./(1+snr_je(q6)));

    sinre2_osj=(snr_s2e./(1+snr_s1e+snr_je(q4)))+((y(q2)*snr_
re(q2))./(1+snr_je(q6)));

    Cs_osj(1,jj)=max(0,0.5*log2(1+sinr1_osj)
-0.5*log2(1+sinre2_osj)
+0.5*log2(1+sinr2_osj)-0.5*log2(1+sinre1_osj));

    outage_osj=outage_osj+(Rs>Cs_osj(1,jj));

  % case:Suboptimal selection with jamming(SSJ)

    [q3 q4]=max(mean(snr_je).*mean(snr_je));      % selecting
jammer in first phase.

    [q1
q2]=max(y.*((snr_rs1./mean(snr_re)).*(snr_rs2./mean(snr_
re))));      % selecting relay in second phase.

    Q1=(q1==0);        %Q1=1==> all the relays fail to decode
    temp=ones(N,1);
    temp(q2,1)=0;

    [q5
q6]=max(temp.*((mean(snr_je)./snr_js1).*(mean(snr_je)./snr_
js2)));      %selecting the jammer in the second phase

    sinr1_ssj=((y(q2)*snr_rs1(q2))./(1+snr_js1(q6)));
    sinr2_ssj=((y(q2)*snr_rs2(q2))./(1+snr_js2(q6)));

    sinre1_ssj=(snr_s1e./(1+snr_s2e+snr_je(q4)))+((y(q2)*snr_
re(q2))./(1+snr_je(q6)));

    sinre2_ssj=(snr_s2e./(1+snr_s1e+snr_je(q4)))+((y(q2)*snr_
re(q2))./(1+snr_je(q6)));

    Cs_ssj(1,jj)=max(0,0.5*log2(1+sinr1_ssj)
-0.5*log2(1+sinre2_ssj)+0.5*log2(1+sinr2_ssj)-
0.5*log2(1+sinre1_ssj));

    outage_ssj=outage_ssj+(Rs>Cs_ssj(1,jj));

  % case:optimal selection with max-min instantaneous secrecy
rate(OSJ-MMISR)

    [q3 q4]=max(min(snr_je,snr_je));      % selecting jammer
in first phase.
```

```
     [q1
q2]=max(min(y.*(snr_rs1./snr_re),y.*(snr_rs2./snr_re)));        %
selecting relay in second phase.

     Q1=(q1==0);        %Q1=1==> all the relays fail to decode
     temp=ones(N,1);
     temp(q2,1)=0;

     [q5
q6]=max(min(temp.*(snr_je./snr_js1),temp.*(snr_je./snr_js2)));
%selecting the jammer in the second phase

     sinr1_osj=((y(q2)*snr_rs1(q2))./(1+snr_js1(q6)));
     sinr2_osj=((y(q2)*snr_rs2(q2))./(1+snr_js2(q6)));

     sinre1_osj=(snr_s1e./(1+snr_s2e+snr_je(q4)))+((y(q2)*snr_
re(q2))./(1+snr_je(q6)));

     sinre2_osj=(snr_s2e./(1+snr_s1e+snr_je(q4)))+((y(q2)*snr_
re(q2))./(1+snr_je(q6)));

     Cs_osjmmisr(1,jj)=max(0,0.5*log2(1+sinr1_osj)-0.5*log2
(1+sinre2_osj)+0.5*log2(1+sinr2_osj)-0.5*log2(1+sinre1_osj));

     outage_osjmmisr=outage_osjmmisr+(Rs>Cs_osjmmisr(1,jj));

  % case:Suboptimal selection with max-min instantaneous
secrecy rate(SSJ-MMISR)

     [q3 q4]=max(min(mean(snr_je),mean(snr_je)));       %
selecting jammer in first phase.

     [q1
q2]=max(min(y.*(snr_rs1./mean(snr_re)),y.*(snr_rs2./mean(snr_
re))));    % selecting relay in second phase.

     Q1=(q1==0);        %Q1=1==> all the relays fail to decode
     temp=ones(N,1);
     temp(q2,1)=0;

     [q5
q6]=max(min(temp.*(mean(snr_je)./snr_js1),temp.*(mean(snr_
je)./snr_js2)));       %selecting the jammer in the second phase

     sinr1_ssj=((y(q2)*snr_rs1(q2))./(1+snr_js1(q6)));
     sinr2_ssj=((y(q2)*snr_rs2(q2))./(1+snr_js2(q6)));

     sinre1_ssj=(snr_s1e./(1+snr_s2e+snr_je(q4)))+((y(q2)*snr_
re(q2))./(1+snr_je(q6)));

     sinre2_ssj=(snr_s2e./(1+snr_s1e+snr_je(q4)))+((y(q2)*snr_
re(q2))./(1+snr_je(q6)));
```

```
      Cs_ssjmmisr(1,jj)=max(0,0.5*log2(1+sinr1_ssj)-0.5*log2
(1+sinre2_ssj)+0.5*log2(1+sinr2_ssj)-0.5*log2(1+sinre1_ssj));

      outage_ssjmmisr=outage_ssjmmisr+(Rs>Cs_ssjmmisr(1,jj));

% case:optimal switching(OW)

      [q3 q4]=max(snr_je.*snr_je);      % selecting jammer in
first phase.

      [q1 q2]=max(y.*((snr_rs1./snr_re).*(snr_rs2./snr_re)));
% selecting relay in second phase.

      Q1=(q1==0);        %Q1=1==> all the relays fail to decode
      temp=ones(N,1);
      temp(q2,1)=0;

      [q5 q6]=max(temp.*((snr_je./snr_js1).*(snr_je./snr_
js2)));       %selecting the jammer in the second phase

      sinr1_os=(y(q2)*snr_rs1(q2));
      sinr2_os=(y(q2)*snr_rs2(q2));
      sinre1_os=(snr_s1e./(1+snr_s2e))+(y(q2)*snr_re(q2));
      sinre2_os=(snr_s2e./(1+snr_s1e))+(y(q2)*snr_re(q2));

      sinr1_osj=((y(q2)*snr_rs1(q2))./(1+snr_js1(q6)));
      sinr2_osj=((y(q2)*snr_rs2(q2))./(1+snr_js2(q6)));

sinre1_osj=(snr_s1e./(1+snr_s2e+snr_je(q4)))+((y(q2)*snr_
re(q2))./(1+snr_je(q6)));

sinre2_osj=(snr_s2e./(1+snr_s1e+snr_je(q4)))+((y(q2)*snr_
re(q2))./(1+snr_je(q6)));

if(((1+sinr1_osj)./(1+sinre2_osj)).*((1+sinr2_osj)./
(1+sinre1_osj))>...

((1+sinr1_os)./(1+sinre2_os)).*((1+sinr2_os)./(1+sinre1_os)))

 Cs_ow(1,jj)=max(0,0.5*log2(1+sinr1_osj)-0.5*log2(1+sinre2_
 osj)+0.5*log2(1+sinr2_osj)-0.5*log2(1+sinre1_osj));

   else

 Cs_ow(1,jj)=max(0,0.5*log2(1+sinr1_os)-0.5*log2(1+sinre2_
 os)+0.5*log2(1+sinr2_os)-0.5*log2(1+sinre1_os));

   end

   outage_ow=outage_ow+(Rs>Cs_ow(1,jj));
```

```
% case:Suboptimal switching(SW)

    [q3 q4]=max(mean(snr_je).*mean(snr_je));     % selecting
jammer in first phase.

    [q1
q2]=max(y.*((snr_rs1./mean(snr_re)).*(snr_rs2./mean(snr_
re)))) ;      % selecting relay in second phase.

    Q1=(q1==0);          %Q1=1==> all the relays fail to decode
    temp=ones(N,1);
    temp(q2,1)=0;

    [q5 q6]=max(temp.*((mean(snr_je)./snr_js1).*(mean(snr_
je)./snr_js2)));      %selecting the jammer in the second phase

    sinr1_ss=y(q2)*snr_s1r(q2);
    sinr2_ss=y(q2)*snr_s2r(q2);
    sinre1_ss=((snr_s1e./(snr_s2e+1))+ y(q2)*snr_re(q2));
    sinre2_ss=((snr_s2e./(snr_s1e+1))+ y(q2)*snr_re(q2));

    sinr1_ssj=((y(q2)*snr_rs1(q2))./(1+snr_js1(q6)));
    sinr2_ssj=((y(q2)*snr_rs2(q2))./(1+snr_js2(q6)));

sinre1_ssj=(snr_s1e./(1+snr_s2e+snr_je(q4)))+((y(q2)*snr_
re(q2))./(1+snr_je(q6)));

sinre2_ssj=(snr_s2e./(1+snr_s1e+snr_je(q4)))+((y(q2)*snr_
re(q2))./(1+snr_je(q6)));

if(((1+sinr1_ssj)./(1+sinre2_ssj)).*((1+sinr2_ssj)./
(1+sinre1_ssj))>...

((1+sinr1_ss)./(1+sinre2_ss)).*((1+sinr2_ss)./(1+sinre1_ss)))

    Cs_sw(1,jj)=max(0,0.5*log2(1+sinr1_ssj)-0.5*log2(1+sinre2_
ssj)+0.5*log2(1+sinr2_ssj)-0.5*log2(1+sinre1_ssj));
  else

  Cs_sw(1,jj)=max(0,0.5*log2(1+sinr1_ss)-0.5*log2(1+sinre2_
ss)+0.5*log2(1+sinr2_ss)-0.5*log2(1+sinre1_ss));

  end
    outage_sw=outage_sw+(Rs>Cs_sw(1,jj));

  % case:optimal selection with controlled jamming(OSCJ)

                F=[];

                for r=1:N
                for j2=1:N
                for j1=1:N

                if (r~=j2)
```

```
    sinr1_oscj=(y(r)*snr_rs1(r));
    sinr2_oscj=(y(r)*snr_rs2(r));

sinre1_oscj=(snr_s1e./(1+snr_s2e+snr_je(j1)))+(y(r)*snr_
re(r)./(1+snr_je(j2)));

sinre2_oscj=(snr_s2e./(1+snr_s1e+snr_je(j1)))+(y(r)*snr_
re(r)./(1+snr_je(j2)));

    C=max(0,0.5*log2(1+sinr1_oscj)
-0.5*log2(1+sinre2_oscj)
+0.5*log2(1+sinr2_oscj)-0.5*log2(1+sinre1_oscj));

        F=[F C];
        end
         end

    end
end

  Cs_oscj(1,jj)=max(F);
  outage_oscj=outage_oscj+(Rs>Cs_oscj(1,jj));

        end

        out_os(ii)=outage_os/Num_Trials
        c_os(ii)=mean(Cs_os)

        out_ss(ii)=outage_ss/Num_Trials
        c_ss(ii)=mean(Cs_ss)

        out_osmmisr(ii)=outage_osmmisr/Num_Trials
        c_osmmisr(ii)=mean(Cs_osmmisr)

         out_ssmmisr(ii)=outage_ssmmisr/Num_Trials
        c_ssmmisr(ii)=mean(Cs_ssmmisr)

        out_osj(ii)=outage_osj/Num_Trials
        c_osj(ii)=mean(Cs_osj)

     out_ssj(ii)=outage_ssj/Num_Trials
        c_ssj(ii)=mean(Cs_ssj)

        out_osjmmisr(ii)=outage_osjmmisr/Num_Trials
        c_osjmmisr(ii)=mean(Cs_osjmmisr)

        out_ssjmmisr(ii)=outage_ssjmmisr/Num_Trials
        c_ssjmmisr(ii)=mean(Cs_ssjmmisr)

        out_ow(ii)=outage_ow/Num_Trials
        c_ow(ii)=mean(Cs_ow)
```

```
     out_sw(ii)=outage_sw/Num_Trials
     c_sw(ii)=mean(Cs_sw)

     out_oscj(ii)=outage_oscj/Num_Trials
     c_oscj(ii)=mean(Cs_oscj)

 end

   plot(P_dB, c_os,'b-x',P_dB, c_ss,'r-s',P_dB, c_osmmisr,'k-
x',P_dB, c_ssmmisr,'k-s',P_dB, c_osj,'b-s',P_dB,
c_ssj,'r-x',...
      P_dB, c_osjmmisr,'g:>',P_dB, c_ssjmmisr,'c:<',P_dB,
c_ow,'b-.+',P_dB, c_sw,'r-.d',P_dB, c_oscj,'m-p')

 xlabel('P[dB]');
 ylabel('secrecy ergodic capacity[BPCU]');
 grid on; box on;
 axis([0 25 0 5]);
 legend ('OS','SS','OS-MMISR','SS-MMISR','OSJ','SSJ','OSJ-
MMISR','SSJ-MMISR','OW','SW','OSCJ',2);
 figure

 semilogy(P_dB, out_os,'b-x',P_dB, out_ss,'r-s',P_dB, out_
osmmisr,'k-x',P_dB,out_ssmmisr,'k-s',P_dB,
out_osj,'b-s',P_dB,out_ssj,'r-x',...
      P_dB, out_osjmmisr,'g:>',P_dB, out_ssjmmisr,'c:<',P_dB,
out_ow,'b-.+',P_dB, out_sw,'r-.d',P_dB, out_oscj,'m-p')

 xlabel('P[dB]');
 ylabel('secrecy outage probability');
 grid on; box on;
 axis([0 40 10^-3 1]);
   legend ('OS','SS','OS-MMISR','SS-MMISR','OSJ','SSJ',
'OSJ-MMISR','SSJ-MMISR','OW','SW','OSCJ',3);
 figure




%****************************************************************
% Presence of M=2 non-cooperating and cooperating eavesdroppers
%****************************************************************

% Joint relay and jammer selection for secure two-way relay
networks
% performance metrics are ergodic secrecy capacity & secrecy
outage
% probability

% Non-cooperative eavesdroppers
 clear all
```

```
clc
%====================
%system parameters
%====================
N=4;                      %number of relay nodes
M=2;                      %number of eavesdroppers
Rs=0.1;                   %target secrecy rate (BPCU)-for outage
probability calculation
R0=2;                     %Transmission Rate-to check which relay
can decode correctly
P_dB=0:5:40;              %Transmitted power (dB)
Num_Trials=3000;          %number of trials
PL_exp=3;                 %path loss exponent
N0=1;                     %noise variance

loc_s1=[0 0];
loc_e=[0.5 1;0 0.5];
loc_s2=[1 0];
loc_r=[0.35 0.3;0.4 0.65;0.6 0.4;0.65 0.7];

z1=(loc_s1(1)-loc_e(:,1)).^2;
z2=(loc_s1(2)-loc_e(:,2)).^2;
z=sqrt(z1+z2);
var_s1e=z.^(-PL_exp);

z1=(loc_s2(1)-loc_e(:,1)).^2;
z2=(loc_s2(2)-loc_e(:,2)).^2;
z=sqrt(z1+z2);
var_s2e=z.^(-PL_exp);

z1=(loc_s1(1)-loc_r(:,1)).^2;
z2=(loc_s1(2)-loc_r(:,2)).^2;
z=sqrt(z1+z2);
var_s1r=z.^(-PL_exp);

z1=(loc_s2(1)-loc_r(:,1)).^2;
z2=(loc_s2(2)-loc_r(:,2)).^2;
z=sqrt(z1+z2);
var_s2r=z.^(-PL_exp);

z1=(loc_r(:,1)-loc_e(1,1)).^2;
z2=(loc_r(:,2)-loc_e(1,2)).^2;
z11=sqrt(z1+z2);
z3=(loc_r(:,1)-loc_e(2,1)).^2;
z4=(loc_r(:,2)-loc_e(2,2)).^2;
z22=sqrt(z3+z4);
z=[z11 z22];
var_re=z.^(-PL_exp);
```

```
%==========================
%simulation start
%==========================
for ii=1:length(P_dB)
    Ps=10^(P_dB(ii)/10);
    Pr=Ps;
    Pj=Ps/100;    %jammer power less than relay node power

    outage_cs=0;
    outage_os=0;
    outage_osmmisr=0;
    outage_ss=0;
    outage_ssmmisr=0;
    outage_osj=0;
    outage_osjmmisr=0;
    outage_ssj=0;
    outage_ssjmmisr=0;
    outage_oscj=0;
    outage_ow=0;
    outage_sw=0;

    for jj=1:Num_Trials

%=============================================================
          %Channel coefficients
%=============================================================
          %broadcasting phase

          hs1r=sqrt(var_s1r/2).*(randn(N,1)+i*randn(N,1));
% S1-->relays
          hs2r=sqrt(var_s2r/2).*(randn(N,1)+i*randn(N,1));
% S2-->relays
          hs1e=sqrt(var_s1e/2).*(randn(M,1)+i*randn(M,1));
% S1-->eavesdropper
          hs2e=sqrt(var_s2e/2).*(randn(M,1)+i*randn(M,1));
% S2-->eavesdropper
          %cooperative links
          %hrs1=sqrt(var_s1r/2).*(randn(N,1)+i*randn(N,1));
% R-->destination1
          %hrs2=sqrt(var_s2r/2).*(randn(N,1)+i*randn(N,1));
% R-->destination2
          hre=sqrt(var_re/2).*(randn(N,M)+i*randn(N,M));
% R-->eavesdropper
```

```
%================================================================
            %instantaneous SNR calculations

%================================================================

        snr_s1r=Ps*abs(hs1r).^2;
        snr_s2r=Ps*abs(hs2r).^2;
        snr_s1e=Ps*abs(hs1e).^2;
        snr_s2e=Ps*abs(hs2e).^2;
        snr_rs1=Pr*abs(hs1r).^2;
        snr_rs2=Pr*abs(hs2r).^2;
      % snr_js1=Pj*abs(hrs1).^2;
      % snr_js2=Pj*abs(hrs2).^2;
        snr_re=Pr*abs(hre).^2;
      % snr_je=Pj*abs(hre).^2;

        cs1r=0.5*log2(1+snr_s1r);
        y1=cs1r>R0;

        cs2r=0.5*log2(1+snr_s2r);
        y2=cs2r>R0;

        y=(y1&y2);
  %========================
      %Non-jamming techniques
  %========================
  % case:optimal selection with maximum sum instantaneous
secrecy
  % rate(OS-MSISR)

              F=[];

              for r=1:N

  sinr1_os=(y(r)*snr_rs1(r));
  sinr2_os=(y(r)*snr_rs2(r));

  sinre1_os=max((snr_s1e./(1+snr_s2e))+(y(r)*snr_re(r,:)'));
  sinre2_os=max((snr_s2e./(1+snr_s1e))+(y(r)*snr_re(r,:)'));

  C=max(0,0.5*log2(1+sinr1_os)
-0.5*log2(1+sinre2_os)+0.5*log2(1+sinr2_os)
-0.5*log2(1+sinre1_os));

        F=[F C];

            end

  Cs_os(1,jj)=max(F);
  outage_os=outage_os+(Rs>Cs_os(1,jj));
```

```
%===========================
   %Jamming techniques
%===========================
% case:optimal selection with jamming(OSJ)

              F=[];
              for j1=1:N
              for r=1:N
              for j2=1:N

              if (r~=j2)

   sinr1_osj=(y(r)*snr_rs1(r)./(1+snr_rs1(j2)/100));
   sinr2_osj=(y(r)*snr_rs2(r)./(1+snr_rs2(j2)/100));

sinre1_osj=max((snr_s1e./(1+snr_s2e+snr_re(j1,:).'/100))+(y(r)
*snr_re(r,:).'./(1+snr_re(j2,:).'/100)));

sinre2_osj=max((snr_s2e./(1+snr_s1e+snr_re(j1,:).'/100))+(y(r)
*snr_re(r,:).'./(1+snr_re(j2,:).'/100)));

   C=max(0,0.5*log2(1+sinr1_osj)
-0.5*log2(1+sinre2_osj)+0.5*log2(1+sinr2_osj)
-0.5*log2(1+sinre1_osj));

      F=[F C];

          end
       end

   end
end

 Cs_osj(1,jj)=max(F);
 outage_osj=outage_osj+(Rs>Cs_osj(1,jj));

 % case:optimal selection with controlled jamming(OSCJ)

              F=[];

              for r=1:N
              for j2=1:N
              for j1=1:N

              if (r~=j2)

   sinr1_oscj=(y(r)*snr_rs1(r));
   sinr2_oscj=(y(r)*snr_rs2(r));
```

```
sinre1_oscj=max((snr_s1e./(1+snr_s2e+snr_re(j1,:).'/100))+
(y(r)*snr_re(r,:).'./(1+snr_re(j2,:).'/100)));
sinre2_oscj=max((snr_s2e./(1+snr_s1e+snr_re(j1,:).'/100))+
(y(r)*snr_re(r,:).'./(1+snr_re(j2,:).'/100)));

   C=max(0,0.5*log2(1+sinr1_oscj)-0.5*log2(1+sinre2_oscj)+0.5*
   log2(1+sinr2_oscj)-0.5*log2(1+sinre1_oscj));

      F=[F C];
      end
       end

   end
end

 Cs_oscj(1,jj)=max(F);
 outage_oscj=outage_oscj+(Rs>Cs_oscj(1,jj));

      end
       out_os(ii)=outage_os/Num_Trials
      c_os(ii)=mean(Cs_os)

      out_osj(ii)=outage_osj/Num_Trials
      c_osj(ii)=mean(Cs_osj)

      out_oscj(ii)=outage_oscj/Num_Trials
      c_oscj(ii)=mean(Cs_oscj)
 end
  plot(P_dB, c_os,'b-x',P_dB, c_osj,'b-s',P_dB, c_oscj,'m-p')
xlabel('P[dB]');
ylabel('secrecy ergodic capacity[BPCU]');
grid on; box on;
axis([0 25 0 5]);
legend ('OS','OSJ','OSCJ',2);
figure

semilogy(P_dB, out_os,'b-x',P_dB,out_osj,'b-s',P_dB,
out_oscj,'m-p')
 xlabel('P[dB]');
ylabel('secrecy outage probability');
grid on; box on;
axis([0 40 10^-3 1]);
  legend ('OS','OSJ','OSCJ',3);
  figure

% cooperative eavesdroppers

clear all
clc
```

```
%===================
%system parameters
%===================
N=4;                    %number of relay nodes
M=2;                    %number of eavesdroppers
Rs=0.1;                 %target secrecy rate (BPCU)-for outage
probability calculation
R0=2;                   %Transmission Rate-to check which relay
can decode correctly
P_dB=0:5:40;            %Transmitted power (dB)
Num_Trials=3000;        %number of trials
PL_exp=3;               %path loss exponent
N0=1;                   %noise variance

loc_s1=[0 0];
loc_e=[0.5 1;0 0.5];
loc_s2=[1 0];
loc_r=[0.35 0.3;0.4 0.65;0.6 0.4;0.65 0.7];

z1=(loc_s1(1)-loc_e(:,1)).^2;
z2=(loc_s1(2)-loc_e(:,2)).^2;
z=sqrt(z1+z2);
var_s1e=z.^(-PL_exp);

z1=(loc_s2(1)-loc_e(:,1)).^2;
z2=(loc_s2(2)-loc_e(:,2)).^2;
z=sqrt(z1+z2);
var_s2e=z.^(-PL_exp);

z1=(loc_s1(1)-loc_r(:,1)).^2;
z2=(loc_s1(2)-loc_r(:,2)).^2;
z=sqrt(z1+z2);
var_s1r=z.^(-PL_exp);

z1=(loc_s2(1)-loc_r(:,1)).^2;
z2=(loc_s2(2)-loc_r(:,2)).^2;
z=sqrt(z1+z2);
var_s2r=z.^(-PL_exp);

z1=(loc_r(:,1)-loc_e(1,1)).^2;
z2=(loc_r(:,2)-loc_e(1,2)).^2;
z11=sqrt(z1+z2);
z3=(loc_r(:,1)-loc_e(2,1)).^2;
z4=(loc_r(:,2)-loc_e(2,2)).^2;
z22=sqrt(z3+z4);
z=[z11 z22];
var_re=z.^(-PL_exp);
```

```
%===========================
%simulation start
%===========================
for ii=1:length(P_dB)
    Ps=10^(P_dB(ii)/10);
    Pr=Ps;
    Pj=Ps/100;    %jammer power less than relay node power

    outage_cs=0;
    outage_os=0;
    outage_osmmisr=0;
    outage_ss=0;
    outage_ssmmisr=0;
    outage_osj=0;
    outage_osjmmisr=0;
    outage_ssj=0;
    outage_ssjmmisr=0;
    outage_oscj=0;
    outage_ow=0;
    outage_sw=0;

    for jj=1:Num_Trials

%================================================================
        %Channel coefficients
%================================================================
        %broadcasting phase

        hs1r=sqrt(var_s1r/2).*(randn(N,1)+i*randn(N,1));
% S1-->relays
        hs2r=sqrt(var_s2r/2).*(randn(N,1)+i*randn(N,1));
% S2-->relays
        hs1e=sqrt(var_s1e/2).*(randn(M,1)+i*randn(M,1));
% S1-->eavesdropper
        hs2e=sqrt(var_s2e/2).*(randn(M,1)+i*randn(M,1));
% S2-->eavesdropper
        %cooperative links
        %hrs1=sqrt(var_s1r/2).*(randn(N,1)+i*randn(N,1));
% R-->destination1
        %hrs2=sqrt(var_s2r/2).*(randn(N,1)+i*randn(N,1));
% R-->destination2
        hre=sqrt(var_re/2).*(randn(N,M)+i*randn(N,M));
% R-->eavesdropper
```

```
%=============================================================
          %instantaneous SNR calculations
%=============================================================

          snr_s1r=Ps*abs(hs1r).^2;
          snr_s2r=Ps*abs(hs2r).^2;
          snr_s1e=Ps*abs(hs1e).^2;
          snr_s2e=Ps*abs(hs2e).^2;
          snr_rs1=Pr*abs(hs1r).^2;
          snr_rs2=Pr*abs(hs2r).^2;
        % snr_js1=Pj*abs(hrs1).^2;
        % snr_js2=Pj*abs(hrs2).^2;
         snr_re=Pr*abs(hre).^2;
        % snr_je=Pj*abs(hre).^2;

          cs1r=0.5*log2(1+snr_s1r);
          y1=cs1r>R0;

          cs2r=0.5*log2(1+snr_s2r);
          y2=cs2r>R0;

          y=(y1&y2);
    %=========================
       %Non-jamming techniques
    %=========================
    % case:optimal selection with maximum sum instantaneous
secrecy
    % rate(OS-MSISR)

                  F=[];

                  for r=1:N

    sinr1_os=(y(r)*snr_rs1(r));
    sinr2_os=(y(r)*snr_rs2(r));

    sinre1_os=sum((snr_s1e./(1+snr_s2e))+(y(r)*snr_re(r,:)'));
    sinre2_os=sum((snr_s2e./(1+snr_s1e))+(y(r)*snr_re(r,:)'));

    C=max(0,0.5*log2(1+sinr1_os)
-0.5*log2(1+sinre2_os)+0.5*log2(1+sinr2_os)
-0.5*log2(1+sinre1_os));

        F=[F C];

            end
```

```
Cs_os(1,jj)=max(F);
outage_os=outage_os+(Rs>Cs_os(1,jj));

  %===========================
    %Jamming techniques
  %===========================
  % case:optimal selection with jamming(OSJ)

               F=[];
               for j1=1:N
               for r=1:N
               for j2=1:N

               if (r~=j2)

   sinr1_osj=(y(r)*snr_rs1(r)./(1+snr_rs1(j2)/100));
   sinr2_osj=(y(r)*snr_rs2(r)./(1+snr_rs2(j2)/100));

sinre1_osj=sum((snr_s1e./(1+snr_s2e+snr_re(j1,:).'/100))+(y(r)
*snr_re(r,:).'./(1+snr_re(j2,:).'/100)));

sinre2_osj=sum((snr_s2e./(1+snr_s1e+snr_re(j1,:).'/100))+(y(r)
*snr_re(r,:).'./(1+snr_re(j2,:).'/100)));

   C=max(0,0.5*log2(1+sinr1_osj)
-0.5*log2(1+sinre2_osj)+0.5*log2(1+sinr2_osj)
-0.5*log2(1+sinre1_osj));

        F=[F C];

            end
        end

   end
end

Cs_osj(1,jj)=max(F);
outage_osj=outage_osj+(Rs>Cs_osj(1,jj));

% case:optimal selection with controlled jamming(OSCJ)

               F=[];

               for r=1:N
               for j2=1:N
               for j1=1:N

               if (r~=j2)
```

```
    sinr1_oscj=(y(r)*snr_rs1(r));
    sinr2_oscj=(y(r)*snr_rs2(r));

sinre1_oscj=sum(((snr_s1e./(1+snr_s2e+snr_re(j1,:).'/100))+
(y(r)*snr_re(r,:).'./(1+snr_re(j2,:).'/100))));

sinre2_oscj=sum(((snr_s2e./(1+snr_s1e+snr_re(j1,:).'/100))+
(y(r)*snr_re(r,:).'./(1+snr_re(j2,:).'/100))));

    C=max(0,0.5*log2(1+sinr1_oscj)-0.5*log2(1+sinre2_
    oscj)+0.5*log2(1+sinr2_oscj)-0.5*log2(1+sinre1_oscj));

      F=[F C];
      end
       end

   end
end

 Cs_oscj(1,jj)=max(F);
 outage_oscj=outage_oscj+(Rs>Cs_oscj(1,jj));

      end
       out_os(ii)=outage_os/Num_Trials
      c_os(ii)=mean(Cs_os)

      out_osj(ii)=outage_osj/Num_Trials
      c_osj(ii)=mean(Cs_osj)

      out_oscj(ii)=outage_oscj/Num_Trials
      c_oscj(ii)=mean(Cs_oscj)
 end
  plot(P_dB, c_os,'b-x',P_dB, c_osj,'b-s',P_dB, c_oscj,'m-p')
 xlabel('P[dB]');
 ylabel('secrecy ergodic capacity[BPCU]');
 grid on; box on;
 axis([0 25 0 5]);
 legend ('OS','OSJ','OSCJ',2);
 figure

 semilogy(P_dB, out_os,'b-x',P_dB,out_osj,'b-s',P_dB,
out_oscj,'m-p')
 xlabel('P[dB]');
 ylabel('secrecy outage probability');
 grid on; box on;
 axis([0 40 10^-3 1]);
  legend ('OS','OSJ','OSCJ',3);
  figure
```

```
%****************************************************************
% Presence of M=3 cooperating eavesdroppers
%****************************************************************

% cooperative eavesdroppers
%Two-way relay networks
clear all
clc
%==================
%system parameters
%==================
N=4;                    %number of relay nodes
M=3;                    %number of eavesdroppers
Rs=0.1;                 %target secrecy rate (BPCU)-for outage
probability calculation
R0=2;                   %Transmission Rate-to check which relay
can decode correctly
P_dB=0:5:40;            %Transmitted power (dB)
Num_Trials=3000;        %number of trials
PL_exp=3;               %path loss exponent
N0=1;                   %noise variance

loc_s1=[0 0];
loc_e=[0.5 1;0 0.5;1 0.5];
loc_s2=[1 0];
loc_r=[0.35 0.3;0.4 0.65;0.6 0.4;0.65 0.7];

z1=(loc_s1(1)-loc_e(:,1)).^2;
z2=(loc_s1(2)-loc_e(:,2)).^2;
z=sqrt(z1+z2);
var_s1e=z.^(-PL_exp);

z1=(loc_s2(1)-loc_e(:,1)).^2;
z2=(loc_s2(2)-loc_e(:,2)).^2;
z=sqrt(z1+z2);
var_s2e=z.^(-PL_exp);

z1=(loc_s1(1)-loc_r(:,1)).^2;
z2=(loc_s1(2)-loc_r(:,2)).^2;
z=sqrt(z1+z2);
var_s1r=z.^(-PL_exp);

z1=(loc_s2(1)-loc_r(:,1)).^2;
z2=(loc_s2(2)-loc_r(:,2)).^2;
z=sqrt(z1+z2);
var_s2r=z.^(-PL_exp);

z1=(loc_r(:,1)-loc_e(1,1)).^2;
z2=(loc_r(:,2)-loc_e(1,2)).^2;
z11=sqrt(z1+z2);
z3=(loc_r(:,1)-loc_e(2,1)).^2;
```

```
z4=(loc_r(:,2)-loc_e(2,2)).^2;
z22=sqrt(z3+z4);
z5=(loc_r(:,1)-loc_e(3,1)).^2;
z6=(loc_r(:,2)-loc_e(3,2)).^2;
z33=sqrt(z5+z6);
z=[z11 z22 z33];
var_re=z.^(-PL_exp);


%========================
%simulation start
%========================
for ii=1:length(P_dB)
    Ps=10^(P_dB(ii)/10);
    Pr=Ps;
    Pj=Ps/100;    %jammer power less than relay node power

    outage_cs=0;
    outage_os=0;
    outage_osmmisr=0;
    outage_ss=0;
    outage_ssmmisr=0;
    outage_osj=0;
    outage_osjmmisr-0;
    outage_ssj=0;
    outage_ssjmmisr=0;
    outage_oscj=0;
    outage_ow=0;
    outage_sw=0;

    for jj=1:Num_Trials

%==================================================================
        %Channel coefficients

%==================================================================
        %broadcasting phase

        hs1r=sqrt(var_s1r/2).*(randn(N,1)+i*randn(N,1));
% S1-->relays
        hs2r=sqrt(var_s2r/2).*(randn(N,1)+i*randn(N,1));
% S2-->relays
        hs1e=sqrt(var_s1e/2).*(randn(M,1)+i*randn(M,1));
% S1-->eavesdropper
        hs2e=sqrt(var_s2e/2).*(randn(M,1)+i*randn(M,1));
% S2-->eavesdropper
        %cooperative links
        %hrs1=sqrt(var_s1r/2).*(randn(N,1)+i*randn(N,1));
% R-->destination1
        %hrs2=sqrt(var_s2r/2).*(randn(N,1)+i*randn(N,1));
% R-->destination2
```

```
            hre=sqrt(var_re/2).*(randn(N,M)+i*randn(N,M));
% R-->eavesdropper

%================================================================
        %instantaneous SNR calculations

%================================================================

        snr_s1r=Ps*abs(hs1r).^2;
        snr_s2r=Ps*abs(hs2r).^2;
        snr_s1e=Ps*abs(hs1e).^2;
        snr_s2e=Ps*abs(hs2e).^2;
        snr_rs1=Pr*abs(hs1r).^2;
        snr_rs2=Pr*abs(hs2r).^2;
      % snr_js1=Pj*abs(hrs1).^2;
      % snr_js2=Pj*abs(hrs2).^2;
        snr_re=Pr*abs(hre).^2;
      % snr_je=Pj*abs(hre).^2;

        cs1r=0.5*log2(1+snr_s1r);
        y1=cs1r>R0;

        cs2r=0.5*log2(1+snr_s2r);
        y2=cs2r>R0;

        y=(y1&y2);
  %========================
     %Non-jamming techniques
  %========================
  % case:optimal selection with maximum sum instantaneous
secrecy
  % rate(OS-MSISR)

            F=[];

            for r=1:N

  sinr1_os=(y(r)*snr_rs1(r));
  sinr2_os=(y(r)*snr_rs2(r));

  sinre1_os=sum((snr_s1e./(1+snr_s2e))+(y(r)*snr_re(r,:)'));
  sinre2_os=sum((snr_s2e./(1+snr_s1e))+(y(r)*snr_re(r,:)'));

  C=max(0,0.5*log2(1+sinr1_os)
-0.5*log2(1+sinre2_os)+0.5*log2(1+sinr2_os)
-0.5*log2(1+sinre1_os));

      F=[F C];

        end
```

```
Cs_os(1,jj)=max(F);
outage_os=outage_os+(Rs>Cs_os(1,jj));

  %=========================
     %Jamming techniques
  %=========================
  % case:optimal selection with jamming(OSJ)

              F=[];
              for j1=1:N
              for r=1:N
              for j2=1:N

              if (r~=j2)

   sinr1_osj=(y(r)*snr_rs1(r)./(1+snr_rs1(j2)/100));
   sinr2_osj=(y(r)*snr_rs2(r)./(1+snr_rs2(j2)/100));

sinre1_osj=sum((snr_s1e./(1+snr_s2e+snr_re(j1,:).'/100))+(y(r)
*snr_re(r,:).'./(1+snr_re(j2,:).'/100)));

sinre2_osj=sum((snr_s2e./(1+snr_s1e+snr_re(j1,:).'/100))+(y(r)
*snr_re(r,:).'./(1+snr_re(j2,:).'/100)));

   C=max(0,0.5*log2(1+sinr1_osj)
   -0.5*log2(1+sinre2_osj)+0.5*log2(1+sinr2_osj)
   -0.5*log2(1+sinre1_osj));

       F=[F C];

           end
       end

   end
end

Cs_osj(1,jj)=max(F);
outage_osj=outage_osj+(Rs>Cs_osj(1,jj));

% case:optimal selection with controlled jamming(OSCJ)

              F=[];

              for r=1:N
              for j2=1:N
              for j1=1:N

              if (r~=j2)

   sinr1_oscj=(y(r)*snr_rs1(r));
   sinr2_oscj=(y(r)*snr_rs2(r));
```

```
sinre1_oscj=sum(((snr_s1e./(1+snr_s2e+snr_re(j1,:).'/100))+
(y(r)*snr_re(r,:).'./(1+snr_re(j2,:).'/100))));

sinre2_oscj=sum(((snr_s2e./(1+snr_s1e+snr_re(j1,:).'/100))+
(y(r)*snr_re(r,:).'./(1+snr_re(j2,:).'/100))));

   C=max(0,0.5*log2(1+sinr1_oscj)
-0.5*log2(1+sinre2_oscj)+0.5*log2(1+sinr2_oscj)
-0.5*log2(1+sinre1_oscj));

      F=[F C];
      end
       end

   end
end

 Cs_oscj(1,jj)=max(F);
 outage_oscj=outage_oscj+(Rs>Cs_oscj(1,jj));

      end
       out_os(ii)=outage_os/Num_Trials
      c_os(ii)=mean(Cs_os)

      out_osj(ii)=outage_osj/Num_Trials
      c_osj(ii)=mean(Cs_osj)

      out_oscj(ii)=outage_oscj/Num_Trials
      c_oscj(ii)=mean(Cs_oscj)
 end
  plot(P_dB, c_os,'b-x',P_dB, c_osj,'b-s',P_dB, c_oscj,'m-p')
 xlabel('P[dB]');
 ylabel('secrecy ergodic capacity[BPCU]');
 grid on; box on;
 axis([0 25 0 5]);
 legend ('OS','OSJ','OSCJ',2);
 figure

 semilogy(P_dB, out_os,'b-x',P_dB,out_osj,'b-s',P_dB,
out_oscj,'m-p')
 xlabel('P[dB]');
 ylabel('secrecy outage probability');
 grid on; box on;
 axis([0 40 10^-3 1]);
  legend ('OS','OSJ','OSCJ',3);
 figure
```

```
%****************************************************************
                % Relays located close to Eavesdroppers
%****************************************************************

% Joint relay and jammer selection for secure two-way relay
networks
% performance metrics are ergodic secrecy capacity & secrecy
outage
% probability
% Intermediate nodes close to E.

 clear all
 clc
 %===================
 %system parameters
 %===================
 N=4;                    %number of relay nodes
 M=1;                    %number of eavesdroppers
 Rs=0.1;                %target secrecy rate (BPCU)-for outage
probability calculation
 R0=2;                  %Transmission Rate-to check which relay
can decode correctly
 P_dB=0:5:40;          %Transmitted power (dB)
 Num_Trials-8000;      %number of trials
 PL_exp=3;             %path loss exponent
 N0=1;                 %noise variance

 loc_s1=[0 0];
 loc_e=[0.5 1];
 loc_s2=[1 0];
 loc_r=[0.4 0.95;0.55 0.9;0.45 0.85;0.6 0.8];

 var_s1e=(sqrt((loc_s1(1)-loc_e(1)).^2+(loc_s1(2)-loc_e(2)).
^2)).^(-PL_exp);
 var_s2e=(sqrt((loc_s2(1)-loc_e(1)).^2+(loc_s2(2)-
loc_e(2)).^2)).^(-PL_exp);

 z1=(loc_s1(1)-loc_r(:,1)).^2;
 z2=(loc_s1(2)-loc_r(:,2)).^2;
 z=sqrt(z1+z2);
 var_s1r=z.^(-PL_exp);

 z1=(loc_s2(1)-loc_r(:,1)).^2;
 z2=(loc_s2(2)-loc_r(:,2)).^2;
 z=sqrt(z1+z2);
 var_s2r=z.^(-PL_exp);
```

```
z1=(loc_e(1)-loc_r(:,1)).^2;
z2=(loc_e(2)-loc_r(:,2)).^2;
z=sqrt(z1+z2);
var_re=z.^(-PL_exp);


%=========================
%simulation start
%=========================
for ii=1:length(P_dB)
    Ps=10^(P_dB(ii)/10);
    Pr=Ps;
    Pj=Ps/100;    %jammer power less than relay node power

    outage_cs=0;
    outage_os=0;
    outage_osmmisr=0;
    outage_ss=0;
    outage_ssmmisr=0;
    outage_osj=0;
    outage_osjmmisr=0;
    outage_ssj=0;
    outage_ssjmmisr=0;
    outage_oscj=0;
    outage_ow=0;
    outage_sw=0;


    for jj=1:Num_Trials


%=====================================================================
        %Channel coefficients
%=====================================================================
        %broadcasting phase

        hs1r=sqrt(var_s1r/2).*(randn(N,1)+i*randn(N,1));
% S1-->relays
        hs2r=sqrt(var_s2r/2).*(randn(N,1)+i*randn(N,1));
% S2-->relays
        hs1e=sqrt(var_s1e/2).*(randn(1,1)+i*randn(1,1));
% S1-->eavesdropper
        hs2e=sqrt(var_s2e/2).*(randn(1,1)+i*randn(1,1));
% S2-->eavesdropper
        %cooperative links
        %hrs1=sqrt(var_s1r/2).*(randn(N,1)+i*randn(N,1));
% R-->destination1
        %hrs2=sqrt(var_s2r/2).*(randn(N,1)+i*randn(N,1));
% R-->destination2
        hre=sqrt(var_re/2).*(randn(N,1)+i*randn(N,1));
% R-->eavesdropper
```

```
%=================================================================
          %instantaneous SNR calculations
%=================================================================

          snr_s1r=Ps*abs(hs1r).^2;
          snr_s2r=Ps*abs(hs2r).^2;
          snr_s1e=Ps*abs(hs1e).^2;
          snr_s2e=Ps*abs(hs2e).^2;
          snr_rs1=Pr*abs(hs1r).^2;
          snr_rs2=Pr*abs(hs2r).^2;
          snr_js1=Pj*abs(hs1r).^2;
          snr_js2=Pj*abs(hs2r).^2;
          snr_re=Pr*abs(hre).^2;
          snr_je=Pj*abs(hre).^2;

          cs1r=0.5*log2(1+snr_s1r);
          y1=cs1r>R0;

          cs2r=0.5*log2(1+snr_s2r);
          y2=cs2r>R0;

          y=(y1&y2);

     %=========================
        %Non-jamming techniques
     %=========================

     %case:optimal selection (OS)

          sinr1_os=(y.*snr_rs1);
          sinr2_os=(y.*snr_rs2);

          sinre1_os=(((snr_s1e)/(snr_s2e+1))+(y.*snr_re));
          sinre2_os=(((snr_s2e)/(snr_s1e+1))+(y.*snr_re));

     [q1 q2]=max(((1+sinr1_os)./(1+sinre2_os)).*((1+sinr2_
os)./(1+sinre1_os)));

          sinr1_osn=(y(q2).*snr_rs1(q2));
          sinr2_osn=(y(q2).*snr_rs2(q2));

sinre1_osn=(((snr_s1e)/(snr_s2e+1))+(y(q2).*snr_re(q2)));

sinre2_osn=(((snr_s2e)/(snr_s1e+1))+(y(q2).*snr_re(q2)));

          Cs_os(1,jj)=max(0,0.5*log2(1+sinr1_osn)-
0.5*log2(1+sinre2_osn)+0.5*log2(1+sinr2_osn)-
0.5*log2(1+sinre1_osn));

     outage_os=outage_os+(Rs>Cs_os(1,jj));
```

```
%case:Suboptimal selection (SS)

        sinr1_ss=(y.*snr_rs1);
        sinr2_ss=(y.*snr_rs2);

        sinre1_ss=((mean(snr_s1e)/(mean(snr_s2e)+1))+
y.*mean(snr_re));
        sinre2_ss=((mean(snr_s2e)/(mean(snr_s1e)+1))+
y.*mean(snr_re));

        [q1 q2]=max((((1+sinr1_ss)./(1+sinre2_ss)).*((1+sinr2_
ss)./(1+sinre1_ss))));

        sinr1_ssn=y(q2)*snr_s1r(q2);
        sinr2_ssn=y(q2)*snr_s2r(q2);
        sinre1_ssn=((snr_s1e./(snr_s2e+1))+ y(q2)*snr_re(q2));
        sinre2_ssn=((snr_s2e./(snr_s1e+1))+ y(q2)*snr_re(q2));

        Cs_ss(1,jj)=max(0,0.5*log2(1+sinr1_ssn)-
0.5*log2(1+sinre2_ssn)+0.5*log2(1+sinr2_ssn)-
0.5*log2(1+sinre1_ssn));

        outage_ss=outage_ss+(Rs>Cs_ss(1,jj));

        % case:optimal selection with max-min instantaneous
secrecy rate(OS-MMISR)

        sinr1_os=(y.*snr_rs1);
        sinr2_os=(y.*snr_rs2);

        sinre1_os=(((snr_s1e)/(snr_s2e+1))+(y.*snr_re));
        sinre2_os=(((snr_s2e)/(snr_s1e+1))+(y.*snr_re));

        [q1 q2]=max(min(((1+sinr1_os)./(1+sinre2_
os)),((1+sinr2_os)./(1+sinre1_os))));

        sinr1_osn=(y(q2).*snr_rs1(q2));
        sinr2_osn=(y(q2).*snr_rs2(q2));

sinre1_osn=(((snr_s1e)/(snr_s2e+1))+(y(q2).*snr_re(q2)));

sinre2_osn=(((snr_s2e)/(snr_s1e+1))+(y(q2).*snr_re(q2)));

        Cs_osmmisr(1,jj)=max(0,0.5*log2(1+sinr1_osn)
-0.5*log2(1+sinre2_osn)+0.5*log2(1+sinr2_osn)-
0.5*log2(1+sinre1_osn));

        outage_osmmisr=outage_osmmisr+(Rs>Cs_osmmisr(1,jj));
```

```
% case:Suboptimal selection with max-min instantaneous
secrecy
    % rate(SS-MMISR)

        sinr1_ss=(y.*snr_rs1);
        sinr2_ss=(y.*snr_rs2);

        sinre1_ss=((mean(snr_s1e)/(mean(snr_s2e)+1))+
y.*mean(snr_re));
        sinre2_ss=((mean(snr_s2e)/(mean(snr_s1e)+1))+
y.*mean(snr_re));

        [q1
q2]=max(min(((1+sinr1_ss)./(1+sinre2_ss)),((1+sinr2_ss)./
(1+sinre1_ss))));

        sinr1_ssn=y(q2)*snr_s1r(q2);
        sinr2_ssn=y(q2)*snr_s2r(q2);
        sinre1_ssn=((snr_s1e./(snr_s2e+1))+ y(q2)*snr_re(q2));
        sinre2_ssn=((snr_s2e./(snr_s1e+1))+ y(q2)*snr_re(q2));

        Cs_ssmmisr(1,jj)=max(0,0.5*log2(1+sinr1_ssn)-
0.5*log2(1+sinre2_ssn)+0.5*log2(1+sinr2_ssn)-
0.5*log2(1+sinre1_ssn));

        outage_ssmmisr=outage_ssmmisr+(Rs>Cs_ssmmisr(1,jj));

    %=========================
     %Jamming techniques
     %=========================
    % case:optimal selection with jamming(OSJ)

        [q3 q4]=max(snr_je.*snr_je);       % selecting jammer in
first phase.

        [q1 q2]=max(y.*((snr_rs1./snr_re).*(snr_rs2./snr_re)));
    % selecting relay in second phase.

        Q1=(q1==0);       %Q1=1==> all the relays fail to decode
        temp=ones(N,1);
        temp(q2,1)=0;

        [q5 q6]=max(temp.*((snr_je./snr_js1).*(snr_je./snr_
js2)));       %selecting the jammer in the second phase

        sinr1_osj=((y(q2)*snr_rs1(q2))./(1+snr_js1(q6)));
        sinr2_osj=((y(q2)*snr_rs2(q2))./(1+snr_js2(q6)));

sinre1_osj=(snr_s1e./(1+snr_s2e+snr_je(q4)))+((y(q2)*snr_
re(q2))./(1+snr_je(q6)));
```

```
sinre2_osj=(snr_s2e./(1+snr_s1e+snr_je(q4)))+((y(q2)*snr_
re(q2))./(1+snr_je(q6)));

    Cs_osj(1,jj)=max(0,0.5*log2(1+sinr1_osj)
-0.5*log2(1+sinre2_osj)+0.5*log2(1+sinr2_osj)-
0.5*log2(1+sinre1_osj));

    outage_osj=outage_osj+(Rs>Cs_osj(1,jj));

% case:Suboptimal selection with jamming(SSJ)

    [q3 q4]=max(mean(snr_je).*mean(snr_je));      % selecting
jammer in first phase.

    [q1
q2]=max(y.*((snr_rs1./mean(snr_re)).*(snr_rs2./mean(snr_
re))));     % selecting relay in second phase.

    Q1=(q1==0);        %Q1=1==> all the relays fail to decode
    temp=ones(N,1);
    temp(q2,1)=0;

    [q5
q6]=max(temp.*((mean(snr_je)./snr_js1).*(mean(snr_je)./snr_
js2)));      %selecting the jammer in the second phase

    sinr1_ssj=((y(q2)*snr_rs1(q2))./(1+snr_js1(q6)));
    sinr2_ssj=((y(q2)*snr_rs2(q2))./(1+snr_js2(q6)));

sinre1_ssj=(snr_s1e./(1+snr_s2e+snr_je(q4)))+((y(q2)*snr_
re(q2))./(1+snr_je(q6)));

sinre2_ssj=(snr_s2e./(1+snr_s1e+snr_je(q4)))+((y(q2)*snr_
re(q2))./(1+snr_je(q6)));

    Cs_ssj(1,jj)=max(0,0.5*log2(1+sinr1_ssj)-
0.5*log2(1+sinre2_ssj)+0.5*log2(1+sinr2_ssj)-
0.5*log2(1+sinre1_ssj));

    outage_ssj=outage_ssj+(Rs>Cs_ssj(1,jj));

% case:optimal selection with max-min instantaneous secrecy
rate(OSJ-MMISR)

    [q3 q4]=max(min(snr_je,snr_je));       % selecting jammer
in first phase.

    [q1
q2]=max(min(y.*(snr_rs1./snr_re),y.*(snr_rs2./snr_re)));
% selecting relay in second phase.

    Q1=(q1==0);        %Q1=1==> all the relays fail to decode
```

```
      temp=ones(N,1);
      temp(q2,1)=0;

      [q5
q6]=max(min(temp.*(snr_je./snr_js1),temp.*(snr_je./snr_js2)));
%selecting the jammer in the second phase

      sinr1_osj=((y(q2)*snr_rs1(q2))./(1+snr_js1(q6)));
      sinr2_osj=((y(q2)*snr_rs2(q2))./(1+snr_js2(q6)));

sinre1_osj=(snr_s1e./(1+snr_s2e+snr_je(q4)))+((y(q2)*snr_
re(q2))./(1+snr_je(q6)));

sinre2_osj=(snr_s2e./(1+snr_s1e+snr_je(q4)))+((y(q2)*snr_
re(q2))./(1+snr_je(q6)));

      Cs_osjmmisr(1,jj)=max(0,0.5*log2(1+sinr1_osj)-
0.5*log2(1+sinre2_osj)+0.5*log2(1+sinr2_osj)-
0.5*log2(1+sinre1_osj));

      outage_osjmmisr=outage_osjmmisr+(Rs>Cs_osjmmisr(1,jj));

   % case:Suboptimal selection with max-min instantaneous
secrecy rate(SSJ-MMISR)

      [q3 q4]=max(min(mean(snr_je),mean(snr_je)));      % selecting
jammer in first phase.
      [q1
q2]=max(min(y.*(snr_rs1./mean(snr_re)),y.*(snr_rs2./mean(snr_
re))));      % selecting relay in second phase.

      Q1=(q1==0);         %Q1=1==> all the relays fail to decode
      temp=ones(N,1);
      temp(q2,1)=0;

      [q5
q6]=max(min(temp.*(mean(snr_je)./snr_js1),temp.*(mean(snr_
je)./snr_js2)));      %selecting the jammer in the second phase

      sinr1_ssj=((y(q2)*snr_rs1(q2))./(1+snr_js1(q6)));
      sinr2_ssj=((y(q2)*snr_rs2(q2))./(1+snr_js2(q6)));

sinre1_ssj=(snr_s1e./(1+snr_s2e+snr_je(q4)))+((y(q2)*snr_
re(q2))./(1+snr_je(q6)));

sinre2_ssj=(snr_s2e./(1+snr_s1e+snr_je(q4)))+((y(q2)*snr_
re(q2))./(1+snr_je(q6)));

      Cs_ssjmmisr(1,jj)=max(0,0.5*log2(1+sinr1_ssj)-
0.5*log2(1+sinre2_ssj)+0.5*log2(1+sinr2_ssj)-
0.5*log2(1+sinre1_ssj));
```

```
    outage_ssjmmisr=outage_ssjmmisr+(Rs>Cs_ssjmmisr(1,jj));

% case:optimal switching(OW)

    [q3 q4]=max(snr_je.*snr_je);      % selecting jammer in
first phase.

    [q1 q2]=max(y.*((snr_rs1./snr_re).*(snr_rs2./snr_re)));
% selecting relay in second phase.

    Q1=(q1==0);        %Q1=1==> all the relays fail to decode
    temp=ones(N,1);
    temp(q2,1)=0;

    [q5 q6]=max(temp.*((snr_je./snr_js1).*(snr_je./snr_
js2)));      %selecting the jammer in the second phase

    sinr1_os=(y(q2)*snr_rs1(q2));
    sinr2_os=(y(q2)*snr_rs2(q2));
    sinre1_os=(snr_s1e./(1+snr_s2e))+(y(q2)*snr_re(q2));
    sinre2_os=(snr_s2e./(1+snr_s1e))+(y(q2)*snr_re(q2));

    sinr1_osj=((y(q2)*snr_rs1(q2))./(1+snr_js1(q6)));
    sinr2_osj=((y(q2)*snr_rs2(q2))./(1+snr_js2(q6)));
sinre1_osj=(snr_s1e./(1+snr_s2e+snr_je(q4)))+((y(q2)*snr_
re(q2))./(1+snr_je(q6)));

sinre2_osj=(snr_s2e./(1+snr_s1e+snr_je(q4)))+((y(q2)*snr_
re(q2))./(1+snr_je(q6)));

if(((1+sinr1_osj)./(1+sinre2_osj)).*((1+sinr2_osj)./
(1+sinre1_osj))>...

((1+sinr1_os)./(1+sinre2_os)).*((1+sinr2_os)./(1+sinre1_os)))

    Cs_ow(1,jj)=max(0,0.5*log2(1+sinr1_osj)-
0.5*log2(1+sinre2_osj)+0.5*log2(1+sinr2_osj)-
0.5*log2(1+sinre1_osj));

  else

  Cs_ow(1,jj)=max(0,0.5*log2(1+sinr1_os)-
0.5*log2(1+sinre2_os)+0.5*log2(1+sinr2_os)-
0.5*log2(1+sinre1_os));

  end

    outage_ow=outage_ow+(Rs>Cs_ow(1,jj));

    % case:Suboptimal switching(SW)
```

```
      [q3 q4]=max(mean(snr_je).*mean(snr_je));     % selecting
jammer in first phase.

      [q1
q2]=max(y.*((snr_rs1./mean(snr_re)).*(snr_rs2./mean(snr_
re))));     % selecting relay in second phase.

      Q1=(q1==0);        %Q1=1==> all the relays fail to decode
      temp=ones(N,1);
      temp(q2,1)=0;

      [q5 q6]=max(temp.*((mean(snr_je)./snr_js1).*(mean(snr_
je)./snr_js2)));     %selecting the jammer in the second phase

      sinr1_ss=y(q2)*snr_s1r(q2);
      sinr2_ss=y(q2)*snr_s2r(q2);
      sinre1_ss=((snr_s1e./(snr_s2e+1))+ y(q2)*snr_re(q2));
      sinre2_ss=((snr_s2e./(snr_s1e+1))+ y(q2)*snr_re(q2));

      sinr1_ssj=((y(q2)*snr_rs1(q2))./(1+snr_js1(q6)));
      sinr2_ssj=((y(q2)*snr_rs2(q2))./(1+snr_js2(q6)));

      sinre1_ssj=(snr_s1e./(1+snr_s2e+snr_je(q4)))+((y(q2)*snr_
re(q2))./(1+snr_je(q6)));

      sinre2_ssj=(snr_s2e./(1+snr_s1e+snr_je(q4)))+((y(q2)*snr_
re(q2))./(1+snr_je(q6)));
if((((1+sinr1_ssj)./(1+sinre2_ssj)).*((1+sinr2_ssj)./
(1+sinre1_ssj)))>...

((1+sinr1_ss)./(1+sinre2_ss)).*((1+sinr2_ss)./(1+sinre1_ss)))

    Cs_sw(1,jj)=max(0,0.5*log2(1+sinr1_ssj)-
0.5*log2(1+sinre2_ssj)+0.5*log2(1+sinr2_ssj)-
0.5*log2(1+sinre1_ssj));
    else

  Cs_sw(1,jj)=max(0,0.5*log2(1+sinr1_ss)
-0.5*log2(1+sinre2_ss)+0.5*log2(1+sinr2_ss)
-0.5*log2(1+sinre1_ss));

    end
      outage_sw=outage_sw+(Rs>Cs_sw(1,jj));

    % case:optimal selection with controlled jamming(OSCJ)
                  F=[];

                  for r=1:N
                  for j2=1:N
                  for j1=1:N

                  if (r~=j2)
```

```
   sinr1_oscj=(y(r)*snr_rs1(r));
   sinr2_oscj=(y(r)*snr_rs2(r));

sinre1_oscj=(snr_s1e./(1+snr_s2e+snr_je(j1)))+(y(r)*snr_
re(r)./(1+snr_je(j2)));

sinre2_oscj=(snr_s2e./(1+snr_s1e+snr_je(j1)))+(y(r)*snr_
re(r)./(1+snr_je(j2)));

   C=max(0,0.5*log2(1+sinr1_oscj)-
0.5*log2(1+sinre2_oscj)+0.5*log2(1+sinr2_oscj)-
0.5*log2(1+sinre1_oscj));

     F=[F C];
     end
      end

  end
end

 Cs_oscj(1,jj)=max(F);
 outage_oscj=outage_oscj+(Rs>Cs_oscj(1,jj));

    end

    out_os(ii)=outage_os/Num_Trials
    c_os(ii)=mean(Cs_os)

    out_ss(ii)=outage_ss/Num_Trials
    c_ss(ii)=mean(Cs_ss)

    out_osmmisr(ii)=outage_osmmisr/Num_Trials
    c_osmmisr(ii)=mean(Cs_osmmisr)

     out_ssmmisr(ii)=outage_ssmmisr/Num_Trials
    c_ssmmisr(ii)=mean(Cs_ssmmisr)

    out_osj(ii)=outage_osj/Num_Trials
    c_osj(ii)=mean(Cs_osj)

   out_ssj(ii)=outage_ssj/Num_Trials
    c_ssj(ii)=mean(Cs_ssj)

    out_osjmmisr(ii)=outage_osjmmisr/Num_Trials
    c_osjmmisr(ii)=mean(Cs_osjmmisr)

    out_ssjmmisr(ii)=outage_ssjmmisr/Num_Trials
    c_ssjmmisr(ii)=mean(Cs_ssjmmisr)

    out_ow(ii)=outage_ow/Num_Trials
```

```
        c_ow(ii)=mean(Cs_ow)

         out_sw(ii)=outage_sw/Num_Trials
        c_sw(ii)=mean(Cs_sw)

        out_oscj(ii)=outage_oscj/Num_Trials
        c_oscj(ii)=mean(Cs_oscj)

  end

    plot(P_dB, c_os,'b-x',P_dB, c_ss,'r-s',P_dB, c_osmmisr,'k-
  x',P_dB, c_ssmmisr,'k-s',P_dB, c_osj,'b-s',P_dB,
  c_ssj,'r-x',...
        P_dB, c_osjmmisr,'g:>',P_dB, c_ssjmmisr,'c:<',P_dB,
  c_ow,'b-.+',P_dB, c_sw,'r-.d',P_dB, c_oscj,'m-p')

  xlabel('P[dB]');
  ylabel('secrecy ergodic capacity[BPCU]');
  grid on; box on;
  axis([0 25 0 5]);
  legend
  ('OS','SS','OS-MMISR','SS-MMISR','OSJ','SSJ','OSJ-MMISR','SSJ-
  MMISR','OW','SW','OSCJ',2);
  figure

  semilogy(P_dB, out_os,'b-x',P_dB, out_ss,'r-s',P_dB, out_
  osmmisr,'k-x',P_dB,out_ssmmisr,'k-s',P_dB,
  out_osj,'b-s',P_dB,out_ssj,'r-x',...
        P_dB, out_osjmmisr,'g:>',P_dB, out_ssjmmisr,'c:<',P_dB,
  out_ow,'b-.+',P_dB, out_sw,'r-.d',P_dB, out_oscj,'m-p')

  xlabel('P[dB]');
  ylabel('secrecy outage probability');
  grid on; box on;
  axis([0 40 10^-3 1]);
  legend
  ('OS','SS','OS-MMISR','SS-MMISR','OSJ','SSJ','OSJ-MMISR','SSJ-
  MMISR','OW','SW','OSCJ',3);
  figure


%*****************************************************************
                    % Relays close to S2
%*****************************************************************

% Joint relay and jammer selection for secure two-way relay
networks
% performance metrics are ergodic secrecy capacity & secrecy
outage
% probability%Intermediate nodes close to S2
```

```
clear all
clc
%====================
%system parameters
%====================
N=4;                    %number of relay nodes
M=1;                    %number of eavesdroppers
Rs=0.1;                 %target secrecy rate (BPCU)-for outage
probability calculation
R0=2;                   %Transmission Rate-to check which relay
can decode correctly
P_dB=0:5:40;            %Transmitted power (dB)
Num_Trials=8000;        %number of trials
PL_exp=3;               %path loss exponent
N0=1;                   %noise variance


loc_s1=[0 0];
loc_e=[0.5 1];
loc_s2=[1 0];
loc_r=[0.95 0.1;0.9 0.2;0.85 0.05;0.8 0.15];

var_s1e=(sqrt((loc_s1(1)-loc_e(1)).^2+(loc_s1(2)-
loc_e(2)).^2)).^(-PL_exp);
var_s2e=(sqrt((loc_s2(1)-loc_e(1)).^2+(loc_s2(2)-
loc_e(2)).^2)).^(-PL_exp);

z1=(loc_s1(1)-loc_r(:,1)).^2;
z2=(loc_s1(2)-loc_r(:,2)).^2;
z=sqrt(z1+z2);
var_s1r=z.^(-PL_exp);


z1=(loc_s2(1)-loc_r(:,1)).^2;
z2=(loc_s2(2)-loc_r(:,2)).^2;
z=sqrt(z1+z2);
var_s2r=z.^(-PL_exp);


z1=(loc_e(1)-loc_r(:,1)).^2;
z2=(loc_e(2)-loc_r(:,2)).^2;
z=sqrt(z1+z2);
var_re=z.^(-PL_exp);


%=========================
%simulation start
%=========================
for ii=1:length(P_dB)
    Ps=10^(P_dB(ii)/10);
    Pr=Ps;
    Pj=Ps/100;   %jammer power less than relay node power
```

```
        outage_cs=0;
        outage_os=0;
        outage_osmmisr=0;
        outage_ss=0;
        outage_ssmmisr=0;
        outage_osj=0;
        outage_osjmmisr=0;
        outage_ssj=0;
        outage_ssjmmisr=0;
        outage_oscj=0;
        outage_ow=0;
        outage_sw=0;

        for jj=1:Num_Trials

%================================================================
        %Channel coefficients
%================================================================
        %broadcasting phase

        hs1r=sqrt(var_s1r/2).*(randn(N,1)+i*randn(N,1));
% S1-->relays
        hs2r=sqrt(var_s2r/2).*(randn(N,1)+i*randn(N,1));
% S2-->relays
        hs1e=sqrt(var_s1e/2).*(randn(1,1)+i*randn(1,1));
% S1-->eavesdropper
        hs2e=sqrt(var_s2e/2).*(randn(1,1)+i*randn(1,1));
% S2-->eavesdropper
        %cooperative links
        %hrs1=sqrt(var_s1r/2).*(randn(N,1)+i*randn(N,1));
% R-->destination1
        %hrs2=sqrt(var_s2r/2).*(randn(N,1)+i*randn(N,1));
% R-->destination2
        hre=sqrt(var_re/2).*(randn(N,1)+i*randn(N,1));
% R-->eavesdropper

%================================================================
        %instantaneous SNR calculations
%================================================================

        snr_s1r=Ps*abs(hs1r).^2;
        snr_s2r=Ps*abs(hs2r).^2;
        snr_s1e=Ps*abs(hs1e).^2;
        snr_s2e=Ps*abs(hs2e).^2;
        snr_rs1=Pr*abs(hs1r).^2;
        snr_rs2=Pr*abs(hs2r).^2;
        snr_js1=Pj*abs(hs1r).^2;
        snr_js2=Pj*abs(hs2r).^2;
        snr_re=Pr*abs(hre).^2;
        snr_je=Pj*abs(hre).^2;
```

```
        cs1r=0.5*log2(1+snr_s1r);
        y1=cs1r>R0;

        cs2r=0.5*log2(1+snr_s2r);
        y2=cs2r>R0;

        y=(y1&y2);

   %=========================
      %Non-jamming techniques
   %=========================

   %case:optimal selection (OS)

        sinr1_os=(y.*snr_rs1);
        sinr2_os=(y.*snr_rs2);

        sinre1_os=(((snr_s1e)/(snr_s2e+1))+(y.*snr_re));
        sinre2_os=(((snr_s2e)/(snr_s1e+1))+(y.*snr_re));

      [q1
q2]=max((((1+sinr1_os)./(1+sinre2_os)).*((1+sinr2_os)./
(1+sinre1_os)));

        sinr1_osn=(y(q2).*snr_rs1(q2));
        sinr2_osn=(y(q2).*snr_rs2(q2));

sinre1_osn=(((snr_s1e)/(snr_s2e+1))+(y(q2).*snr_re(q2)));

sinre2_osn=(((snr_s2e)/(snr_s1e+1))+(y(q2).*snr_re(q2)));

      Cs_os(1,jj)=max(0,0.5*log2(1+sinr1_osn)-
0.5*log2(1+sinre2_osn)+0.5*log2(1+sinr2_osn)-
0.5*log2(1+sinre1_osn));

      outage_os=outage_os+(Rs>Cs_os(1,jj));

  %case:Suboptimal selection (SS)

        sinr1_ss=(y.*snr_rs1);
        sinr2_ss=(y.*snr_rs2);

        sinre1_ss=((mean(snr_s1e)/(mean(snr_s2e)+1))+
y.*mean(snr_re));
        sinre2_ss=((mean(snr_s2e)/(mean(snr_s1e)+1))+
y.*mean(snr_re));

      [q1
q2]=max((((1+sinr1_ss)./(1+sinre2_ss)).*((1+sinr2_ss)./
(1+sinre1_ss)));
```

```
        sinr1_ssn=y(q2)*snr_s1r(q2);
        sinr2_ssn=y(q2)*snr_s2r(q2);
        sinre1_ssn=((snr_s1e./(snr_s2e+1))+
y(q2)*snr_re(q2));
        sinre2_ssn=((snr_s2e./(snr_s1e+1))+
y(q2)*snr_re(q2));

Cs_ss(1,jj)=max(0,0.5*log2(1+sinr1_ssn)-
0.5*log2(1+sinre2_ssn)+0.5*log2(1+sinr2_ssn)-
0.5*log2(1+sinre1_ssn));

        outage_ss=outage_ss+(Rs>Cs_ss(1,jj));

        % case:optimal selection with max-min instantaneous
secrecy rate(OS-MMISR)

        sinr1_os=(y.*snr_rs1);
        sinr2_os=(y.*snr_rs2);

        sinre1_os=(((snr_s1e)/(snr_s2e+1))+(y.*snr_re));
        sinre2_os=(((snr_s2e)/(snr_s1e+1))+(y.*snr_re));

        [q1
q2]=max(min(((1+sinr1_os)./(1+sinre2_os)),((1+sinr2_os)./
(1+sinre1_os))));

        sinr1_osn=(y(q2).*snr_rs1(q2));
        sinr2_osn=(y(q2).*snr_rs2(q2));

sinre1_osn=(((snr_s1e)/(snr_s2e+1))+(y(q2).*snr_re(q2)));

sinre2_osn=(((snr_s2e)/(snr_s1e+1))+(y(q2).*snr_re(q2)));

        Cs_osmmisr(1,jj)=max(0,0.5*log2(1+sinr1_osn)-
0.5*log2(1+sinre2_osn)+0.5*log2(1+sinr2_osn)-
0.5*log2(1+sinre1_osn));

        outage_osmmisr=outage_osmmisr+(Rs>Cs_osmmisr(1,jj));

        % case:Suboptimal selection with max-min instantaneous
secrecy
        % rate(SS-MMISR)

        sinr1_ss=(y.*snr_rs1);
        sinr2_ss=(y.*snr_rs2);

        sinre1_ss=((mean(snr_s1e)/(mean(snr_s2e)+1))+
y.*mean(snr_re));
        sinre2_ss=((mean(snr_s2e)/(mean(snr_s1e)+1))+
y.*mean(snr_re));
```

```
        [q1
q2]=max(min(((1+sinr1_ss)./(1+sinre2_ss)),((1+sinr2_ss)./
(1+sinre1_ss))));

        sinr1_ssn=y(q2)*snr_s1r(q2);
        sinr2_ssn=y(q2)*snr_s2r(q2);
        sinre1_ssn=((snr_s1e./(snr_s2e+1))+ y(q2)*snr_re(q2));
        sinre2_ssn=((snr_s2e./(snr_s1e+1))+ y(q2)*snr_re(q2));

        Cs_ssmmisr(1,jj)=max(0,0.5*log2(1+sinr1_ssn)-
0.5*log2(1+sinre2_ssn)+0.5*log2(1+sinr2_ssn)-
0.5*log2(1+sinre1_ssn));

        outage_ssmmisr=outage_ssmmisr+(Rs>Cs_ssmmisr(1,jj));

    %==========================
     %Jamming techniques
     %==========================
    % case:optimal selection with jamming(OSJ)

        [q3 q4]=max(snr_je.*snr_je);      % selecting jammer in
first phase.

        [q1 q2]=max(y.*((snr_rs1./snr_re).*(snr_rs2./snr_re)));
% selecting relay in second phase.

        Q1=(q1==0);        %Q1=1==> all the relays fail to decode
        temp=ones(N,1);
        temp(q2,1)=0;

        [q5 q6]=max(temp.*((snr_je./snr_js1).*(snr_je./snr_js2)));
%selecting the jammer in the second phase

        sinr1_osj=((y(q2)*snr_rs1(q2))./(1+snr_js1(q6)));
        sinr2_osj=((y(q2)*snr_rs2(q2))./(1+snr_js2(q6)));
        sinre1_osj=(snr_s1e./(1+snr_s2e+snr_je(q4)))+((y(q2)*snr_
re(q2))./(1+snr_je(q6)));
        sinre2_osj=(snr_s2e./(1+snr_s1e+snr_je(q4)))+((y(q2)*snr_
re(q2))./(1+snr_je(q6)));

        Cs_osj(1,jj)=max(0,0.5*log2(1+sinr1_osj)-
0.5*log2(1+sinre2_osj)+0.5*log2(1+sinr2_osj)-
0.5*log2(1+sinre1_osj));

        outage_osj=outage_osj+(Rs>Cs_osj(1,jj));

    % case:Suboptimal selection with jamming(SSJ)

        [q3 q4]=max(mean(snr_je).*mean(snr_je));      % selecting
jammer in first phase.
```

```
        [q1
q2]=max(y.*((snr_rs1./mean(snr_re)).*(snr_rs2./mean(snr_
re))));      % selecting relay in second phase.

    Q1=(q1==0);        %Q1=1==> all the relays fail to decode
    temp=ones(N,1);
    temp(q2,1)=0;

        [q5 q6]=max(temp.*((mean(snr_je)./snr_js1).*(mean(snr_
je)./snr_js2)));      %selecting the jammer in the second phase

    sinr1_ssj=((y(q2)*snr_rs1(q2))./(1+snr_js1(q6)));
    sinr2_ssj=((y(q2)*snr_rs2(q2))./(1+snr_js2(q6)));

    sinre1_ssj=(snr_s1e./(1+snr_s2e+snr_je(q4)))+((y(q2)*snr_
re(q2))./(1+snr_je(q6)));
    sinre2_ssj=(snr_s2e./(1+snr_s1e+snr_je(q4)))+((y(q2)*snr_
re(q2))./(1+snr_je(q6)));

    Cs_ssj(1,jj)=max(0,0.5*log2(1+sinr1_ssj)-
0.5*log2(1+sinre2_ssj)+0.5*log2(1+sinr2_ssj)-
0.5*log2(1+sinre1_ssj));

    outage_ssj=outage_ssj+(Rs>Cs_ssj(1,jj));

 % case:optimal selection with max-min instantaneous secrecy
rate(OSJ-MMISR)

        [q3 q4]=max(min(snr_je,snr_je));      % selecting jammer
in first phase.

        [q1
q2]=max(min(y.*(snr_rs1./snr_re),y.*(snr_rs2./snr_re)));      %
selecting relay in second phase.

    Q1=(q1==0);        %Q1=1==> all the relays fail to decode
    temp=ones(N,1);
    temp(q2,1)=0;

        [q5
q6]=max(min(temp.*(snr_je./snr_js1),temp.*(snr_je./snr_js2)));
%selecting the jammer in the second phase

    sinr1_osj=((y(q2)*snr_rs1(q2))./(1+snr_js1(q6)));
    sinr2_osj=((y(q2)*snr_rs2(q2))./(1+snr_js2(q6)));

sinre1_osj=(snr_s1e./(1+snr_s2e+snr_je(q4)))+((y(q2)*snr_
re(q2))./(1+snr_je(q6)));
sinre2_osj=(snr_s2e./(1+snr_s1e+snr_je(q4)))+((y(q2)*snr_
re(q2))./(1+snr_je(q6)));
```

```
     Cs_osjmmisr(1,jj)=max(0,0.5*log2(1+sinr1_osj)-
0.5*log2(1+sinre2_osj)+0.5*log2(1+sinr2_osj)-
0.5*log2(1+sinre1_osj));

     outage_osjmmisr=outage_osjmmisr+(Rs>Cs_osjmmisr(1,jj));

  % case:Suboptimal selection with max-min instantaneous
secrecy rate(SSJ-MMISR)

     [q3 q4]=max(min(mean(snr_je),mean(snr_je)));     %
selecting jammer in first phase.

     [q1
q2]=max(min(y.*(snr_rs1./mean(snr_re)),y.*(snr_rs2./mean(snr_
re))));     % selecting relay in second phase.

     Q1=(q1==0);     %Q1=1==> all the relays fail to decode
     temp=ones(N,1);
     temp(q2,1)=0;

     [q5
q6]=max(min(temp.*(mean(snr_je)./snr_js1),temp.*(mean(snr_
je)./snr_js2)));     %selecting the jammer in the second phase

     sinr1_ssj=((y(q2)*snr_rs1(q2))./(1+snr_js1(q6)));
     sinr2_ssj=((y(q2)*snr_rs2(q2))./(1+snr_js2(q6)));

sinre1_ssj=(snr_s1e./(1+snr_s2e+snr_je(q4)))+((y(q2)*snr_
re(q2))./(1+snr_je(q6)));

sinre2_ssj=(snr_s2e./(1+snr_s1e+snr_je(q4)))+((y(q2)*snr_
re(q2))./(1+snr_je(q6)));

     Cs_ssjmmisr(1,jj)=max(0,0.5*log2(1+sinr1_ssj)-
0.5*log2(1+sinre2_ssj)+0.5*log2(1+sinr2_ssj)-
0.5*log2(1+sinre1_ssj));

     outage_ssjmmisr=outage_ssjmmisr+(Rs>Cs_ssjmmisr(1,jj));

  % case:optimal switching(OW)

     [q3 q4]=max(snr_je.*snr_je);     % selecting jammer in
first phase.

     [q1
q2]=max(y.*((snr_rs1./snr_re).*(snr_rs2./snr_re)));     %
selecting relay in second phase.

     Q1=(q1==0);     %Q1=1==> all the relays fail to decode
     temp=ones(N,1);
     temp(q2,1)=0;
```

```
      [q5
q6]=max(temp.*((snr_je./snr_js1).*(snr_je./snr_js2)));
%selecting the jammer in the second phase

      sinr1_os=(y(q2)*snr_rs1(q2));
      sinr2_os=(y(q2)*snr_rs2(q2));

sinre1_os=(snr_s1e./(1+snr_s2e))+(y(q2)*snr_re(q2));

sinre2_os=(snr_s2e./(1+snr_s1e))+(y(q2)*snr_re(q2));

      sinr1_osj=((y(q2)*snr_rs1(q2))./(1+snr_js1(q6)));
      sinr2_osj=((y(q2)*snr_rs2(q2))./(1+snr_js2(q6)));

sinre1_osj=(snr_s1e./(1+snr_s2e+snr_je(q4)))+((y(q2)*snr_
re(q2))./(1+snr_je(q6)));
sinre2_osj=(snr_s2e./(1+snr_s1e+snr_je(q4)))+((y(q2)*snr_
re(q2))./(1+snr_je(q6)));

if((((1+sinr1_osj)./(1+sinre2_osj)).*((1+sinr2_osj)./
(1+sinre1_osj)))>...

((1+sinr1_os)./(1+sinre2_os)).*((1+sinr2_os)./(1+sinre1_os)))

  Cs_ow(1,jj)=max(0,0.5*log2(1+sinr1_osj)-
0.5*log2(1+sinre2_osj)+0.5*log2(1+sinr2_osj)-
0.5*log2(1+sinre1_osj));

    else

  Cs_ow(1,jj)=max(0,0.5*log2(1+sinr1_os)-
0.5*log2(1+sinre2_os)+0.5*log2(1+sinr2_os)-
0.5*log2(1+sinre1_os));

    end

    outage_ow=outage_ow+(Rs>Cs_ow(1,jj));

    % case:Suboptimal switching(SW)

      [q3 q4]=max(mean(snr_je).*mean(snr_je));      % selecting
jammer in first phase.

      [q1
q2]=max(y.*((snr_rs1./mean(snr_re)).*(snr_rs2./mean(snr_
re))));      % selecting relay in second phase.

      Q1=(q1==0);          %Q1=1==> all the relays fail to decode
      temp=ones(N,1);
      temp(q2,1)=0;
```

```
      [q5
q6]=max(temp.*((mean(snr_je)./snr_js1).*(mean(snr_je)./snr_
js2)));     %selecting the jammer in the second phase

      sinr1_ss=y(q2)*snr_s1r(q2);
      sinr2_ss=y(q2)*snr_s2r(q2);
      sinre1_ss=((snr_s1e./(snr_s2e+1))+ y(q2)*snr_re(q2));
      sinre2_ss=((snr_s2e./(snr_s1e+1))+ y(q2)*snr_re(q2));

      sinr1_ssj=((y(q2)*snr_rs1(q2))./(1+snr_js1(q6)));
      sinr2_ssj=((y(q2)*snr_rs2(q2))./(1+snr_js2(q6)));

sinre1_ssj=(snr_s1e./(1+snr_s2e+snr_je(q4)))+((y(q2)*snr_
re(q2))./(1+snr_je(q6)));

sinre2_ssj=(snr_s2e./(1+snr_s1e+snr_je(q4)))+((y(q2)*snr_
re(q2))./(1+snr_je(q6)));

if(((1+sinr1_ssj)./(1+sinre2_ssj)).*((1+sinr2_ssj)./
(1+sinre1_ssj))>...

((1+sinr1_ss)./(1+sinre2_ss)).*((1+sinr2_ss)./(1+sinre1_ss)))


      Cs_sw(1,jj)=max(0,0.5*log2(1+sinr1_ssj)-
0.5*log2(1+sinre2_ssj)+0.5*log2(1+sinr2_ssj)-
0.5*log2(1+sinre1_ssj));
   else

  Cs_sw(1,jj)=max(0,0.5*log2(1+sinr1_ss)-
0.5*log2(1+sinre2_ss)+0.5*log2(1+sinr2_ss)-
0.5*log2(1+sinre1_ss));

   end
    outage_sw=outage_sw+(Rs>Cs_sw(1,jj));

    % case:optimal selection with controlled jamming(OSCJ)

              F=[];

              for r=1:N
              for j2=1:N
              for j1=1:N

              if (r~=j2)

  sinr1_oscj=(y(r)*snr_rs1(r));
  sinr2_oscj=(y(r)*snr_rs2(r));
```

```
sinre1_oscj=(snr_s1e./(1+snr_s2e+snr_je(j1)))+(y(r)*snr_
re(r)./(1+snr_je(j2)));

sinre2_oscj=(snr_s2e./(1+snr_s1e+snr_je(j1)))+(y(r)*snr_
re(r)./(1+snr_je(j2)));

   C=max(0,0.5*log2(1+sinr1_oscj)-
0.5*log2(1+sinre2_oscj)+0.5*log2(1+sinr2_oscj)-
0.5*log2(1+sinre1_oscj));

     F=[F C];
     end
      end

  end
end

 Cs_oscj(1,jj)=max(F);
 outage_oscj=outage_oscj+(Rs>Cs_oscj(1,jj));

    end

    out_os(ii)=outage_os/Num_Trials
    c_os(ii)=mean(Cs_os)

    out_ss(ii)=outage_ss/Num_Trials
    c_ss(ii)=mean(Cs_ss)

    out_osmmisr(ii)=outage_osmmisr/Num_Trials
    c_osmmisr(ii)=mean(Cs_osmmisr)

     out_ssmmisr(ii)=outage_ssmmisr/Num_Trials
    c_ssmmisr(ii)=mean(Cs_ssmmisr)

    out_osj(ii)=outage_osj/Num_Trials
    c_osj(ii)=mean(Cs_osj)

   out_ssj(ii)=outage_ssj/Num_Trials
    c_ssj(ii)=mean(Cs_ssj)

    out_osjmmisr(ii)=outage_osjmmisr/Num_Trials
    c_osjmmisr(ii)=mean(Cs_osjmmisr)

    out_ssjmmisr(ii)=outage_ssjmmisr/Num_Trials
    c_ssjmmisr(ii)=mean(Cs_ssjmmisr)

    out_ow(ii)=outage_ow/Num_Trials
    c_ow(ii)=mean(Cs_ow)

     out_sw(ii)=outage_sw/Num_Trials
    c_sw(ii)=mean(Cs_sw)
```

```
    out_oscj(ii)=outage_oscj/Num_Trials
    c_oscj(ii)=mean(Cs_oscj)

end

  plot(P_dB, c_os,'b-x',P_dB, c_ss,'r-s',P_dB, c_osmmisr,'k-
x',P_dB, c_ssmmisr,'k-s',P_dB, c_osj,'b-s',P_dB, c_ssj,'r-x',...
      P_dB, c_osjmmisr,'g:>',P_dB, c_ssjmmisr,'c:<',P_dB,
c_ow,'b-.+',P_dB, c_sw,'r-.d',P_dB, c_oscj,'m-p')

 xlabel('P[dB]');
 ylabel('secrecy ergodic capacity[BPCU]');
 grid on; box on;
 axis([0 25 0 5]);
 legend('OS','SS','OS-MMISR','SS-MMISR','OSJ','SSJ','OSJ-
MMISR','SSJ-MMISR','OW','SW','OSCJ',2);
 figure

 semilogy(P_dB, out_os,'b-x',P_dB, out_ss,'r-s',P_dB,
out_osmmisr,'k-x',P_dB,out_ssmmisr,'k-s',P_dB,out_osj,'b-
s',P_dB,out_ssj,'r-x',...
      P_dB, out_osjmmisr,'g:>',P_dB, out_ssjmmisr,'c:<',P_dB,
out_ow,'b-.+',P_dB, out_sw,'r-.d',P_dB, out_oscj,'m-p')

 xlabel('P[dB]');
 ylabel('secrecy outage probability');
 grid on; box on;
 axis([0 40 10^-3 1]);
 legend('OS','SS','OS-MMISR','SS-MMISR','OSJ','SSJ','OSJ-
MMISR','SSJ- MMISR','OW','SW','OSCJ',3);
 figure


%****************************************************************
%    Relays distributed dispersedly between S1,S2,E nodes
%****************************************************************
% Joint relay and jammer selection for secure two-way relay
networks
% performance metrics are ergodic secrecy capacity & secrecy
outage
% probability
% Intermediate nodes have comparable links with S1,S2 and E
nodes.

 clear all
 clc
 %===================
 %system parameters
 %===================
 N=4;                %number of relay nodes
```

```
    M=1;                    %number of eavesdroppers
    Rs=0.1;                 %target secrecy rate (BPCU)-for outage
probability calculation
    R0=2;                   %Transmission Rate-to check which relay
can decode correctly
    P_dB=0:5:40;            %Transmitted power (dB)
    Num_Trials=5000;        %number of trials
    PL_exp=3;               %path loss exponent
    N0=1;                   %noise variance

    loc_s1=[0 0];
    loc_e=[0.5 1];
    loc_s2=[1 0];
    loc_r=[0.35 0.3;0.4 0.65;0.6 0.4;0.65 0.7];

    var_s1e=(sqrt((loc_s1(1)-loc_e(1)).^2+(loc_s1(2)-
loc_e(2)).^2)).^(-PL_exp);
    var_s2e=(sqrt((loc_s2(1)-loc_e(1)).^2+(loc_s2(2)-
loc_e(2)).^2)).^(-PL_exp);

    z1=(loc_s1(1)-loc_r(:,1)).^2;
    z2=(loc_s1(2)-loc_r(:,2)).^2;
    z=sqrt(z1+z2);
    var_s1r=z.^(-PL_exp);

    z1=(loc_s2(1)-loc_r(:,1)).^2;
    z2=(loc_s2(2)-loc_r(:,2)).^2;
    z=sqrt(z1+z2);
    var_s2r=z.^(-PL_exp);

    z1=(loc_e(1)-loc_r(:,1)).^2;
    z2=(loc_e(2)-loc_r(:,2)).^2;
    z=sqrt(z1+z2);
    var_re=z.^(-PL_exp);

    %========================
    %simulation start
    %========================
    for ii=1:length(P_dB)
        Ps=10^(P_dB(ii)/10);
        Pr=Ps;
        Pj=Ps/100;   %jammer power less than relay node power

        outage_cs=0;
        outage_os=0;
        outage_osmmisr=0;
        outage_ss=0;
        outage_ssmmisr=0;
        outage_osj=0;
        outage_osjmmisr=0;
```

```
        outage_ssj=0;
        outage_ssjmmisr=0;
        outage_oscj=0;
        outage_ow=0;
        outage_sw=0;

        for jj=1:Num_Trials

%==============================================================
        %Channel coefficients
%==============================================================
        %broadcasting phase

        hs1r=sqrt(var_s1r/2).*(randn(N,1)+i*randn(N,1));
% S1-->relays
        hs2r=sqrt(var_s2r/2).*(randn(N,1)+i*randn(N,1));
% S2-->relays
        hs1e=sqrt(var_s1e/2).*(randn(1,1)+i*randn(1,1));
% S1-->eavesdropper
        hs2e=sqrt(var_s2e/2).*(randn(1,1)+i*randn(1,1));
% S2-->eavesdropper
        %cooperative links
        %hrs1=sqrt(var_s1r/2).*(randn(N,1)+i*randn(N,1));
% R-->destination1
        %hrs2=sqrt(var_s2r/2).*(randn(N,1)+i*randn(N,1));
% R-->destination2
        hre=sqrt(var_re/2).*(randn(N,1)+i*randn(N,1));
% R-->eavesdropper

%==============================================================
        %instantaneous SNR calculations
%==============================================================

        snr_s1r=Ps*abs(hs1r).^2;
        snr_s2r=Ps*abs(hs2r).^2;
        snr_s1e=Ps*abs(hs1e).^2;
        snr_s2e=Ps*abs(hs2e).^2;
        snr_rs1=Pr*abs(hs1r).^2;
        snr_rs2=Pr*abs(hs2r).^2;
        snr_js1=Pj*abs(hs1r).^2;
        snr_js2=Pj*abs(hs2r).^2;
        snr_re=Pr*abs(hre).^2;
        snr_je=Pj*abs(hre).^2;

        cs1r=0.5*log2(1+snr_s1r);
        y1=cs1r>R0;

        cs2r=0.5*log2(1+snr_s2r);
        y2=cs2r>R0;
```

```
        y=(y1&y2);

   %Two-way relay networks
   %=========================
      %Non-jamming techniques
   %=========================

   %case:optimal selection (OS)

        sinr1_os=(y.*snr_rs1);
        sinr2_os=(y.*snr_rs2);

        sinre1_os=(((snr_s1e)/(snr_s2e+1))+(y.*snr_re));
        sinre2_os=(((snr_s2e)/(snr_s1e+1))+(y.*snr_re));

        [q1

q2]=max((((1+sinr1_os)./(1+sinre2_os)).*((1+sinr2_os)./
(1+sinre1_os)));

        sinr1_osn=(y(q2).*snr_rs1(q2));
        sinr2_osn=(y(q2).*snr_rs2(q2));

sinre1_osn=(((snr_s1e)/(snr_s2e+1))+(y(q2).*snr_re(q2)));

sinre2_osn=(((snr_s2e)/(snr_s1e+1))+(y(q2).*snr_re(q2)));

        Cs_os(1,jj)=max(0,0.5*log2(1+sinr1_osn)-
0.5*log2(1+sinre2_osn)+0.5*log2(1+sinr2_osn)-
0.5*log2(1+sinre1_osn));

        outage_os=outage_os+(Rs>Cs_os(1,jj));

   %case:Suboptimal selection (SS)

        sinr1_ss=(y.*snr_rs1);
        sinr2_ss=(y.*snr_rs2);

        sinre1_ss=((mean(snr_s1e)/(mean(snr_s2e)+1))+
y.*mean(snr_re));
        sinre2_ss=((mean(snr_s2e)/(mean(snr_s1e)+1))+
y.*mean(snr_re));

        [q1
q2]=max((((1+sinr1_ss)./(1+sinre2_ss)).*((1+sinr2_ss)./
(1+sinre1_ss)));

        sinr1_ssn=y(q2)*snr_s1r(q2);
        sinr2_ssn=y(q2)*snr_s2r(q2);
```

```
        sinre1_ssn=((snr_s1e./(snr_s2e+1))+ y(q2)*snr_re(q2));
        sinre2_ssn=((snr_s2e./(snr_s1e+1))+ y(q2)*snr_re(q2));

        Cs_ss(1,jj)=max(0,0.5*log2(1+sinr1_ssn)-
0.5*log2(1+sinre2_ssn)+0.5*log2(1+sinr2_ssn)-
0.5*log2(1+sinre1_ssn));

        outage_ss=outage_ss+(Rs>Cs_ss(1,jj));

        % case:optimal selection with max-min instantaneous
secrecy rate(OS-MMISR)

        sinr1_os=(y.*snr_rs1);
        sinr2_os=(y.*snr_rs2);

        sinre1_os=(((snr_s1e)/(snr_s2e+1))+(y.*snr_re));
        sinre2_os=(((snr_s2e)/(snr_s1e+1))+(y.*snr_re));

        [q1
q2]=max(min(((1+sinr1_os)./(1+sinre2_os)),((1+sinr2_os)./
(1+sinre1_os))));

        sinr1_osn=(y(q2).*snr_rs1(q2));
        sinr2_osn=(y(q2).*snr_rs2(q2));

sinre1_osn=(((snr_s1e)/(snr_s2e+1))+(y(q2).*snr_re(q2)));

sinre2_osn=(((snr_s2e)/(snr_s1e+1))+(y(q2).*snr_re(q2)));

      Cs_osmmisr(1,jj)=max(0,0.5*log2(1+sinr1_osn)-
0.5*log2(1+sinre2_osn)+0.5*log2(1+sinr2_osn)-
0.5*log2(1+sinre1_osn));

      outage_osmmisr=outage_osmmisr+(Rs>Cs_osmmisr(1,jj));

    % case:Suboptimal selection with max-min instantaneous
secrecy
    % rate(SS-MMISR)

        sinr1_ss=(y.*snr_rs1);
        sinr2_ss=(y.*snr_rs2);

        sinre1_ss=((mean(snr_s1e)/(mean(snr_s2e)+1))+
y.*mean(snr_re));
        sinre2_ss=((mean(snr_s2e)/(mean(snr_s1e)+1))+
y.*mean(snr_re));
```

```
        [q1
q2]=max(min(((1+sinr1_ss)./(1+sinre2_ss)),((1+sinr2_ss)./
(1+sinre1_ss)))));

        sinr1_ssn=y(q2)*snr_s1r(q2);
        sinr2_ssn=y(q2)*snr_s2r(q2);
        sinre1_ssn=((snr_s1e./(snr_s2e+1))+ y(q2)*snr_re(q2));
        sinre2_ssn=((snr_s2e./(snr_s1e+1))+ y(q2)*snr_re(q2));

        Cs_ssmmisr(1,jj)=max(0,0.5*log2(1+sinr1_ssn)-
0.5*log2(1+sinre2_ssn)+0.5*log2(1+sinr2_ssn)-
0.5*log2(1+sinre1_ssn));

        outage_ssmmisr=outage_ssmmisr+(Rs>Cs_ssmmisr(1,jj));

    %==========================
     %Jamming techniques
    %==========================
  % case:optimal selection with jamming(OSJ)

    [q3 q4]=max(snr_je.*snr_je);       % selecting jammer in
first phase.

    [q1 q2]=max(y.*((snr_rs1./snr_re).*(snr_rs2./snr_re)));
% selecting relay in second phase.

    Q1=(q1==0);       %Q1=1==> all the relays fail to decode
    temp=ones(N,1);
    temp(q2,1)=0;

    [q5 q6]=max(temp.*((snr_je./snr_js1).*(snr_je./snr_
js2)));       %selecting the jammer in the second phase

    sinr1_osj=((y(q2)*snr_rs1(q2))./(1+snr_js1(q6)));
    sinr2_osj=((y(q2)*snr_rs2(q2))./(1+snr_js2(q6)));

sinre1_osj=(snr_s1e./(1+snr_s2e+snr_je(q4)))+((y(q2)*snr_
re(q2))./(1+snr_je(q6)));

sinre2_osj=(snr_s2e./(1+snr_s1e+snr_je(q4)))+((y(q2)*snr_
re(q2))./(1+snr_je(q6)));

    Cs_osj(1,jj)=max(0,0.5*log2(1+sinr1_osj)-
0.5*log2(1+sinre2_osj)+0.5*log2(1+sinr2_osj)-
0.5*log2(1+sinre1_osj));

    outage_osj=outage_osj+(Rs>Cs_osj(1,jj));

  % case:Suboptimal selection with jamming(SSJ)
```

```
    [q3 q4]=max(mean(snr_je).*mean(snr_je));    % selecting
jammer in first phase.

    [q1
q2]=max(y.*((snr_rs1./mean(snr_re)).*(snr_rs2./mean(snr_
re))));     % selecting relay in second phase.

    Q1=(q1==0);        %Q1=1==> all the relays fail to decode
    temp=ones(N,1);
    temp(q2,1)=0;

    [q5
q6]=max(temp.*((mean(snr_je)./snr_js1).*(mean(snr_je)./snr_
js2)));     %selecting the jammer in the second phase

    sinr1_ssj=((y(q2)*snr_rs1(q2))./(1+snr_js1(q6)));
    sinr2_ssj=((y(q2)*snr_rs2(q2))./(1+snr_js2(q6)));

sinre1_ssj=(snr_s1e./(1+snr_s2e+snr_je(q4)))+((y(q2)*snr_
re(q2))./(1+snr_je(q6)));

sinre2_ssj=(snr_s2e./(1+snr_s1e+snr_je(q4)))+((y(q2)*snr_
re(q2))./(1+snr_je(q6)));

    Cs_ssj(1,jj)=max(0,0.5*log2(1+sinr1_ssj)-
0.5*log2(1+sinre2_ssj)+0.5*log2(1+sinr2_ssj)-
0.5*log2(1+sinre1_ssj));

    outage_ssj=outage_ssj+(Rs>Cs_ssj(1,jj));

  % case:optimal selection with max-min instantaneous secrecy
rate(OSJ-MMISR)

     [q3 q4]=max(min(snr_je,snr_je));    % selecting jammer
in first phase.

    [q1
q2]=max(min(y.*(snr_rs1./snr_re),y.*(snr_rs2./snr_re)));    %
selecting relay in second phase.

    Q1=(q1==0);        %Q1=1==> all the relays fail to decode
    temp=ones(N,1);
    temp(q2,1)=0;

    [q5
q6]=max(min(temp.*(snr_je./snr_js1),temp.*(snr_je./snr_js2)));
%selecting the jammer in the second phase

    sinr1_osj=((y(q2)*snr_rs1(q2))./(1+snr_js1(q6)));
    sinr2_osj=((y(q2)*snr_rs2(q2))./(1+snr_js2(q6)));
```

```
sinre1_osj=(snr_s1e./(1+snr_s2e+snr_je(q4)))+((y(q2)*snr_
re(q2))./(1+snr_je(q6)));

sinre2_osj=(snr_s2e./(1+snr_s1e+snr_je(q4)))+((y(q2)*snr_
re(q2))./(1+snr_je(q6)));

    Cs_osjmmisr(1,jj)=max(0,0.5*log2(1+sinr1_osj)-
0.5*log2(1+sinre2_osj)+0.5*log2(1+sinr2_osj)-
0.5*log2(1+sinre1_osj));

    outage_osjmmisr=outage_osjmmisr+(Rs>Cs_osjmmisr(1,jj));

  % case:Suboptimal selection with max-min instantaneous
secrecy rate(SSJ-MMISR)

      [q3 q4]=max(min(mean(snr_je),mean(snr_je)));     %
selecting jammer in first phase.

    [q1
q2]=max(min(y.*(snr_rs1./mean(snr_re)),y.*(snr_rs2./mean(snr_
re))));     % selecting relay in second phase.

    Q1=(q1==0);      %Q1=1==> all the relays fail to decode
    temp=ones(N,1);
    temp(q2,1)=0;

    [q5
q6]=max(min(temp.*(mean(snr_je)./snr_js1),temp.*(mean(snr_je)./
snr_js2)));
%selecting the jammer in the second phase

    sinr1_ssj=((y(q2)*snr_rs1(q2))./(1+snr_js1(q6)));
    sinr2_ssj=((y(q2)*snr_rs2(q2))./(1+snr_js2(q6)));

sinre1_ssj=(snr_s1e./(1+snr_s2e+snr_je(q4)))+((y(q2)*snr_
re(q2))./(1+snr_je(q6)));

sinre2_ssj=(snr_s2e./(1+snr_s1e+snr_je(q4)))+((y(q2)*snr_
re(q2))./(1+snr_je(q6)));

    Cs_ssjmmisr(1,jj)=max(0,0.5*log2(1+sinr1_ssj)-
0.5*log2(1+sinre2_ssj)+0.5*log2(1+sinr2_ssj)-
0.5*log2(1+sinre1_ssj));

    outage_ssjmmisr=outage_ssjmmisr+(Rs>Cs_ssjmmisr(1,jj));

  % case:optimal switching(OW)

    [q3 q4]=max(snr_je.*snr_je);      % selecting jammer in
first phase.
```

```
    [q1 q2]=max(y.*((snr_rs1./snr_re).*(snr_rs2./snr_re)));
% selecting relay in second phase.

    Q1=(q1==0);        %Q1=1==> all the relays fail to decode
    temp=ones(N,1);
    temp(q2,1)=0;

    [q5 q6]=max(temp.*((snr_je./snr_js1).*(snr_je./snr_
js2)));     %selecting the jammer in the second phase

    sinr1_os=(y(q2)*snr_rs1(q2));
    sinr2_os=(y(q2)*snr_rs2(q2));
    sinre1_os=(snr_s1e./(1+snr_s2e))+(y(q2)*snr_re(q2));
    sinre2_os=(snr_s2e./(1+snr_s1e))+(y(q2)*snr_re(q2));

    sinr1_osj=((y(q2)*snr_rs1(q2))./(1+snr_js1(q6)));
    sinr2_osj=((y(q2)*snr_rs2(q2))./(1+snr_js2(q6)));

sinre1_osj=(snr_s1e./(1+snr_s2e+snr_je(q4)))+((y(q2)*snr_
re(q2))./(1+snr_je(q6)));

sinre2_osj=(snr_s2e./(1+snr_s1e+snr_je(q4)))+((y(q2)*snr_
re(q2))./(1+snr_je(q6)));

if(((1+sinr1_osj)./(1+sinre2_osj)).*((1+sinr2_osj)./
(1+sinre1_osj))>...

((1+sinr1_os)./(1+sinre2_os)).*((1+sinr2_os)./(1+sinre1_os)))

  Cs_ow(1,jj)=max(0,0.5*log2(1+sinr1_osj)-
0.5*log2(1+sinre2_osj)+0.5*log2(1+sinr2_osj)-
0.5*log2(1+sinre1_osj));

  else

  Cs_ow(1,jj)=max(0,0.5*log2(1+sinr1_os)-
0.5*log2(1+sinre2_os)+0.5*log2(1+sinr2_os)-
0.5*log2(1+sinre1_os));

  end

    outage_ow=outage_ow+(Rs>Cs_ow(1,jj));

    % case:Suboptimal switching(SW)

    [q3 q4]=max(mean(snr_je).*mean(snr_je));      % selecting
jammer in first phase.

    [q1
q2]=max(y.*((snr_rs1./mean(snr_re)).*(snr_rs2./mean(snr_
re))));      % selecting relay in second phase.
```

```
      Q1=(q1==0);        %Q1=1==> all the relays fail to decode
      temp=ones(N,1);
      temp(q2,1)=0;

      [q5
q6]=max(temp.*((mean(snr_je)./snr_js1).*(mean(snr_je)./snr_
js2)));       %selecting the jammer in the second phase

      sinr1_ss=y(q2)*snr_s1r(q2);
      sinr2_ss=y(q2)*snr_s2r(q2);
      sinre1_ss=((snr_s1c./(snr_s2c+1))+ y(q2)*snr_re(q2)),
      sinre2_ss=((snr_s2e./(snr_s1e+1))+ y(q2)*snr_re(q2));

      sinr1_ssj=((y(q2)*snr_rs1(q2))./(1+snr_js1(q6)));
      sinr2_ssj=((y(q2)*snr_rs2(q2))./(1+snr_js2(q6)));

sinre1_ssj=(snr_s1e./(1+snr_s2e+snr_je(q4)))+((y(q2)*snr_
re(q2))./(1+snr_je(q6)));

sinre2_ssj=(snr_s2e./(1+snr_s1e+snr_je(q4)))+((y(q2)*snr_
re(q2))./(1+snr_je(q6)));

if((((1+sinr1_ssj)./(1+sinre2_ssj)).*((1+sinr2_ssj)./
(1+sinre1_ssj))>...

((1+sinr1_ss)./(1+sinre2_ss)).*((1+sinr2_ss)./(1+sinre1_ss)))

      Cs_sw(1,jj)=max(0,0.5*log2(1+sinr1_ssj)-
0.5*log2(1+sinre2_ssj)+0.5*log2(1+sinr2_ssj)-
0.5*log2(1+sinre1_ssj));
   else

  Cs_sw(1,jj)=max(0,0.5*log2(1+sinr1_ss)-
0.5*log2(1+sinre2_ss)+0.5*log2(1+sinr2_ss)-
0.5*log2(1+sinre1_ss));

  end
    outage_sw=outage_sw+(Rs>Cs_sw(1,jj));

    % case:optimal selection with controlled jamming(OSCJ)

            F=[];

            for r=1:N
            for j2=1:N
            for j1=1:N

            if (r~=j2)

  sinr1_oscj=(y(r)*snr_rs1(r));
  sinr2_oscj=(y(r)*snr_rs2(r));
```

```
sinre1_oscj=(snr_s1e./(1+snr_s2e+snr_je(j1)))+(y(r)*snr_
re(r)./(1+snr_je(j2)));

sinre2_oscj=(snr_s2e./(1+snr_s1e+snr_je(j1)))+(y(r)*snr_
re(r)./(1+snr_je(j2)));

    C=max(0,0.5*log2(1+sinr1_oscj)-
0.5*log2(1+sinre2_oscj)+0.5*log2(1+sinr2_oscj)-
0.5*log2(1+sinre1_oscj));

    F=[F C];
    end
     end

  end
end

 Cs_oscj(1,jj)=max(F);
 outage_oscj=outage_oscj+(Rs>Cs_oscj(1,jj));

    end

    out_os(ii)=outage_os/Num_Trials
    c_os(ii)=mean(Cs_os)

    out_ss(ii)=outage_ss/Num_Trials
    c_ss(ii)=mean(Cs_ss)

    out_osmmisr(ii)=outage_osmmisr/Num_Trials
    c_osmmisr(ii)=mean(Cs_osmmisr)

     out_ssmmisr(ii)=outage_ssmmisr/Num_Trials
    c_ssmmisr(ii)=mean(Cs_ssmmisr)

    out_osj(ii)=outage_osj/Num_Trials
    c_osj(ii)=mean(Cs_osj)

   out_ssj(ii)=outage_ssj/Num_Trials
    c_ssj(ii)=mean(Cs_ssj)

    out_osjmmisr(ii)=outage_osjmmisr/Num_Trials
    c_osjmmisr(ii)=mean(Cs_osjmmisr)

    out_ssjmmisr(ii)=outage_ssjmmisr/Num_Trials
    c_ssjmmisr(ii)=mean(Cs_ssjmmisr)

    out_ow(ii)=outage_ow/Num_Trials
    c_ow(ii)=mean(Cs_ow)

     out_sw(ii)=outage_sw/Num_Trials
```

```
        c_sw(ii)=mean(Cs_sw)

        out_oscj(ii)=outage_oscj/Num_Trials
        c_oscj(ii)=mean(Cs_oscj)

    end

    plot(P_dB, c_os,'b-x',P_dB, c_ss,'r-s',P_dB, c_osmmisr,'k-
x',P_dB, c_ssmmisr,'k-s',P_dB, c_osj,'b-s',P_dB, c_ssj,'r-x',...
        P_dB, c_osjmmisr,'g:>',P_dB, c_ssjmmisr,'c:<',P_dB,
c_ow,'b-.+',P_dB, c_sw,'r-.d',P_dB, c_oscj,'m-p')

    xlabel('P[dB]');
    ylabel('secrecy ergodic capacity[BPCU]');
    grid on; box on;
    axis([0 25 0 5]);
    legend ('OS','SS','OS-MMISR','SS-MMISR','OSJ','SSJ',
'OSJ-MMISR','SSJ-MMISR','OW','SW','OSCJ',2);
    figure

    semilogy(P_dB, out_os,'b-x',P_dB, out_ss,'r-s',P_dB, out_
osmmisr,'k-x',P_dB,out_ssmmisr,'k-s',P_dB,
out_osj,'b-s',P_dB,out_ssj,'r-x',...
        P_dB, out_osjmmisr,'g:>',P_dB, out_ssjmmisr,'c:<',P_dB,
out_ow,'b-.+',P_dB, out_sw,'r-.d',P_dB, out_oscj,'m-p')

    xlabel('P[dB]');
    ylabel('secrecy outage probability');
    grid on; box on;
    axis([0 40 10^-3 1]);
    legend ('OS','SS','OS-MMISR','SS-MMISR','OSJ','SSJ',
'OSJ-MMISR','SSJ-MMISR','OW','SW','OSCJ',3);
    figure
```

Appendix C: MATLAB® Simulation Codes for Chapter 7

```
%*************************************************************
       % ------- Theoretical Results -----------
%*************************************************************
% Fig. 7.5
clc
clear all
n=500;
p=0.2;    %Adversary probability of compromising
rounds=25;    % max no. of rounds
copromise=100;    % max no. of compromised sensors
t1=6;            %proactive peers alone
t2=6;            %reactive peers alone
t1_pro=3;        %proactive peers proposed
t2_pro=3;        %reactive peers proposed
hold on
%############ PROACTIVE  ALONE PROACTIVE  ALONE ############
psp(1,1)=0;pgp(1,1)=1;
for r=2:rounds
    for k=2:copromise
        %   disp(['r= ',num2str(r),'   k=   ',num2str(k) '
t=  ',num2str(t) ' i=   ',num2str(i),'  a=  ',num2str(a),'
b=  ',num2str(b)]);
        pgp(r,k)=1-k/n-psp(r-1,k-1)*(1-t1*(1-p)/(n-1))^(n*pgp
(r-1,k-1)-k);
        psp(r,k)=1-pgp(r,k);
    end
end
```

307

```
pgp(1,:)=1;pgp(:,1)=1;
pgp=[ones(1,k);pgp];
pgp=[ones(1,r+1)',pgp];

psp=[zeros(1,k);psp];
psp=[zeros(1,r+1)',psp];
%save proactive


%############## REACTIVE ALONE REACTIVE ALONE ##############3
psr(1,1)=0;
for r=2:rounds
    for k=2:copromise
        summ=0;
        for i=0:t2
            a=combnk_as(n*(1-psr(r-1,k-1))-k,i);
            b=combnk_as(n*psr(r-1,k-1)+k-1,t2-i);
            c=combnk_as(n-1,t2);
            %disp(['r= ',num2str(r),'   k=   ',num2str(k) '
t2=  ',num2str(t2) '  i=   ',num2str(i),'  a=  ',num2str(a),'
b=  ',num2str(b)]);
            xx=p^i*a*b/c;
            summ=summ+xx;
        end
        pgr(r,k)=1-k/n-psr(r-1,k-1)*summ;
        psr(r,k)=1-pgr(r,k);
    end
end
pgr(1,:)=1;pgr(:,1)=1;
pgr=[ones(1,k);pgr];
pgr=[ones(1,r+1)',pgr];
psr=[zeros(1,k);psr];
psr=[zeros(1,r+1)',psr];
%save reactive


%############### PROPOSED PROPOSED PROPOSED ###############
%###################### 1-PROACTIVE ######################
psp_pro(1,1)=0;
pgp_pro(1,1)=1;
for r=2:rounds
    for k=2:copromise
        %   disp(['r= ',num2str(r),'   k=   ',num2str(k) '
t=  ',num2str(t) '  i=   ',num2str(i),'  a=  ',num2str(a),'
b=  ',num2str(b)]);
            pgp_pro(r,k)=1-k/n-psp_pro(r-1,k-1)*(1-t1_pro*(1-p)/
(n-1))^(n*pgp_pro(r-1,k-1)-k);
            psp_pro(r,k)=1-pgp_pro(r,k);
    end
end
pgp_pro(1,:)=1;pgp_pro(:,1)=1;
pgp_pro=[ones(1,k);pgp_pro];
```

```
pgp_pro=[ones(1,r+1)',pgp_pro];
psp_pro=[zeros(1,k);psp_pro];
psp_pro=[zeros(1,r+1)',psp_pro];

%######################## 2-REACTIVE ########################3
psr_pro(1,1)=0;
for r=2:rounds
    for k=2:copromise
        summ=0;
        for i=0:t2_pro
            a=combnk_as(n*(1-psr_pro(i-1,k-1))-k,i);
            b=combnk_as(n*psr_pro(r-1,k-1)+k-1,t2_pro-i);
            c=combnk_as(n-1,t2_pro);
            %disp(['r= ',num2str(r),'  k=   ',num2str(k) '
t2=  ',num2str(t2) '  i=   ',num2str(i),'  a=  ',num2str(a),'
b=  ',num2str(b)]);
            xx=p^i*a*b/c;
            summ=summ+xx;
        end
        pgr_pro(r,k)=1-k/n-psr_pro(r-1,k-1)*summ;
        psr_pro(r,k)=1-pgr_pro(r,k);
    end
end
pgr_pro(1,:)=1;pgr_pro(:,1)=1;
pgr_pro=[ones(1,k);pgr_pro];
pgr_pro=[ones(1,r+1)',pgr_pro];
psr_pro=[zeros(1,k);psr_pro];
psr_pro=[zeros(1,r+1)',psr_pro];

pstotal=psr_pro.*psp_pro;
pgtotal=1-pstotal;
pstotal_ratio_1=pstotal;
%save proactive_reactive_proposed_ratio_1
plot(0:r,pstotal_ratio_1(:,101),':go',
'LineWidth',2,'MarkerEdgeColor','k','MarkerFaceColor','b',
'MarkerSize',2)

clear all
n=500;
p=0.2;    %Adversary probability of compromising
rounds=25;    % max no. of rounds
copromise=100;   % max no. of compromised sensors
t1=6;           %proactive peers alone
t2=6;           %reactive peers alone
t1_pro=4;     %proactive peers proposed
t2_pro=2;      %reactive peers proposed

%############ PROACTIVE  ALONE PROACTIVE  ALONE ############
psp(1,1)=0;pgp(1,1)=1;
for r=2:rounds
```

```
     for k=2:copromise
         %   disp(['r= ',num2str(r),'   k=    ',num2str(k) '
t=   ',num2str(t) '  i=    ',num2str(i),'  a=  ',num2str(a),'
b=   ',num2str(b)]);
           pgp(r,k)=1-k/n-psp(r-1,k-1)*(1-t1*(1-p)/
(n-1))^(n*pgp(r-1,k-1)-k);
           psp(r,k)=1-pgp(r,k);
     end
end
pgp(1,:)=1;pgp(:,1)=1;
pgp=[ones(1,k);pgp];
pgp=[ones(1,r+1)',pgp];

psp=[zeros(1,k);psp];
psp=[zeros(1,r+1)',psp];
%save proactive

%############## REACTIVE ALONE REACTIVE ALONE ##############3
psr(1,1)=0;
for r=2:rounds
     for k=2:copromise
          summ=0;
          for i=0:t2
              a=combnk_as(n*(1-psr(r-1,k-1))-k,i);
              b=combnk_as(n*psr(r-1,k-1)+k-1,t2-i);
              c=combnk_as(n-1,t2);
              %disp(['r= ',num2str(r),'   k=    ',num2str(k) '
t2=  ',num2str(t2) '  i=    ',num2str(i),'  a=  ',num2str(a),'
b=   ',num2str(b)]);
              xx=p^i*a*b/c;
              summ=summ+xx;
          end
          pgr(r,k)=1-k/n-psr(r-1,k-1)*summ;
          psr(r,k)=1-pgr(r,k);
     end
end
pgr(1,:)=1;pgr(:,1)=1;
pgr=[ones(1,k);pgr];
pgr=[ones(1,r+1)',pgr];
psr=[zeros(1,k);psr];
psr=[zeros(1,r+1)',psr];
%save reactive

%############### PROPOSED PROPOSED PROPOSED ###############
%####################### 1-PROACTIVE #######################
psp_pro(1,1)=0;
pgp_pro(1,1)=1;
for r=2:rounds
```

```
    for k=2:copromise
        %    disp(['r= ',num2str(r),'    k=    ',num2str(k) '
t=  ',num2str(t) '  i=    ',num2str(i),'  a=  ',num2str(a),'
b=  ',num2str(b)]);
        pgp_pro(r,k)=1-k/n-psp_pro(r-1,k-1)*(1-t1_pro*(1-p)/
(n-1))^(n*pgp_pro(r-1,k-1)-k);
        psp_pro(r,k)=1-pgp_pro(r,k);
    end
end
pgp_pro(1,:)=1;pgp_pro(:,1)=1;
pgp_pro=[ones(1,k);pgp_pro];
pgp_pro=[ones(1,r+1)',pgp_pro];
psp_pro=[zeros(1,k);psp_pro];
psp_pro=[zeros(1,r+1)',psp_pro];

%####################### 2-REACTIVE #######################3
psr_pro(1,1)=0;
for r=2:rounds
    for k=2:copromise
        summ=0;
        for i=0:t2_pro
            a=combnk_as(n*(1-psr_pro(r-1,k-1))-k,i);
            b=combnk_as(n*psr_pro(r-1,k-1)+k-1,t2_pro-i);
            c=combnk_as(n-1,t2_pro);
            %disp(['r= ',num2str(r),'    k=    ',num2str(k) '
t2=  ',num2str(t2) '  i=    ',num2str(i),'  a=  ',num2str(a),'
b=  ',num2str(b)]);
            xx=p^i*a*b/c;
            summ=summ+xx;
        end
        pgr_pro(r,k)=1-k/n-psr_pro(r-1,k-1)*summ;
        psr_pro(r,k)=1-pgr_pro(r,k);
    end
end
pgr_pro(1,:)=1;pgr_pro(:,1)=1;
pgr_pro=[ones(1,k);pgr_pro];
pgr_pro=[ones(1,r+1)',pgr_pro];
psr_pro=[zeros(1,k);psr_pro];
psr_pro=[zeros(1,r+1)',psr_pro];

pstotal=psr_pro.*psp_pro;
pgtotal=1-pstotal;
pstotal_ratio_2=pstotal;
%save proactive_reactive_proposed_ratio_2
plot(0:r,pstotal_ratio_2(:,101),'m*:',
'LineWidth',2,'MarkerEdgeColor','k','MarkerFaceColor','b',
'MarkerSize',2)
```

```
clc
clear all
n=500;
p=0.2;    %Adversary probability of compromising
rounds=25;   % max no. of rounds
copromise=100;   % max no. of compromised sensors
t1=6;            %proactive peers alone
t2=6;            %reactive peers alone
t1_pro=2;      %proactive peers proposed
t2_pro=4;       %reactive peers proposed

%############ PROACTIVE  ALONE PROACTIVE  ALONE ############
psp(1,1)=0;pgp(1,1)=1;
for r=2:rounds
    for k=2:copromise
        %   disp(['r= ',num2str(r),'    k=    ',num2str(k) '
t= ',num2str(t) ' i=    ',num2str(i),' a=   ',num2str(a),'
b=   ',num2str(b)]);
        pgp(r,k)=1-k/n-psp(r-1,k-1)*(1-t1*(1-p)/(n-1))^(n*pgp
(r-1,k-1)-k);
        psp(r,k)=1-pgp(r,k);
    end
end
pgp(1,:)=1;pgp(:,1)=1;
pgp=[ones(1,k);pgp];
pgp=[ones(1,r+1)',pgp];

psp=[zeros(1,k);psp];
psp=[zeros(1,r+1)',psp];
%save proactive

%############# REACTIVE ALONE REACTIVE ALONE #############3
psr(1,1)=0;
for r=2:rounds
    for k=2:copromise
        summ=0;
        for i=0:t2
            a=combnk_as(n*(1-psr(r-1,k-1))-k,i);
            b=combnk_as(n*psr(r-1,k-1)+k-1,t2-i);
            c=combnk_as(n-1,t2);
            %disp(['r= ',num2str(r),'    k=    ',num2str(k) '
t2= ',num2str(t2) ' i=    ',num2str(i),' a=   ',num2str(a),'
b=   ',num2str(b)]);
            xx=p^i*a*b/c;
            summ=summ+xx;
        end
        pgr(r,k)=1-k/n-psr(r-1,k-1)*summ;
        psr(r,k)=1-pgr(r,k);
    end
end
```

```
pgr(1,:)=1;pgr(:,1)=1;
pgr=[ones(1,k);pgr];
pgr=[ones(1,r+1)',pgr];
psr=[zeros(1,k);psr];
psr=[zeros(1,r+1)',psr];
%save reactive


%############### PROPOSED PROPOSED PROPOSED ###############
%####################### 1-PROACTIVE #######################
psp_pro(1,1)=0;
pgp_pro(1,1)=1;
for r=2:rounds
    for k=2:copromise
        %  disp(['r= ',num2str(r),'   k=   ',num2str(k) '
t= ',num2str(t) ' i=   ',num2str(i),' a= ',num2str(a),'
b= ',num2str(b)]);
        pgp_pro(r,k)=1-k/n-psp_pro(r-1,k-1)*(1-t1_pro*(1-p)/
(n-1))^(n*pgp_pro(r-1,k-1)-k);
        psp_pro(r,k)=1-pgp_pro(r,k);
    end
end
pgp_pro(1,:)=1;pgp_pro(:,1)=1;
pgp_pro=[ones(1,k);pgp_pro];
pgp_pro=[oncs(1,r+1)',pgp_pro];
psp_pro=[zeros(1,k);psp_pro];
psp_pro=[zeros(1,r+1)',psp_pro];


%####################### 2-REACTIVE #######################3
psr_pro(1,1)=0;
for r=2:rounds
    for k=2:copromise
        summ=0;
        for i=0:t2_pro
            a=combnk_as(n*(1-psr_pro(r-1,k-1))-k,i);
            b=combnk_as(n*psr_pro(r-1,k-1)+k-1,t2_pro-i);
            c=combnk_as(n-1,t2_pro);
            %disp(['r= ',num2str(r),'   k=   ',num2str(k) '
t2= ',num2str(t2) ' i=   ',num2str(i),' a=   ',num2str(a),'
b= ',num2str(b)]);
            xx=p^i*a*b/c;
            summ=summ+xx;
        end
        pgr_pro(r,k)=1-k/n-psr_pro(r-1,k-1)*summ;
        psr_pro(r,k)=1-pgr_pro(r,k);
    end
end
pgr_pro(1,:)=1;pgr_pro(:,1)=1;
pgr_pro=[ones(1,k);pgr_pro];
pgr_pro=[ones(1,r+1)',pgr_pro];
psr_pro=[zeros(1,k);psr_pro];
```

```
psr_pro=[zeros(1,r+1)',psr_pro];

pstotal=psr_pro.*psp_pro;
pgtotal=1-pstotal;
pstotal_ratio_half=pstotal;

%save proactive_reactive_proposed_ratio_half



plot(0:r,pstotal_ratio_half(:,101),'c+:',
'LineWidth',2,'MarkerEdgeColor','k','MarkerFaceColor','b',
'MarkerSize',2)

xlabel('Rounds (r)','FontSize',7)
ylabel('Probability of Compromised Sensors','FontSize',7)
legend('ratio=1','ratio=2','ratio=0.5')

%#############################%###########################%#######
##############
%####%##############################%#########################%##
##############
% Fig. 7.6
clc
clear all
n=500;
p=0.2;   %Adversary probability of compromising
rounds=25;   % max no. of rounds
copromise=100;   % max no. of compromised sensors
t1=6;           %proactive peers alone
t2=6;           %reactive peers alone
t1_pro=3;       %proactive peers proposed
t2_pro=3;        %reactive peers proposed

%############# PROACTIVE  ALONE PROACTIVE  ALONE #############
psp(1,1)=0;pgp(1,1)=1;
for r=2:rounds
    for k=2:copromise
        %   disp(['r= ',num2str(r),'   k=   ',num2str(k) '
t= ',num2str(t) '  i=   ',num2str(i),'  a= ',num2str(a),'
b= ',num2str(b)]);
        pgp(r,k)=1-k/n-psp(r-1,k-1)*(1-t1*(1-p)/(n-1))^(n*pgp
(r-1,k-1)-k);
        psp(r,k)=1-pgp(r,k);
    end
end
pgp(1,:)=1;pgp(:,1)=1;
pgp=[ones(1,k);pgp];
pgp=[ones(1,r+1)',pgp];
```

```
psp=[zeros(1,k);psp];
psp=[zeros(1,r+1)',psp];
save proactive

%############### REACTIVE ALONE REACTIVE ALONE ###############3
psr(1,1)=0;
for r=2:rounds
    for k=2:copromise
        summ=0;
        for i=0:t2
            a=combnk_as(n*(1 psr(r-1,k-1))-k,i);
            b=combnk_as(n*psr(r-1,k-1)+k-1,t2-i);
            c=combnk_as(n-1,t2);
            %disp(['r= ',num2str(r),'   k=   ',num2str(k) '
t2=  ',num2str(t2) '  i=   ',num2str(i),'  a=  ',num2str(a),'
b=  ',num2str(b)]);
            xx=p^i*a*b/c;
            summ=summ+xx;
        end
        pgr(r,k)=1-k/n-psr(r-1,k-1)*summ;
        psr(r,k)=1-pgr(r,k);
    end
end
pgr(1,:)=1;pgr(:,1)-1;
pgr=[ones(1,k);pgr];
pgr=[ones(1,r+1)',pgr];
psr=[zeros(1,k);psr];
psr=[zeros(1,r+1)',psr];
save reactive

%############### PROPOSED PROPOSED PROPOSED ###############
%####################### 1-PROACTIVE #######################
psp_pro(1,1)=0;
pgp_pro(1,1)=1;
for r=2:rounds
    for k=2:copromise
        %   disp(['r= ',num2str(r),'   k=   ',num2str(k) '
t=  ',num2str(t) '  i=   ',num2str(i),'  a=  ',num2str(a),'
b=  ',num2str(b)]);
            pgp_pro(r,k)=1-k/n-psp_pro(r-1,k-1)*(1-t1_pro*(1-p)/
(n-1))^(n*pgp_pro(r-1,k-1)-k);
            psp_pro(r,k)=1-pgp_pro(r,k);
    end
end
pgp_pro(1,:)=1;pgp_pro(:,1)=1;
pgp_pro=[ones(1,k);pgp_pro];
pgp_pro=[ones(1,r+1)',pgp_pro];
psp_pro=[zeros(1,k);psp_pro];
psp_pro=[zeros(1,r+1)',psp_pro];
```

```
%######################## 2-REACTIVE ########################3
psr_pro(1,1)=0;
for r=2:rounds
    for k=2:copromise
        summ=0;
        for i=0:t2_pro
            a=combnk_as(n*(1-psr_pro(r-1,k-1))-k,i);
            b=combnk_as(n*psr_pro(r-1,k-1)+k-1,t2_pro-i);
            c=combnk_as(n-1,t2_pro);
            %disp(['r= ',num2str(r),'   k=   ',num2str(k) '
t2=  ',num2str(t2) '  i=   ',num2str(i),'  a=  ',num2str(a),'
b=  ',num2str(b)]);
            xx=p^i*a*b/c;
            summ=summ+xx;
        end
        pgr_pro(r,k)=1-k/n-psr_pro(r-1,k-1)*summ;
        psr_pro(r,k)=1-pgr_pro(r,k);
    end
end
pgr_pro(1,:)=1;pgr_pro(:,1)=1;
pgr_pro=[ones(1,k);pgr_pro];
pgr_pro=[ones(1,r+1)',pgr_pro];
psr_pro=[zeros(1,k);psr_pro];
psr_pro=[zeros(1,r+1)',psr_pro];

pstotal=psr_pro.*psp_pro;
pgtotal=1-pstotal;
save proactive_reactive_proposed

figure
box on
hold
plot(0:r,pgp(:,101),'-r*',
'LineWidth',2,'MarkerEdgeColor','k','MarkerFaceColor','b',
'MarkerSize',2)
plot(0:r,pgr(:,101),':gd',
'LineWidth',2,'MarkerEdgeColor','k','MarkerFaceColor','b',
'MarkerSize',2)
plot(0:r,pgtotal(:,101),'--bv',
'LineWidth',2,'MarkerEdgeColor','k','MarkerFaceColor','b',
'MarkerSize',2)
xlabel('Rounds (r)','FontSize',7)
ylabel('Normalized Healthy Sensors','FontSize',7)
legend('Proactive','Reactive','CHSHRD','FontSize',7)
figure
box on
hold
plot(0:r,psp(:,101),'-r*',
'LineWidth',2,'MarkerEdgeColor','k','MarkerFaceColor','b',
'MarkerSize',2)
```

```
plot(0:r,psr(:,101),':gd',
'LineWidth',2,'MarkerEdgeColor','k','MarkerFaceColor','b',
'MarkerSize',2)
plot(0:r,pstotal(:,101),'--bv',
'LineWidth',2,'MarkerEdgeColor','k','MarkerFaceColor','b',
'MarkerSize',2)
xlabel('Rounds (r)','FontSize',7)
ylabel('Normalized Compromised Sensors','FontSize',7)
legend('Proactive','Reactive','CHSHRD','FontSize',7)
figure
box on
hold
plot(0:k,pgp(20,:),'-r*',
'LineWidth',2,'MarkerEdgeColor','k','MarkerFaceColor','b',
'MarkerSize',2)
plot(0:k,pgr(20,:),':gd', 'LineWidth',2,'MarkerEdgeColor','k',
'MarkerFaceColor','b','MarkerSize',2)
plot(0:k,pgtotal(20,:),'--bv',
'LineWidth',2,'MarkerEdgeColor','k','MarkerFaceColor','b',
'MarkerSize',2)
xlabel('Adversary Capapility (k)','FontSize',7)
ylabel('Normalized Healthy Sensors','FontSize',7)
legend('Proactive','Reactive','CHSHRD','FontSize',7)
figure
box on
hold
plot(0:k,psp(20,:),'-r*',
'LineWidth',2,'MarkerEdgeColor','k','MarkerFaceColor','b',
'MarkerSize',2)
plot(0:k,psr(20,:),':gd',
'LineWidth',2,'MarkerEdgeColor','k','MarkerFaceColor','b',
'MarkerSize',2)
plot(0:k,pstotal(20,:),'--bv',
'LineWidth',2,'MarkerEdgeColor','k','MarkerFaceColor','b',
'MarkerSize',2)
xlabel('Adversary Capapility (k)','FontSize',7)
ylabel('Normalized Compromised Sensors','FontSize',7)
legend('Proactive','Reactive','CHSHRD','FontSize',7)


figure
box on
hold
plot(0:r,pgtotal(:,41),'-r',
'LineWidth',2,'MarkerEdgeColor','k','MarkerFaceColor','b',
'MarkerSize',2)
plot(0:r,pgtotal(:,61),':gd',
'LineWidth',2,'MarkerEdgeColor','k','MarkerFaceColor','b',
'MarkerSize',2)
plot(0:r,pgtotal(:,81),'--bv',
```

```
'LineWidth',2,'MarkerEdgeColor','k','MarkerFaceColor','b',
'MarkerSize',2)
plot(0:r,pgtotal(:,101),'-.kp',
'LineWidth',2,'MarkerEdgeColor','k','MarkerFaceColor','b',
'MarkerSize',2)
xlabel('Rounds','FontSize',7)
ylabel('Probability of healthy Sensors','FontSize',7)
legend('k=40','k=60','k=80','k=100','FontSize',7)

figure
box on
hold
plot(0:r,pstotal(:,41),'-r',
'LineWidth',2,'MarkerEdgeColor','k','MarkerFaceColor','b',
'MarkerSize',2)
plot(0:r,pstotal(:,61),':gd',
'LineWidth',2,'MarkerEdgeColor','k','MarkerFaceColor','b',
'MarkerSize',2)
plot(0:r,pstotal(:,81),'--bv',
'LineWidth',2,'MarkerEdgeColor','k','MarkerFaceColor','b',
'MarkerSize',2)
plot(0:r,pstotal(:,101),'-.kp',
'LineWidth',2,'MarkerEdgeColor','k','MarkerFaceColor','b',
'MarkerSize',2)
xlabel('Rounds','FontSize',7)
ylabel('Probability of compromised Sensors','FontSize',7)
legend('k=40','k=60','k=80','k=100','FontSize',7)
%****************************************************************
%****************************************************************

function res=combnk_as(n,r)
summm=1;
w=n;
for i=1:r
    summm=summm*w;
    w=w-1;
end
res=summm/factorial(r);


%****************************************************************
%****************************************************************


%****************************************************************
%        % ------- Simulation Results -----------
%****************************************************************

% MAIN DECLARATION
% "DF:" refer to a definition
```

```
% 1st modification instead of using a predefined compromised
probability, i
% select it totally random
clc;
clear all;
start=clock;
tic
%###################################################################
##########################
%################### PARAMETERS DEFINITION ###################
for adv_TR=[ 70    ];
figure
nTime=100; % number of iteration to reduce randomness
node_no=500;% node number
maxx=500;    % width of the network
maxy=500;    % length of the network
TR=60;       % Sensor Transmission range  and ADV  range
adv_vol=10; % ADV speed
% the ADV velocity must be 2*(ADV range) so that we guarantee
that each time
% the ADV occupy a new set of health sensors in our case the
ADV velocity
% must be 2*60=120 meter/round
update_t-1;  % Updating time
Pr_t=0.3; % Threshold probability of compromising
Pr_up=0.001;% Upbound threshold probability
Iter_node_set=zeros(node_no,21);  % 20 refer to the number of
columns in which data measured stored
Iter_node_dataset=[];  % this matrix changes every nTime
iteration, it contains the data of the matrix (node),
                            % and it is updated every
(nTime) iteration
self_healing=[];
%###################################################################
##########################
%##################### THE BEGINNING ###################

for iTime=1:nTime
    disp(['iteration number', num2str(iTime)]);
    clear node;    % To clear the previous data stored in
matrix "node"
    sick_node_befor_heal=0;
    sick_node_after_heal=0;
    node =  topo_AS(node_no, maxx, maxy);
    % DF "topo_AS()": is a function used to generate three
columns in the node matrix the first and second columns are
      % the X and Y axis of a sensor  the third column
contains the compromised probability of each si
      % this function run only one time each iteration to
distribute the nodes at their location coordinates (X,Y)
```

```
node =[node,zeros(node_no,18)];
%XXXXXXXXXXXXXXXXXXXXX  The formation of matrix "node"  XXX
XXXXXXXXXXXXXXXXXXXXXXXXXXXXX
% 1st  Col, node(:,1): x axis
% 2nd  Col, node(:,2): y axis

%XXXXXXXXXXXXXXXXXXXXXXXXXXXXXXXXXXXXXXXXXXXXXXXXXXXXXXXXXXXXXX
XXXXXXXXXXXXXXXXXXXXXXXXXX
% 3rd  Col, node(:,3): Pi the Backward secrecy of si (the
compromised probability of si)

%XXXXXXXXXXXXXXXXXXXXXXXXXXXXXXXXXXXXXXXXXXXXXXXXXXXXXXXXXXXXXX
XXXXXXXXXXXXXXXXXXXXXXXXXX
% 4th  Col, node(:,4): NBi, the number of neighbor nodes
of si

%XXXXXXXXXXXXXXXXXXXXXXXXXXXXXXXXXXXXXXXXXXXXXXXXXXXXXXXXXXXXXX
XXXXXXXXXXXXXXXXXXXXXXXXXX
% 5th  Col, node(:,5): t, the number of qualified neighbor
nodes which have a compromised probability greater than
% the threshold probability (Pr_t) note that (t <= NBi)
%XXXXXXXXXXXXXXXXXXXXXXXXXXX  Naive scheme  XXXXXXXXXXXXXXXX
XXXXXXXXXXXXXXXXXXXXXXXXXXXX
% 6th  Col, node(:,6): Pr_BSe_comp_naive, the same as
node(:,3)
% 10th Col, node(:,10): Pr_reliab_naive, data reliability
%XXXXXXXXXXXXXXXXXXXXXXXXXXX Wang's Infocom 09 scheme XXXXXXXXX
XXXXXXXXXXXXXXXXXXXXXXXXXX
% 7th  Col, node(:,7):  Pr_BSe_comp_wang_eq1, tau=1, the
BSe of si
% 14th Col, node(:,14): Pr_BSe_comp_wang_less1, tau>1, the
BSe of si
% 11th Col, node(:,11): Pr_reliab_wang_eq1, tau=1, data
reliability
% 15th Col, node(:,15): Pr_reliab_wang_less1, tau>1, data
reliability
%XXXXXXXXXXXXXXXXXXXXXXXXXXXXXX  Proposed scheme  XXXXXXXXXXXXX
XXXXXXXXXXXXXXXXXXXXXXXXX
% 8th  Col, node(:,8): Pr_BSe_comp_proposed_eq1, tau=1,
the BSe of s_i to be compromised
% 9th  Col, node(:,9): Pr_BSe_comp_proposed_less1, tau>1,
the BSe of s_i to be compromised
% 12th Col, node(:,12): Pr_reliab_proposed_eq1, tau=1,
data reliability
% 13th Col, node(:,13): Pr_reliab_proposed_less1, tau>1,
data reliability

%XXXXXXXXXXXXXXXXXXXXXXXXXXXXXXXXXXXXXXXXXXXXXXXXXXXXXXXXXXXXXX
XXXXXXXXXXXXXXXXXXXXXX
```

```
     ro=1;

     node=advr_AS_1(maxx,maxy,node,node_no,adv_TR,adv_vol,
update_t,ro);

     %######### NODES ARE NOT COMPROMISED  NOT HEALED#########
     not_sick_node=find(node(:,3)<0.3);
%    not_sick_node=not_sick_node';
     %sick_node_befor_heal(healing)=length(sick_node);
     %??? ?? ????? ?? ?? ??? ????? ??????? ??? ?? ???? ??
????????
%    for i= not_sick_node
            %disp([' lower than 0.3  ' num2str(node(i,3))])
%[nb_set,nb_no,qlf_candd_set,qlf_candd_no,not_qlf_candd_
set,not_qlf_candd_no] =...
        %nb_qlf_node_AS(i,Pr_t,TR,node,node_no);

%    end
     %######### NODES ARE NOT COMPROMISED  NOT HEALED#########


     %######### NODES ARE COMPROMISED  WILL BE HEALED#########
     for healing=1:10
     if healing==1
     else
         node(:,3)=node(:,8);
         if healing<43
             ro=healing;

node=advr_AS_1(maxx,maxy,node,node_no,adv_TR,adv_vol,update_t,
ro);
         end
     end
     sick_node=find(node(:,3)>=0.3);
     sick_node=sick_node';
     sick_node_befor_heal(healing)=length(sick_node);

     %disp(['sick_node before   ' num2str(length(sick_node))  '
Healing round   ' num2str(healing)])

     %for i =sick_node
     for i =1:node_no
            %disp([' greater than 0.3  ' num2str(node(i,3))
'   #########################'])

        %
node=advr_AS_2(maxx,maxy,node,node_no,adv_TR,
i,adv_vol,update_t);
```

```
%           node=advr_AS_rand_1(maxx,maxy,node,node_no,adv_TR,i);
%           node=advr_AS_rand_2(maxx,maxy,node,node_no,adv_TR,i);
          % repeat the function mean that there are another
adversary
          % "advr_AS_1" "advr_AS_2" these functions runs for
each si to compute the following :
          % 1-location of  ADV (x,y), which depend on prevous
location, spread, the update time and direction of ADV
          % 2-the distance between the ADV and all sensors in
the area
          % 3-the compromised probability of all sensors
according to their positions from the ADV "node(:,3)"

          %%%disp(['adversary    11  ', num2str(sum(node(:,3)))])
          %%%disp(['adversary    22  ',
num2str(sum(node(:,18)))])

          %for jjj=1:node_no,
          %   if node(jjj,3)>node(jjj,18)
             %else
             %node(jjj,3)=node(jjj,18);
             %end
          %end
          % the above lines guarantee that if si is compromised
by adv1 and adv2 , then the final compromising probability is
the greater
          % one; for example if si compromising probability is
.5 due to adv1 and it is .6 due to the adv2, then the
effective compromising
          % probability is the greater one (0.6)

[nb_set,nb_no,qlf_candd_set,qlf_candd_no,not_qlf_candd_set,
not_qlf_candd_no] =...
          nb_qlf_node_AS(i,Pr_t,TR,node,node_no);
          % DF "nb_qlf_node_AS": it is a function used to
compute many parameters such as:
          % 1- Compute neighbor node set "nb_set" and neighbor
node number "nb_no" which
             % are in the transmission range of si
          % 2- Compute the qualified candidate set "qlf_candd_
set" and the qualified
             % candidate number "qlf_candd_no", this is a part
of the neighbor set but
             % there compromised probability not exceed the
threshold probability

          node(i,4)=nb_no;        % total number of neighbors for
each sensor i
          node(i,5)=qlf_candd_no;% number of qualified neighbor
nodes for each sensor i
```

```
%################################################################
#################################

    %################################################################
#############################
            %#################### Naive scheme Naive scheme Naive
scheme #####################
            %-----------------------------------------------------------
------------------%
            node(i,6)=node(i,3);     % Pr_BSe_comp_naive, Naive
scheme, the BSe of s_i
            %disp(['1- node(i,3)=   ',num2str(node(i,3)),'
node(i,6)=   ',num2str(node(i,6)),   '     ' num2str(i)])
            node(i,10)=1-node(i,3); % Pr_reliab_naive, Naive
scheme, data reliability

%@@@@@@@@@@@@@@@@@@@@@@@@@@@@@@@@@@@@@@@@@@@@@@@@@@@@@@@@@@@@@@
    %@@@@@@@@@@@@@@@@@@@@@@@@@@@@@@@@@@@@@@@@@@@@@@@@@@@@@@@@@@
            %@@@@@@ Proposed Scheme Proposed Scheme  Proposed
Scheme  @@@@@@@@@@


            %-------------------------------------------------------
--------------%
            % data cannot be recovered, if one data part is lost.
            % If there is enough qualified candidates, data is
distributed stored. Otherwise, data is stored locally.
            if qlf_candd_no > 0
                % tau=1, that is m = n = t, then Pr_BSe_comp
=prod_{i=1}^{m}P_{i,j}*P_i
                sort_qlf_candd_set=sort(qlf_candd_set);
                % "sort_qlf_candd_set" is an ordered version of
"qlf_candd_set"
                randperm_qlf_candd_set= qlf_candd_set(randperm
(qlf_candd_no));
                % "randperm_qlf_candd_set" is a different order
version of "qlf_candd_set"
                randperm_nb_set = nb_set(randperm(nb_no));
                % "randperm_nb_set" is a different order version
of "nb_set"
                % in other words the same values exist in both
matrices but
                % with different order


                sort_qlf_candd_set_reliab = 1-sort_qlf_candd_
set;     % AMIRAMIRAMIR   ADDED
```

```
                   randperm_qlf_candd_set_reliab =
1-randperm_qlf_candd_set;
                   randperm_nb_set_reliab = 1-randperm_nb_set;
                   % note that (qlf_candd_set) &  (nb_set) contain
the compromised probability
                   % and (randperm_qlf_candd_set_reliab) &
(randperm_nb_set_reliab) contain the reliable probability
which
                   % equal to (1-compromised probability)

                   for  temp_n = 1:length(sort_qlf_candd_set)
                       % here we select the first t peers  from
"sort_qlf_candd_set"
                       % so that the product of the compromising
probability of these
                       % t peers is lower than the upperbound
threshold probability
                       if  prod(sort_qlf_candd_set(1:temp_n))>
Pr_up
                           %disp(['temp_n 1    ',num2str(temp_n),'
the product    ', num2str(prod(sort_qlf_candd_set
(1:temp_n)))])
                       else
                           % "else" mean that  the "prod(sort_
qlf_candd_set(1:temp_n))" is lower than "Pr_up"
                           % disp(['temp_n 2    ',num2str
(temp_n),'    the product    ', num2str(prod(sort_qlf_
candd_set(1:temp_n)))])
                           break;  % this break will terminate the for
loop
                       end
                       %disp(['temp_n 3    ',num2str(temp_n),'
the product    ',
num2str(prod(sort_qlf_candd_set(1:temp_n)))])
                   end
                   %disp(['temp_n 4    ',num2str(temp_n),'    the
product    ', num2str(prod(sort_qlf_candd_set(1:temp_n)))])

                   node(i,8)=prod(sort_qlf_candd_set(1:temp_n))
*node(i,3);  % tau=1. comp. prob.
                   % chose the first "temp_n" nodes from the
"sort_qlf_candd_set"
                   %which have the lower compromising probability
and multiply
                   %them with each other and with the compromising
probability
                   %of the sick node itself..............WE MUST
USE THE SAME
                   %NODES WHICH USED TO ENHANCE THE COMPROMISING
PROBABILITY
```

```
                %OF THE SICK NODE ..... TO CACULATE THE DATA
RELIABILITY  OF
                %THIS SICK NODE      SO THAT THE SUM OF COLUMNS 8
AND 12 =1
                % THIS IS THE SAME DONE BY YIREN FOT TAU>1




                %node(i,12)=prod(randperm_qlf_candd_set_
reliab(1:temp_n));  %YIREN tau=1. reliability prob.
                %node(i,12)=prod(sort_qlf_candd_set_
reliab(1:temp_n));  %AMIRAMIR tau=1. reliability prob.
                node(i,12)=1-prod(sort_qlf_candd_set(1:temp_n));
%AMIRAMIR tau=1. reliability prob.


                % Proposed scheme, m>(nt)/(t+2), n=t
                m=ceil((temp_n*temp_n)/(temp_n+2)); % this is
law exist in [58]
                node(i,9)=prod(sort_qlf_candd_set(1:m))
*node(i,3);      % tau>1. comp. prob.
                nodc(i,13)=1 prod(sort_qlf_candd_set(1:m));
% tau>1. reliability prob.



%&&&&&&&&&&&&&&&&&&&&&&&&&&&&&&&&&&&&&&&&&&&&&&&&&&&&&&&&&&&&
&&&&&&&&&&&&&&&&&&&&&&&&
    %&&&&&&&&&&&&&&&&&&&&&&&&&&&&&&&&&&&&&&&&&&&&&&&&&&&&&&&&&&&&
&&&&&&&&&&&&&&
        %&&&&&&&&&& Wang's Infocom 09 scheme  Wang's Infocom
09 scheme &&&&&&&&&&&&&&&&
        %----------------------------------------------------
----------------%


                node(i,7)=prod(nb_set(1:temp_n))*node(i,3);
% tau=1   YIREN comp. prob
                %node(i,7)=prod(randperm_nb_set(1:temp_n))
*node(i,3);     % tau=1   AMIRAMIR comp. prob

                node(i,11)=prod(randperm_nb_set_reliab
(1:temp_n));            % tau=1 reliability prob.
                %  Randomly select m nodes
                node(i,14)=prod(randperm_nb_set(1:m))*node(i,3);
% tau>1 comp. prob
                node(i,15)=1-prod(randperm_nb_set(1:m));
% tau>1 reliability prob.
        else
```

```
                % this mean that the (qlf_candd_no = 0) which
mean that there is no qualified neighbor nodes
                % and hence there is no self-healing
(cooperation) and any compromised probability will be
                % the defined compromised probability defined in
columns 3. And any reliable probability will be
                % (1-compromised probability)
                node(i,8)=node(i,3);node(i,7)=node(i,3);
                node(i,9)=node(i,3);node(i,14)=node(i,3);
                %disp(['2- node(i,3)=   ',num2str(node(i,3)),'
node(i,7)=   ',num2str(node(i,7))   '   '   num2str(i)])
                %disp(['3- node(i,3)=   ',num2str(node(i,3)),'
node(i,8)=   ',num2str(node(i,8))   '   '   num2str(i)])
                %disp(['4- node(i,3)=   ',num2str(node(i,3)),'
node(i,9)=   ',num2str(node(i,9))   '   '   num2str(i)])
                %disp(['5- node(i,3)=   ',num2str(node(i,3)),'
node(i,14)=   ',num2str(node(i,14))   '   '   num2str(i)])

                node(i,12)=1-node(i,3);node(i,11)=1-node(i,3);
                node(i,13)=1-node(i,3);node(i,15)=1-node(i,3);
        end                             %  end of ( if qlf_
candd_no > 0)




            %disp(['final- node(i,3)=   ',num2str(node(i,3))  '
'   num2str(i)])
%disp('************************************************************
********')

    end                                 %  end of (for i =
1:node_no)


    sick_node=find(node(:,3)>=0.3);
    sick_node=sick_node';
    sick_node_befor_heal(healing)=length(sick_node);
    %disp(['sick_node after1    ' num2str(length(sick_node))   '
Healing round    ' num2str(healing)])

    sick_node=find(node(:,8)>=0.3);
    sick_node_after_heal(healing)=length(sick_node);
    %disp(['sick_node after2    ' num2str(length(sick_node))   '
Healing round    ' num2str(healing)])
    %disp('***********************************')
    end
```

```
%###############################################################
####################
%    node(not_sick_node,3)=0;
 %        node(not_sick_node,4)=0;%nb_no;        % total number
of neighbors for each sensor i
  %       node(not_sick_node,5)=0;%qlf_candd_no;% number of
qualified neighbor nodes for each sensor i
  %       node(not_sick_node,6)=0;%node(i,3);     % Pr_BSe_comp_
naive, Naive scheme, the BSe of s_i
   %      node(not_sick_node,10)=0;%1-node(i,3); % Pr_reliab_
naive, Naive scheme, data reliability
   %      node(not_sick_node,8)=0;%node(i,3);
   %   node(not_sick_node,7)=0;%node(i,3);
     % node(not_sick_node,9)=0;%node(i,3);
%        node(not_sick_node,14)=0;%node(i,3);
 %       node(not_sick_node,12)=0;%1-node(i,3);
  %      node(not_sick_node,11)=0;%1-node(i,3);
   %     node(not_sick_node,13)=0;%1-node(i,3);
    %    node(not_sick_node,15)=0;%1-node(i,3);



%###############################################################
##############


    %sum(node(:,3))  % the sum of all compromised probability
of all nodes
    %length(find(node(:,4)==node(:,5)))

    Iter_node_set=Iter_node_set+node;                % sum
all the node matrices nTime


%    node(not_sick_node,:)=[];    % Amir

    Iter_node_dataset=[Iter_node_dataset;node];     % store
all the node matrices
    %size(Iter_node_dataset)
    self_healing=[self_healing;sick_node_befor_heal,0,sick_
node_after_heal];
end                             %  end of (for
iTime=1:nTime)
net_healing=[self_healing(:,1) self_healing(:,(2+healing):(2*h
ealing+1))];
average_net_healing=mean(net_healing);
%***************************************************************
********************
```

```
      %**********************************************************
**********************************
            %********************** PLOTTING   PLOTTING
PLOTTING **********************
%-------------------------------------------------------------------
% mean of Iter_node_set
Iter_node_set=Iter_node_set/nTime;

nb_no_comp_reliab_set=unique(Iter_node_dataset(:,4));



No_of_quif_set=unique(Iter_node_dataset(:,5));   %  AMIEAMIR



%size(nb_no_comp_reliab_set)
% nb_no_comp_reliab_set(:,1): the number of neighbor nodes
% column number 4 contains the number of neighbor for each
sensor i while
% the (nb_no_comp_reliab_set) contains the same values as in
(Iter_node_dataset(:,4)) but
% with no repetitions (unique). it will also be sorted
nb_no_comp_reliab_set=[nb_no_comp_reliab_set,zeros(length(nb_
no_comp_reliab_set),10)];



No_of_quif_set=[No_of_quif_set,zeros(length(No_of_quif_
set),10)];   %  AMIEAMIR


% the above line to make "nb_no_comp_reliab_set" has a 11
columns

for temp_nb_no = 1:length(nb_no_comp_reliab_set)
    %if nb_no_comp_reliab_set(1,1)==0
    %    temp_nb_no=temp_nb_no-1;
     %   iiiii=1;
    %else
        iiiii=0;
    %end
    nb_no_comp_reliab_set(temp_nb_no+iiiii,2)=
mean(Iter_node_dataset(find(Iter_node_dataset(:,4)==temp_nb_
no),8));
    % nb_no_comp_reliab_set(:,2): mean_Pr_BSe_comp.
Proposed scheme, tau=1
    % 8th  Col, node(:,8): Pr_BSe_comp_proposed_eq1, tau=1,
the BSe of s_i to be compromised
```

```
        nb_no_comp_reliab_set(temp_nb_no+iiiii,3)=
mean(Iter_node_dataset(find(Iter_node_dataset(:,4)==temp_nb_
no),12));
        % nb_no_comp_reliab_set(:,3): mean_Pr_reliab.
Proposed scheme, tau=1
        % 12th Col, node(:,12): Pr_reliab_proposed_eq1, tau=1,
data reliability

        nb_no_comp_reliab_set(temp_nb_no+iiiii,4)=
mean(Iter_node_dataset(find(Iter_node_dataset(:,4)==temp_nb_
no),9));
        % nb_no_comp_reliab_set(:,4): mean_Pr_BSe_comp.
Proposed scheme, tau>1
        % 9th  Col, node(:,9): Pr_BSe_comp_proposed_less1, tau>1,
the BSe of s_i to be compromised

        nb_no_comp_reliab_set(temp_nb_no+iiiii,5)=
mean(Iter_node_dataset(find(Iter_node_dataset(:,4)==temp_nb_
no),13));
        % nb_no_comp_reliab_set(:,5): mean_Pr_reliab.
Proposed scheme, tau>1
        % 13th Col, node(:,13): Pr_reliab_proposed_less1, tau>1,
data reliability

        nb_no_comp_reliab_set(temp_nb_no+iiiii,6)=
mean(Iter_node_dataset(find(Iter_node_dataset(:,4)==temp_nb_
no),7));
        % nb_no_comp_reliab_set(:,6): mean_Pr_BSe_comp.
wang, tau=1
        % 7th  Col, node(:,7):   Pr_BSe_comp_wang_eq1, tau=1, the
BSe of si

        nb_no_comp_reliab_set(temp_nb_no+iiiii,7)=
mean(Iter_node_dataset(find(Iter_node_dataset(:,4)==temp_nb_
no),11));
        % nb_no_comp_reliab_set(:,7): mean_Pr_reliab.
wang, tau=1
        % 11th Col, node(:,11): Pr_reliab_wang_eq1, tau=1, data
reliability

        nb_no_comp_reliab_set(temp_nb_no+iiiii,8)=
mean(Iter_node_dataset(find(Iter_node_dataset(:,4)==temp_nb_
no),14));
        % nb_no_comp_reliab_set(:,8): mean_Pr_BSe_comp.
wang, tau>1
        % 14th Col, node(:,14): Pr_BSe_comp_wang_less1, tau>1,
the BSe of si
```

```
    nb_no_comp_reliab_set(temp_nb_no+iiiii,9)=
mean(Iter_node_dataset(find(Iter_node_dataset(:,4)==temp_nb_
no),15));
    % nb_no_comp_reliab_set(:,9): mean_Pr_reliab.
wang, tau>1
    % 15th Col, node(:,15): Pr_reliab_wang_less1, tau>1, data
reliability

    nb_no_comp_reliab_set(temp_nb_no+iiiii,10)=
mean(Iter_node_dataset(find(Iter_node_dataset(:,4)==temp_nb_
no),6));
    % nb_no_comp_reliab_set(:,10):mean_Pr_BSe_comp.
naive
    % 6th  Col, node(:,6): Pr_BSe_comp_naive, the same as
node(:,3)

    nb_no_comp_reliab_set(temp_nb_no+iiiii,11)=
mean(Iter_node_dataset(find(Iter_node_dataset(:,4)==temp_nb_
no),10));
    % nb_no_comp_reliab_set(:,11):mean_Pr_reliab.
naive
    % 10th Col, node(:,10): Pr_reliab_naive, data reliability


end




%AMIRAMIRAMIRAMIRAMIRAMIRAMIRAMIRAMIRAMIRAMIRAMIRAMIRAMIR
AMIRAMIRAMIR
%AMIRAMIRAMIRAMIRAMIRAMIRAMIRAMIRAMIRAMIRAMIRAMIRAMIRAMIR
AMIRAMIRAMIR
%AMIRAMIRAMIRAMIRAMIRAMIRAMIRAMIRAMIRAMIRAMIRAMIRAMIRAMIR
AMIRAMIRAMIR
for temp_qf_no = 1:length(No_of_quif_set)
    % if No_of_quif_set(1,1)==0
     %     temp_qf_no=temp_qf_no-1;
     %    iii=1;
    % else
        iii=0;
     %end
        temp_qf_no
        temp_qf_no+iii
    No_of_quif_set(temp_qf_no+iii,2)=
mean(Iter_node_dataset(find(Iter_node_dataset(:,5)==temp_qf_
no),8));
```

```
      % No_of_quif_set(:,2): mean_Pr_BSe_comp.          Proposed
scheme, tau=1
      % 8th  Col, node(:,8): Pr_BSe_comp_proposed_eq1, tau=1,
the BSe of s_i to be compromised

      No_of_quif_set(temp_qf_no+iii,3)= mean(Iter_node_dataset
(find(Iter_node_dataset(:,5)==temp_qf_no),12));
      % No_of_quif_set(:,3): mean_Pr_reliab.
Proposed scheme, tau=1
      % 12th Col, node(:,12): Pr_reliab_proposed_eq1, tau=1,
data reliability

      No_of_quif_set(temp_qf_no+iii,4)= mean(Iter_node_dataset
(find(Iter_node_dataset(:,5)==temp_qf_no),9));
      % No_of_quif_set(:,4): mean_Pr_BSe_comp.          Proposed
scheme, tau>1
      % 9th  Col, node(:,9): Pr_BSe_comp_proposed_less1, tau>1,
the BSe of s_i to be compromised

      No_of_quif_set(temp_qf_no+iii,5)=
mean(Iter_node_dataset(find(Iter_node_dataset(:,5)==temp_qf_
no),13));
      % No_of_quif_set(:,5): mean_Pr_reliab.
Proposed scheme, tau>1
      % 13th Col, node(:,13): Pr_reliab_proposed_less1, tau>1,
data reliability

      No_of_quif_set(temp_qf_no+iii,6)=
mean(Iter_node_dataset(find(Iter_node_dataset(:,5)==temp_qf_
no),7));
      % No_of_quif_set(:,6): mean_Pr_BSe_comp.          wang,
tau=1
      % 7th  Col, node(:,7):   Pr_BSe_comp_wang_eq1, tau=1, the
BSe of si

      No_of_quif_set(temp_qf_no+iii,7)=
mean(Iter_node_dataset(find(Iter_node_dataset(:,5)==temp_qf_
no),11));
      % No_of_quif_set(:,7): mean_Pr_reliab.
wang, tau=1
      % 11th Col, node(:,11): Pr_reliab_wang_eq1, tau=1, data
reliability

      No_of_quif_set(temp_qf_no+iii,8)=
mean(Iter_node_dataset(find(Iter_node_dataset(:,5)==temp_qf_
no),14));
      % No_of_quif_set(:,8): mean_Pr_BSe_comp.          wang,
tau>1
      % 14th Col, node(:,14): Pr_BSe_comp_wang_less1, tau>1,
the BSe of si
```

```
    No_of_quif_set(temp_qf_no+iii,9)=
mean(Iter_node_dataset(find(Iter_node_dataset(:,5)==temp_qf_
no),15));
    % No_of_quif_set(:,9): mean_Pr_reliab.
wang, tau>1
    % 15th Col, node(:,15): Pr_reliab_wang_less1, tau>1, data
reliability


    No_of_quif_set(temp_qf_no+iii,10)=
mean(Iter_node_dataset(find(Iter_node_dataset(:,5)==temp_qf_
no),6));
    % No_of_quif_set(:,10):mean_Pr_BSe_comp.         naive
    % 6th  Col, node(:,6): Pr_BSe_comp_naive, the same as
node(:,3)


    No_of_quif_set(temp_qf_no+iii,11)=
mean(Iter_node_dataset(find(Iter_node_dataset(:,5)==temp_qf_
no),10));
    % No_of_quif_set(:,11):mean_Pr_reliab.             naive
    % 10th Col, node(:,10): Pr_reliab_naive, data reliability

end
%AMIRAMIRAMIRAMIRAMIRAMIRAMIRAMIRAMIRAMIRAMIRAMIRAMIRAMIR
AMIRAMIRAMIR
%AMIRAMIRAMIRAMIRAMIRAMIRAMIRAMIRAMIRAMIRAMIRAMIRAMIRAMIR
AMIRAMIRAMIR
%AMIRAMIRAMIRAMIRAMIRAMIRAMIRAMIRAMIRAMIRAMIRAMIRAMIRAMIR
AMIRAMIRAMIR


%$$$$$$$$$$$$$$$$$$$$$$$$$$$$$$$$$$$$$$$$$$$$$$$$$$$$$$$$$$$$
$$$$$$$$$$$$$$$$$$$$$$$$$$$$$$$$$$
        %$$$$$$$$$$$$$$$$$$$$$$$$$$$$$$$$$$$$$$ THE PLOTTING
$$$$$$$$$$$$$$$$$$$$$$$$$$$$$$$$$$
%----------------------------------------------------------------
-----%


        %----------------- Pr_BSe_comp,
tau =1----------------%
subplot(2,3,1);plot(Iter_node_set(:,6),'ro-','MarkerSize',
4);hold on;  % naive scheme
subplot(2,3,1);plot(Iter_node_set(:,7),'gd-','MarkerSize',4);
hold on;  % Wang's scheme - randomly select
subplot(2,3,1);plot(Iter_node_set(:,8),'bx-','MarkerSize',4);
% Ours - optimized select
title('(A) \tau=1');xlabel('Index of nodes');ylabel('Prob. of
BSe to be compromised');
legend('\fontsize{7} Naive scheme','\fontsize {7} Wang \it{et
al.}  [10]','\fontsize {7} Ours','Location','NorthEast');
```

```
            %----------------- Pr_BSe_comp,  tau >1
----------------%
subplot(2,3,4);plot(Iter_node_set(:,6),'ro-','MarkerSize',
4);hold on;    % naive scheme
subplot(2,3,4);plot(Iter_node_set(:,14),'gd-','MarkerSize',
4);hold on;  % Wang's scheme - randomly select
subplot(2,3,4);plot(Iter_node_set(:,9),'bx-','MarkerSize',4);
% proposed scheme - optimized select
title('(D) \tau<1');xlabel('Index of nodes');ylabel('Prob. of
BSe to be compromised');
legend('\fontsize{7} Naive scheme','\fontsize {7} Wang \it{et
al.}  [10]','\fontsize {7} Ours','Location','NorthEast');

            %----------------- Pr_reliab, tau = 1
----------------%
subplot(2,3,2);plot(Iter_node_set(:,10),'ro-','MarkerSize',4);
hold on;    % naive scheme
subplot(2,3,2);plot(Iter_node_set(:,11),'gd-','MarkerSize',4);
hold on;     % Wang's scheme - randomly select
subplot(2,3,2);plot(Iter_node_set(:,12),'bx-','MarkerSize',4);
% proposed scheme - optimized select
title('(B) \tau=1');xlabel('Index of nodes');ylabel('Prob. of
data reliability');
legend('\fontsize{7} Naive scheme','\fontsize {7} Wang \it{et
al.}  [10]','\fontsize {7} Ours','Location','NorthEast');

            %----------------- Pr_reliab, tau > 1
----------------%
subplot(2,3,5);plot(Iter_node_set(:,10),'ro-','MarkerSize',4);
hold on;    % naive scheme
subplot(2,3,5);plot(Iter_node_set(:,15),'gd-','MarkerSize',4);
hold on;    % Wang's scheme - randomly select
subplot(2,3,5);plot(Iter_node_set(:,13),'bx-','MarkerSize',4);
% proposed scheme - optimized select
title('(E) \tau<1');xlabel('Index of nodes');ylabel('Prob. of
data reliability');
legend('\fontsize{7} Naive scheme','\fontsize {7} Wang \it{et
al.}  [10]','\fontsize {7} Ours','Location','NorthEast');

            %--------------- Pr_BSe_comp compare with mean
neighbor node--------------%
subplot(2,3,3);plot(nb_no_comp_reliab_set(1:15,10),'ro-',
'MarkerSize',4);hold on;  % naive scheme
subplot(2,3,3);plot(nb_no_comp_reliab_set(1:15,6),'gd-',
'MarkerSize',4);hold on;   % Wang's scheme randomly select,
tau=1
subplot(2,3,3);plot(nb_no_comp_reliab_set(1:15,8),'g.-',
'MarkerSize',4);hold on;% Wang's scheme randomly select, tau>1
```

```
subplot(2,3,3);plot(nb_no_comp_reliab_set(1:15,2),'bx-
','MarkerSize',4);hold on;% proposed scheme - optimized
select, \tau=1
subplot(2,3,3);plot(nb_no_comp_reliab_set(1:15,4),'b+-
','MarkerSize',4);
title('(C)');% proposed scheme - optimized select, \tau<1
xlabel('The number of neighbor nodes');ylabel('Prob. of BSe to
be compromised');
legend('\fontsize{7} Naive scheme','\fontsize {7} Wang \it{et
al.}  [10],\tau=1',...
    '\fontsize {7} Wang \it{et al.}  [10],\tau<1','\fontsize
{7} Ours, \tau=1','\fontsize {7} Ours, \tau<1','Location',
'NorthEast');


            %--------------- Pr_reliab compare with mean
neighbor node --------------%

subplot(2,3,6);plot(nb_no_comp_reliab_set(1:15,11),'ro-',
'MarkerSize',4);hold on;% naive scheme
subplot(2,3,6);plot(nb_no_comp_reliab_set(1:15,7),'gd-',
'MarkerSize',4);hold on;% Wang's Infocom 09 scheme - randomly
select, \tau=1
subplot(2,3,6);plot(nb_no_comp_reliab_set(1:15,9),'g.-',
'MarkerSize',4);hold on;% Wang's Infocom 09 scheme - randomly
select, \tau<1
subplot(2,3,6);plot(nb_no_comp_reliab_set(1:15,3),'bx-',
'MarkerSize',4);hold on;% proposed scheme - optimized select,
\tau=1
subplot(2,3,6);plot(nb_no_comp_reliab_set(1:15,5),'b+-',
'MarkerSize',4);% proposed scheme - optimized select, \tau<1
title('(F)');xlabel('The number of neighbor nodes');
ylabel('Prob. of data reliability');
legend('\fontsize{7} Naive scheme','\fontsize {7} Wang \it{et
al.}  [10],\tau=1',...
    '\fontsize {7} Wang \it{et al.}  [10],\tau<1','\fontsize
{7} Ours, \tau=1','\fontsize {7} Ours, \tau<1','Location',
'NorthEast');

if adv_TR==30
    title({['(F)'],['n=',num2str(node_no),'
itr=',num2str(nTime),'  spd=',num2str(adv_vol),'
rang=',num2str(adv_TR)]})
elseif adv_TR==60
    title({['(F)'],['n=',num2str(node_no),'
itr=',num2str(nTime),'   spd=',num2str(adv_vol),'
rang=',num2str(adv_TR)]})
elseif adv_TR==90
    title({['(F)'],['n=',num2str(node_no),'
itr=',num2str(nTime),'   spd=',num2str(adv_vol),'
rang=',num2str(adv_TR)]})
```

```
else
     title({['(F)'],['n=',num2str(node_no),'
itr=',num2str(nTime),'    spd=',num2str(adv_vol),'
rang=',num2str(adv_TR)]})
end


%AMIRAMIRAMIRAMIRAMIRAMIRAMIRAMIRAMIRAMIRAMIRAMIRAMIR
AMIRAMIRAMIR
%AMIRAMIRAMIRAMIRAMIRAMIRAMIRAMIRAMIRAMIRAMIRAMIRAMIR
AMIRAMIRAMIR
%AMIRAMIRAMIRAMIRAMIRAMIRAMIRAMIRAMIRAMIRAMIRAMIRAMIR
AMIRAMIRAMIR
                %--------------- Pr_BSe_comp compare with mean
neighbor node--------------%
figure
subplot(1,2,1);plot(No_of_quif_set(1:15,10),'ro-',
'MarkerSize',4);hold on;  % naive scheme
subplot(1,2,1);plot(No_of_quif_set(1:15,6),'gd-',
'MarkerSize',4);hold on;   % Wang's scheme randomly select,
tau=1
subplot(1,2,1);plot(No_of_quif_set(1:15,8),'g.-',
'MarkerSize',4);hold on;% Wang's scheme randomly select, tau>1
subplot(1,2,1);plot(No_of_quif_set(1:15,2),'bx-',
'MarkerSize',4);hold on;% proposed scheme - optimized select,
\tau=1
subplot(1,2,1);plot(No_of_quif_set(1:15,1),'bi-',
'MarkerSize',4);
title('(A)');% proposed scheme - optimized select, \tau<1
xlabel('The number of qualified nodes');ylabel('Prob. of BSe
to be compromised');
legend('\fontsize{7} Naive scheme','\fontsize {7} Wang \it{et
al.}  [10],\tau=1',...
    '\fontsize {7} Wang \it{et al.}  [10],\tau<1','\fontsize
{7} Ours, \tau=1','\fontsize {7} Ours, \tau<1','Location',
'NorthEast');


                %--------------- Pr_reliab compare with mean
neighbor node --------------%

subplot(1,2,2);plot(No_of_quif_set(1:15,11),'ro-',
'MarkerSize',4);hold on;% naive scheme
subplot(1,2,2);plot(No_of_quif_set(1:15,7),'gd-',
'MarkerSize',4);hold on;% Wang's Infocom 09 scheme - randomly
select, \tau=1
subplot(1,2,2);plot(No_of_quif_set(1:15,9),'g.-',
'MarkerSize',4);hold on;% Wang's Infocom 09 scheme - randomly
select, \tau<1
subplot(1,2,2);plot(No_of_quif_set(1:15,3),'mx-',
'MarkerSize',4);hold on;% proposed scheme - optimized select,
\tau=1
```

```
subplot(1,2,2);plot(No_of_quif_set(1:15,5),'b+-',
'MarkerSize',4);% proposed scheme - optimized select, \tau<1
title('(B)');xlabel('The number of qualified
nodes');ylabel('Prob. of data reliability');
legend('\fontsize{7} Naive scheme','\fontsize {7} Wang \it{et
al.}  [10],\tau=1',...
     '\fontsize {7} Wang \it{et al.}  [10],\tau<1','\fontsize
{7} Ours, \tau=1','\fontsize {7} Ours, \tau<1','Location',
'NorthEast');


figure
plot(1:healing+1,average_net_healing)


%AMIRAMIRAMIRAMIRAMIRAMIRAMIRAMIRAMIRAMIRAMIRAMIRAMIRAMIR
AMIRAMIRAMIR
%AMIRAMIRAMIRAMIRAMIRAMIRAMIRAMIRAMIRAMIRAMIRAMIRAMIRAMIR
AMIRAMIRAMIR
%AMIRAMIRAMIRAMIRAMIRAMIRAMIRAMIRAMIRAMIRAMIRAMIRAMIRAMIR
AMIRAMIRAMIR

 if adv_TR==30
    save adv_tr30
     title({['(F)'],['n=',num2str(node_no),'
itr=',num2str(nTime),'   spd=',num2str(adv_vol),'
rang=',num2str(adv_TR)]})
elseif adv_TR==60
    save adv_TR60
     title({['(F)'],['n=',num2str(node_no),'
itr=',num2str(nTime),'   spd=',num2str(adv_vol),'
rang=',num2str(adv_TR)]})
elseif adv_TR==90
    save adv_TR90
     title({['(F)'],['n=',num2str(node_no),'
itr=',num2str(nTime),'   spd=',num2str(adv_vol),'
rang=',num2str(adv_TR)]})
else
    save adv_TR120
     title({['(F)'],['n=',num2str(node_no),'
itr=',num2str(nTime),'   spd=',num2str(adv_vol),'
rang=',num2str(adv_TR)]})
end
toc
end
endtime=clock;
disp(['start time  =  '   num2str(start)])
disp(['end time    =  '   num2str(endtime)])
beep,beep,beep,beep,beep,beep,beep,beep,beep,beep,beep,beep,
beep,beep;
```

```
length(find(Iter_node_dataset(:,4)==1))+length(find(Iter_node_
dataset(:,4)==2))+length(find(Iter_node_dataset(:,4)==3))+leng
th(find(Iter_node_dataset(:,4)==4))+length(find(Iter_node_data
set(:,4)==5))+length(find(Iter_node_dataset(:,4)==6))+length
(find(Iter_node_dataset(:,4)==7))+length(find(Iter_node_
dataset(:,4)==8))+length(find(Iter_node_dataset(:,4)==9))+
length(find(Iter_node_dataset(:,4)==10))+length(find(Iter_
node_dataset(:,4)==11))+length(find(Iter_node_dataset(:,4)
==12))+length(find(Iter_node_dataset(:,4)==13))+length(find
(Iter_node_dataset(:,4)==14))+length(find(Iter_node_dataset(:,
4)==15))+length(find(Iter_node_dataset(:,4)==16))+length(find
(Iter_node_dataset(:,4)==17))+length(find(Iter_node_dataset(:,
4)==18))+length(find(Iter_node_dataset(:,4)==19))+length(find
(Iter_node_dataset(:,4)==20))+length(find(Iter_node_dataset(:,
4)==21))+length(find(Iter_node_dataset(:,4)==22))+length(find
(Iter_node_dataset(:,4)==23))+length(find(Iter_node_dataset(:,
4)==24))+length(find(Iter_node_dataset(:,4)==25))+length(find
(Iter_node_dataset(:,4)==26))+length(find(Iter_node_dataset(:,
4)==27))+length(find(Iter_node_dataset(:,4)==28))+length(find
(Iter_node_dataset(:,4)==29))+length(find(Iter_node_dataset(:,
4)==30))+length(find(Iter_node_dataset(:,4)==31))+length(find
(Iter_node_dataset(:,4)==32))+length(find(Iter_node_dataset(:,
4)==33))+length(find(Iter_node_dataset(:,4)==34))+length(find
(Iter_node_dataset(:,4)==35))+length(find(Iter_node_dataset(:,
4)==36))+length(find(Iter_node_dataset(:,4)==37))+length(find
(Iter_node_dataset(:,4)==38))+length(find(Iter_node_dataset(:,
4)==39))+length(find(Iter_node_dataset(:,4)==40))+length(find
(Iter_node_dataset(:,4)==41))+length(find(Iter_node_dataset(:,
4)==42))+length(find(Iter_node_dataset(:,4)==43))+length(find
(Iter_node_dataset(:,4)==44))+length(find(Iter_node_dataset(:,
4)==45))+length(find(Iter_node_dataset(:,4)==46))+length(find
(Iter_node_dataset(:,4)==47))+length(find(Iter_node_dataset(:,
4)==48))+length(find(Iter_node_dataset(:,4)==49))+length(find
(Iter_node_dataset(:,4)==50))+length(find(Iter_node_dataset(:,
4)==51))+length(find(Iter_node_dataset(:,4)==52))+length(find
(Iter_node_dataset(:,4)==53))

%###########################################################
##############
%###########################################################
##############
% This function has the following assumption
%-------------------------------------------------
% 1- the adversary can compromise the nodes in its
transmission range
%-------------------------------------------------
% This function used to give the following outputs
%-------------------------------------------------
% 1- the "x_loc" of the adversary
% 2- the "y_loc" of the adversary
```

```
%----------------------------------------
% This function take the following inputs
%----------------------------------------
% 1- maxmumx: the maxmum distance in x direction
% 2- maxmumy: the maxmum distance in y direction
% 3- node_number: the total number of node
% 4- node: the "node" matrix which hold sensor data
% 3- trans_rang: the transmission range of normal sensor node
% 4- node_index_i: the index of the current sensor Si
% 5- adv_vol: the adversary velocity
% 6- update_t: the update time of adversary motion
function
[node]=advr_AS_1(maxmumx,maxmumy,node,node_number,trans_
rang,adv_vol,update_t,rond)
%size(node)
%node(:,1) = node(:,1)*maxmumx;  % Here it generates the X
coordinate of the node
%node(:,2) = node(:,2)*maxmumy;  % Here it generates the Y
coordinate of the node
    node(rond,16)=rand*maxmumx;
    node(rond,17)=rand*maxmumy;


    % the direction of motion of adversary range from 0:2*pi



    for jj=1:node_number
        adv_dist=sqrt((node(jj,1)-node(rond,16))^2+
(node(jj,2)-node(rond,17))^2);
        node(jj,21)=adv_dist;
        % adv_dist: is the distance between the adv location
to the location
        % of the current sensor node Si
        if  adv_dist<=trans_rang
            % node(jj,3)=0.6*rand;
            node(jj,3)=max(0.6+0.4*rand,node(jj,3));     %
0.6:1
            %node(jj,3)=0.6*rand+0.4;
            %plot(node(jj, 1), node(jj, 2), 'k.',
'MarkerSize', 7);
            % node(i, 1), node(i, 2) refer to X and Y
coordinate of node i
            %text(node(jj, 1),node(jj, 2),num2str(jj));
        elseif (adv_dist>trans_rang &&
adv_dist<=1.5*trans_rang)
            node(jj,3)=max(0.4+0.2*rand,node(jj,3));     %
0.4:0.6
```

```
                % node(jj,3)=0.2+0.2*rand;
                  %plot(node(jj, 1), node(jj, 2), 'b.',
'MarkerSize', 7);
                  %text(node(jj, 1),node(jj, 2),num2str(jj));
            elseif (adv_dist>1.5*trans_rang &&
adv_dist<=2*trans_rang)
                  node(jj,3)=max(0.2+0.2*rand,node(jj,3));      %
0.2:0.4
                  %node(jj,3)=0.2*rand;
                  %plot(node(jj, 1), node(jj, 2), 'r.',
'MarkerSize', 17);
                  %text(node(jj, 1),node(jj, 2),num2str(jj));
            else
                   node(jj,3)=max(0.2*rand,node(jj,3));     % 0 : 0.2
                  %node(jj,3)=0.001;
                  %plot(node(jj, 1), node(jj, 2), 'g.',
'MarkerSize', 7);
                  %text(node(jj, 1),node(jj, 2),num2str(jj));
            end

%disp('&&&&&&&&&&&&&&&&&&&&&&&&&&&&&&&&&&&&&&&&&&&&&&&&&&&
&&&&&&&&')
            %disp(['adversary- node(jj,3)-   ',num2str(node
(jj,3))  '     num2str(jj)])

%disp('&&&&&&&&&&&&&&&&&&&&&&&&&&&&&&&&&&&&&&&&&&&&&&&&&&&
&&&&&&&&&')
      end

    %title({['Network Topology: ',num2str(node_number), '
sensor nodes '];[num2str(black_node), '% black nodes, '  ,
num2str(blue_node),'% blue nodes, ',num2str(red_node), '% red
nodes, ',num2str(green_node),'% green nodes.'];['with
different compromise probability'];['black='    ,
num2str(black_node/node_number),'%,blue=',num2str(blue_node/
node_number),'%,red=',num2str(red_node/node_number),'%,
green=',num2str(green_node/node_number),'%.']});
    %xlabel('The Horizontal X');
    %ylabel('The Vertical Y');
    %axis([0, maxmumx, 0, maxmumy]);
    %set(gca, 'XTick', [0;maxmumx]);
    %set(gca, 'YTick', [0;maxmumy]);


    % DF "gca":  Get handle to Current Axis
    % DF "'XTick', [0; maxx]": make X axis begin with 0 and
ends with maxx
    % if we write "set(gca, 'XTick', 0:20:maxx)" thus will
divide the X
```

```
    % axis as 0,20,40,60,80.........,maxx


return;



%############################################################
##############
%############################################################
##############
% This function has the following assumption
%-------------------------------------------------
% 1- the adversary can compromise the nodes in its
transmission range
%----------------------------------------------------
% This function used to give the following outputs
%----------------------------------------------------
% 1- the "x_loc" of the adversary
% 2- the "y_loc" of the adversary
%----------------------------------------------
% This function take the following inputs
%---------------------------------------------
% 1- maxmumx: the maxmum distance in x direction
% 2- maxmumy: the maxmum distance in y direction
% 3- node_number: the total number of node
% 4- node: the "node" matrix which hold sensor data
% 3- trans_rang: the transmission range of normal sensor node
% 4- node_index_i: the index of the current sensor Si
% 5- adv_vol: the adversary velocity
% 6- update_t: the update time of adversary motion
function
[node]=advr_AS_2(maxmumx,maxmumy,node,node_number,trans_rang,
node_index_i,adv_vol,update_t,drawFigure)
direction=2*pi*rand;
%size(node)
%node(:,1) = node(:,1)*maxmumx;  % Here it generates the X
coordinate of the node
%node(:,2) = node(:,2)*maxmumy;  % Here it generates the Y
coordinate of the node
if  node_index_i==1
    node(node_index_i,19)=rand*maxmumx;
    node(node_index_i,20)=rand*maxmumy;
    % the direction of motion of adversary range from 0:2*pi
else
    % "else" refer to node_index_i>1
    d_travel=adv_vol*update_t;
    size(node);
    node((node_index_i-1),19);
```

```
      node(node_index_i,19)=node((node_index_i-1),19)
+d_travel*cos(direction);
      node(node_index_i,20)=node((node_index_i-1),20)
+d_travel*sin(direction);
      if (node(node_index_i,19)<0 ||node(node_index_i,19)>
maxmumx)
          direction=direction+pi;
          node(node_index_i,19)=node((node_index_i-1),19)
+d_travel*cos(direction);
      end
      if (node(node_index_i,20)<0 ||node(node_index_i,20)>
maxmumy)
          direction=direction+pi;

          node(node_index_i,20)=node((node_index_i-1),20)
+d_travel*cos(direction);
      end
end

if   drawFigure >= 1
      colordef white;   % DF "colordef": Set the figure
background defaults color to white
      %whitebg;          % DF "whitebg": complements the colors
in the current figure
      %figure(1);
      axis equal
      hold on;
      box on;           % DF "box on": displays the boundary of
the current axes.


      for jj=1:node_number

          adv_dist=sqrt((node(jj,1)-node(node_index_i,19))^2+
(node(jj,2)-node(node_index_i,20))^2);
          % adv_dist: is the distance between the adv location
to the location
          % of the current sensor node Si
          if  adv_dist<=trans_rang
              node(jj,18)=0.6*rand;
              %plot(node(jj, 1), node(jj, 2), 'k.',
'MarkerSize', 7);
              % node(i, 1), node(i, 2) refer to X and Y
coordinate of node i
              %text(node(jj, 1),node(jj, 2),num2str(jj));
          elseif (adv_dist>trans_rang &&
adv_dist<=2*trans_rang)
              node(jj,18)=0.4+0.2*rand;
              %plot(node(jj, 1), node(jj, 2), 'b.',
'MarkerSize', 7);
```

```
            %text(node(jj, 1),node(jj, 2),num2str(jj));
        elseif (adv_dist>2*trans_rang &&
adv_dist<=3*trans_rang)
            node(jj,18)=0.2+0.2*rand;
            %plot(node(jj, 1), node(jj, 2), 'r.',
'MarkerSize', 20);
            %text(node(jj, 1),node(jj, 2),num2str(jj));
        else
            node(jj,18)=0.2*rand;
            %plot(node(jj, 1), node(jj, 2), 'g.',
'MarkerSize', 7);
            %text(node(jj, 1),node(jj, 2),num2str(jj));
        end
    end
    %black_node=length(find(node(:,3)>=0.6));
    %blue_node=length(find((node(:,3)>=0.4 &
node(:,3)<0.6)));
    %red_node=length(find((node(:,3)>=0.2 & node(:,3)<0.4)));
    %green_node=length(find(node(:,3)<0.2));

    %title({['Network Topology: ',num2str(node_number), '
sensor nodes '];[num2str(black_node), '% black nodes, '  ,
num2str(blue_node),'% blue nodes, ',num2str(red_node), '% red
nodes, ',num2str(green_node),'% green nodes.'];['with
different compromise probability'];['black='     ,
num2str(black_node/node_number),'%,blue=',num2str(blue_node/
node_number),'%,red=',num2str(red_node/node_number),'%,
green=',num2str(green_node/node_number),'%.']});
    %xlabel('The Horizontal X');
    %ylabel('The Vertical Y');
    %axis([0, maxmumx, 0, maxmumy]);
    %set(gca, 'XTick', [0;maxmumx]);
    %set(gca, 'YTick', [0;maxmumy]);


    % DF "gca":  Get handle to Current Axis
    % DF "'XTick', [0; maxx]": make X axis begin with 0 and
ends with maxx
    % if we write "set(gca, 'XTick', 0:20:maxx)" thus will
divide the X
    % axis as 0,20,40,60,80.........,maxx
end


return;


%##############################################################
##############
```

```
%###############################################################
###############
% This function has the following assumption
%--------------------------------------------
% 1- the adversary can compromise the nodes in its
transmission range
%----------------------------------------------------
% This function used to give the following outputs
%----------------------------------------------------
% 1- the "x_loc" of the adversary
% 2- the "y_loc" of the adversary
%------------------------------------------
% This function take the following inputs
%------------------------------------
% 1- maxmumx: the maxmum distance in x direction
% 2- maxmumy: the maxmum distance in y direction
% 3- node_number: the total number of node
% 4- node: the "node" matrix which hold sensor data
% 3- trans_rang: the transmission range of normal sensor node
% 4- node_index_i: the index of the current sensor Si
% 5- adv_vol: the adversary velocity
% 6- update_t: the update time of adversary motion
function
[node]=advr_AS_rand_1(maxmumx,maxmumy,node,node_number,trans_
rang,node_index_i)
%size(node)
%node(:,1) = node(:,1)*maxmumx;   % Here it generates the X
coordinate of the node
%node(:,2) = node(:,2)*maxmumy;   % Here it generates the Y
coordinate of the node
    node(node_index_i,16)=rand*maxmumx;
    node(node_index_i,17)=rand*maxmumy;
    % the direction of motion of adversary range from 0:2*pi


    for jj=1:node_number
        adv_dist=sqrt((node(jj,1)-node
(node_index_i,16))^2+(node(jj,2)-node(node_index_i,17))^2);
        % adv_dist: is the distance between the adv location
to the location
        % of the current sensor node Si
        if  adv_dist<=trans_rang
            node(jj,3)=0.6*rand;
            %plot(node(jj, 1), node(jj, 2), 'k.',
'MarkerSize', 7);
            % node(i, 1), node(i, 2) refer to X and Y
coordinate of node i
            %text(node(jj, 1),node(jj, 2),num2str(jj));
        elseif (adv_dist>trans_rang &&
adv_dist<=2*trans_rang)
```

```
                node(jj,3)=0.4+0.2*rand;
                %plot(node(jj, 1), node(jj, 2), 'b.',
'MarkerSize', 7);
                %text(node(jj, 1),node(jj, 2),num2str(jj));
          elseif (adv_dist>2*trans_rang &&
adv_dist<=3*trans_rang)
                node(jj,3)=0.2+0.2*rand;
                %plot(node(jj, 1), node(jj, 2), 'r.',
'MarkerSize', 17);
                %text(node(jj, 1),node(jj, 2),num2str(jj));
          else
                node(jj,3)=0.2*rand;
                %plot(node(jj, 1), node(jj, 2), 'g.',
'MarkerSize', 7);
                %text(node(jj, 1),node(jj, 2),num2str(jj));
          end
     end
     %black_node=length(find(node(:,3)>=0.6));
     %blue_node=length(find((node(:,3)>=0.4 &
node(:,3)<0.6)));
     %red_node=length(find((node(:,3)>=0.2 & node(:,3)<0.4)));
     %green_node=length(find(node(:,3)<0.2));

     %title({['Network Topology: ',num2str(node_number), '
sensor nodes '];[num2str(black_node), '% black nodes, '  ,
num2str(blue_node),'% blue nodes, ',num2str(red_node), '% red
nodes, ',num2str(green_node),'% green nodes.'];['with
different compromise probability'];['black='        ,
num2str(black_node/node_number),'%,blue=',num2str(blue_node/
node_number),'%,red=',num2str(red_node/node_number),'%,
green=',num2str(green_node/node_number),'%.']});
     %xlabel('The Horizontal X');
     %ylabel('The Vertical Y');
     %axis([0, maxmumx, 0, maxmumy]);
     %set(gca, 'XTick', [0;maxmumx]);
     %set(gca, 'YTick', [0;maxmumy]);


     % DF "gca":  Get handle to Current Axis
     % DF "'XTick', [0; maxx]": make X axis begin with 0 and
ends with maxx
     % if we write "set(gca, 'XTick', 0:20:maxx)" thus will
divide the X
     % axis as 0,20,40,60,80.........,maxx


return;
```

```
%#################################################################
##############
%#################################################################
##############
% This function has the following assumption
%-------------------------------------------------
% 1- the adversary can compromise the nodes in its
transmission range
%-------------------------------------------------
% This function used to give the following outputs
%-------------------------------------------------
% 1- the "x_loc" of the adversary
% 2- the "y_loc" of the adversary
%-----------------------------------------
% This function take the following inputs
%-----------------------------------------
% 1- maxmumx: the maxmum distance in x direction
% 2- maxmumy: the maxmum distance in y direction
% 3- node_number: the total number of node
% 4- node: the "node" matrix which hold sensor data
% 3- trans_rang: the transmission range of normal sensor node
% 4- node_index_i: the index of the current sensor Si
% 5- adv_vol: the adversary velocity
% 6  update_t: the update time of adversary motion
function [node]=advr_AS_rand_2(maxmumx,maxmumy,node,no
de_number,trans_rang,node_index_i)
%size(node)
%node(:,1) = node(:,1)*maxmumx;   % Here it generates the X
coordinate of the node
%node(:,2) = node(:,2)*maxmumy;   % Here it generates the Y
coordinate of the node
    node(node_index_i,19)=rand*maxmumx;
    node(node_index_i,20)=rand*maxmumy;
    % the direction of motion of adversary range from 0:2*pi

    for jj=1:node_number

        adv_dist=sqrt((node(jj,1)-node
(node_index_i,19))^2+(node(jj,2)-node(node_index_i,20))^2);
        % adv_dist: is the distance between the adv location
to the location
        % of the current sensor node Si
        if  adv_dist<=trans_rang
            node(jj,18)=0.6*rand;
            %plot(node(jj, 1), node(jj, 2), 'k.',
'MarkerSize', 7);
            % node(i, 1), node(i, 2) refer to X and Y
coordinate of node i
            %text(node(jj, 1),node(jj, 2),num2str(jj));
```

```
            elseif (adv_dist>trans_rang &&
adv_dist<=2*trans_rang)
                node(jj,18)=0.4+0.2*rand;
                %plot(node(jj, 1), node(jj, 2), 'b.',
'MarkerSize', 7);
                %text(node(jj, 1),node(jj, 2),num2str(jj));
            elseif (adv_dist>2*trans_rang &&
adv_dist<=3*trans_rang)
                node(jj,18)=0.2+0.2*rand;
                %plot(node(jj, 1), node(jj, 2), 'r.',
'MarkerSize', 20);
                %text(node(jj, 1),node(jj, 2),num2str(jj));
            else
                node(jj,18)=0.2*rand;
                %plot(node(jj, 1), node(jj, 2), 'g.',
'MarkerSize', 7);
                %text(node(jj, 1),node(jj, 2),num2str(jj));
            end
        end
    %black_node=length(find(node(:,3)>=0.6));
    %blue_node=length(find((node(:,3)>=0.4 & node(:,3)<
0.6)));
    %red_node=length(find((node(:,3)>=0.2 & node(:,3)<0.4)));
    %green_node=length(find(node(:,3)<0.2));

    %title({['Network Topology: ',num2str(node_number), '
sensor nodes '];[num2str(black_node), '% black nodes, '  ,
num2str(blue_node),'% blue nodes, ',num2str(red_node), '% red
nodes, ',num2str(green_node),'% green nodes.'];['with
different compromise probability'];['black='       ,
num2str(black_node/node_number),'%,blue=',num2str(blue_node/
node_number),'%,red=',num2str(red_node/node_number),'%,
green=',num2str(green_node/node_number),'%.']});
    %xlabel('The Horizontal X');
    %ylabel('The Vertical Y');
    %axis([0, maxmumx, 0, maxmumy]);
    %set(gca, 'XTick', [0;maxmumx]);
    %set(gca, 'YTick', [0;maxmumy]);

    % DF "gca":  Get handle to Current Axis
    % DF "'XTick', [0; maxx]": make X axis begin with 0 and
ends with maxx
    % if we write "set(gca, 'XTick', 0:20:maxx)" thus will
divide the X
    % axis as 0,20,40,60,80.........,maxx
return;


%###############################################################
##############
```

```
%###############################################################
###############
% This function used to get the following outputs
%-------------------------------------------------
    % 1- nb_set:              neighbor node set,
    % 2- nb_no:               the number of neighbor node,
    % 3- qlf_candd_set:       qualified candidate set,
    % 4- qlf_candd_no:        the number of qualified
candidates.
    % 5- not_qlf_candd_set:   the not qualified candidate set
    % 6- not_qlf_candd_no:    the number of not qualified
candidates
%-----------------------------------------
% This function take the following inputs:
%-----------------------------------------
    % 1- node_index_i:        refer to the index of Si
    % 2- Pr_t:                threshold compromised probability
    % 3- trans_range:         transmission range
    % 4- node:                the matrix "node" of size (500*20)
    % 5- node_no:             the total number of nodes=sensors

function
[nb_set,nb_no,qlf_candd_set,qlf_candd_no,not_qlf_candd_set,
not_qlf_candd_no]...
        = nb_qlf_node_AS(node_index_i,Pr_t,trans_range,
node,node_no)

% Intializing the 6 outputs variables
nb_set = zeros(node_no,1);qlf_candd_set = zeros(node_no,1);
not_qlf_candd_set= zeros(node_no,1);
nb_no = 0;qlf_candd_no = 0;not_qlf_candd_no=0;
for j = 1:node_no
    dis_i_j = topo_distance_AS(node_index_i,j,node_no,node);
    % DF "topo_distance_AS": function to calculate the
distance between node i and node j
    if  (dis_i_j>0 && dis_i_j< trans_range)
        % this condition to check if the node j within the
transmitting range of node i
        nb_no = nb_no+1;     % increase the total number of
neighbor nodes
        nb_set(nb_no) = node(j,3);
        % Generate neighbor set which contain the compromised
probability of node j
        if  node(j,3)<=Pr_t  % this condition to check that
the node j has
                             % a compromised probability lower
than
                             % the threshold compromised
probability (Pr_t)
            %disp(['neighbor node id = ' num2str(j)...
```

```
                %       ';   distance between node ' num2str(node_
index_i) '   and node  ' num2str(j) '   is=  '
num2str(dis_i_j)...
                %    '; Compromised probability = '
num2str(node(j,3)*100) '%' '; Qualified']);
            qlf_candd_no = qlf_candd_no+1;
            % increase the qualified by 1
            qlf_candd_set(qlf_candd_no) = node(j,3);
        else
            % "else" mean that the compromized probability of
node j is
            % greater the threshold compromised probability
(Pr_t)
            not_qlf_candd_no=not_qlf_candd_no+1;
            % increase the not qualified by 1
            not_qlf_candd_set(not_qlf_candd_no) = node(j,3);
            %disp(['neighbor node id = ' num2str(j)...
            %       ';   distance between node ' num2str(node_
index_i) '   and node  ' num2str(j) '   is=  '
num2str(dis_i_j)...
                %    '; Compromised probability = '
num2str(node(j,3)*100) '%' '; none']);
        end
    end
end
%disp([num2str(node_index_i),'  ',num2str(j),' total neighbors
nodes
%',num2str(nb_no),'  qualified  ',num2str(qlf_candd_no),'  not
qualified
%',num2str(not_qlf_candd_no) ]) %AS
nb_set=nb_set(1:nb_no);
qlf_candd_set=qlf_candd_set(1:qlf_candd_no);
not_qlf_candd_set=not_qlf_candd_set(1:not_qlf_candd_no);
% the above three line to reduce the size of the three
matrices
return




%################################################################
##############
%################################################################
##############
load adv_TR60


                %------------------ Pr_BSe_comp,  tau =1
----------------%
subplot(2,3,1);plot(Iter_node_set(:,6),'ro-','MarkerSize',4);
hold on;  % naive scheme
```

```
subplot(2,3,1);plot(Iter_node_set(:,7),'gd-','MarkerSize',4);
hold on;  % Wang's scheme - randomly select
subplot(2,3,1);plot(Iter_node_set(:,8),'bx-','MarkerSize',4);
% Ours - optimized select
title('(A) \tau=1');xlabel('Index of nodes');ylabel('Prob. of
BSe to be compromised');
legend('\fontsize{7} Naive scheme','\fontsize {7} Wang \it{et
al.}  [10]','\fontsize {7} Ours','Location','NorthEast');


            %----------------- Pr_BSe_comp,  tau >1
---------------%
subplot(2,3,4);plot(Iter_node_set(:,6),'ro-','MarkerSize',4);
hold on;    % naive scheme
subplot(2,3,4);plot(Iter_node_set(:,14),'gd-','MarkerSize',4);
hold on;  % Wang's scheme - randomly select
subplot(2,3,4);plot(Iter_node_set(:,9),'bx-','MarkerSize',4);
% proposed scheme - optimized select
title('(D) \tau<1');xlabel('Index of nodes');ylabel('Prob. of
BSe to be compromised');
legend('\fontsize{7} Naive scheme','\fontsize {7} Wang \it{et
al.}  [10]','\fontsize {7} Ours','Location','NorthEast');


            %----------------- Pr_reliab, tau =
1----------------%
subplot(2,3,2);plot(Iter_node_set(:,10),'ro-','MarkerSize',4);
hold on;    % naive scheme
subplot(2,3,2);plot(Iter_node_set(:,11),'gd-','MarkerSize',4);
hold on;    % Wang's scheme - randomly select
subplot(2,3,2);plot(Iter_node_set(:,12),'bx-','MarkerSize',4);
% proposed scheme - optimized select
title('(B) \tau=1');xlabel('Index of nodes');ylabel('Prob. of
data reliability');
legend('\fontsize{7} Naive scheme','\fontsize {7} Wang \it{et
al.}  [10]','\fontsize {7} Ours','Location','NorthEast');


            %----------------- Pr_reliab, tau >
1---------------%
subplot(2,3,5);plot(Iter_node_set(:,10),'ro-','MarkerSize',4);
hold on;    % naive scheme
subplot(2,3,5);plot(Iter_node_set(:,15),'gd-','MarkerSize',4);
hold on;    % Wang's scheme - randomly select
subplot(2,3,5);plot(Iter_node_set(:,13),'bx-','MarkerSize',4);
% proposed scheme - optimized select
title('(E) \tau<1');xlabel('Index of nodes');ylabel('Prob. of
data reliability');
legend('\fontsize{7} Naive scheme','\fontsize {7} Wang \it{et
al.}  [10]','\fontsize {7} Ours','Location','NorthEast');


            %--------------- Pr_BSe_comp compare with mean
neighbor node--------------%
```

```
subplot(2,3,3);plot(nb_no_comp_reliab_set(1:15,10),'ro-',
'MarkerSize',4);hold on;   % naive scheme
subplot(2,3,3);plot(nb_no_comp_reliab_set(1:15,6),'gd-',
'MarkerSize',4);hold on;   % Wang's scheme randomly select,
tau=1
subplot(2,3,3);plot(nb_no_comp_reliab_set(1:15,8),'g.-',
'MarkerSize',4);hold on;% Wang's scheme randomly select, tau>1
subplot(2,3,3);plot(nb_no_comp_reliab_set(1:15,2),'bx-
','MarkerSize',4);hold on;% proposed scheme - optimized
select, \tau=1
subplot(2,3,3);plot(nb_no_comp_reliab_set(1:15,4),'b+-',
'MarkerSize',4);
title('(C)');% proposed scheme - optimized select, \tau<1
xlabel('The number of neighbor nodes');ylabel('Prob. of BSe to
be compromised');
legend('\fontsize{7} Naive scheme','\fontsize {7} Wang \it{et
al.}   [10],\tau=1',...
    '\fontsize {7} Wang \it{et al.}   [10],\tau<1','\fontsize
{7} Ours, \tau=1','\fontsize {7} Ours, \
tau<1','Location','NorthEast');

            %--------------- Pr_reliab compare with mean
neighbor node -------------%

subplot(2,3,6);plot(nb_no_comp_reliab_set(1:15,11),'ro-',
'MarkerSize',4);hold on;% naive scheme
subplot(2,3,6);plot(nb_no_comp_reliab_set(1:15,7),'gd-',
'MarkerSize',4);hold on;% Wang's Infocom 09 scheme - randomly
select, \tau=1
subplot(2,3,6);plot(nb_no_comp_reliab_set(1:15,9),'g.-',
'MarkerSize',4);hold on;% Wang's Infocom 09 scheme - randomly
select, \tau<1
subplot(2,3,6);plot(nb_no_comp_reliab_set(1:15,3),'bx-',
'MarkerSize',4);hold on;% proposed scheme - optimized select,
\tau=1
subplot(2,3,6);plot(nb_no_comp_reliab_set(1:15,5),'b+-
','MarkerSize',4);% proposed scheme - optimized select, \tau<1
title('(F)');xlabel('The number of neighbor
nodes');ylabel('Prob. of data reliability');
legend('\fontsize{7} Naive scheme','\fontsize {7} Wang \it{et
al.}   [10],\tau=1',...
    '\fontsize {7} Wang \it{et al.}   [10],\tau<1','\fontsize
{7} Ours, \tau=1','\fontsize {7} Ours, \tau<1','Location',
'NorthEast');

if adv_TR==30
    title({['(F)'],['n=',num2str(node_no),'
itr=',num2str(nTime),'    spd=',num2str(adv_vol),'
rang=',num2str(adv_TR)]})
```

```
elseif adv_TR==60
    title({['(F)'],['n=',num2str(node_no),'
itr=',num2str(nTime),'    spd=',num2str(adv_vol),'
rang=',num2str(adv_TR)]})
elseif adv_TR==90
    title({['(F)'],['n=',num2str(node_no),'
itr=',num2str(nTime),'    spd=',num2str(adv_vol),'
rang=',num2str(adv_TR)]})
else
    title({['(F)'],['n=',num2str(node_no),'
itr=',num2str(nTime),'    spd=',num2str(adv_vol),'
rang=',num2str(adv_TR)]})
end


%AMIRAMIRAMIRAMIRAMIRAMIRAMIRAMIRAMIRAMIRAMIRAMIRAMIRAMIR
AMIRAMIRAMIR
%AMIRAMIRAMIRAMIRAMIRAMIRAMIRAMIRAMIRAMIRAMIRAMIRAMIRAMIR
AMIRAMIRAMIR
%AMIRAMIRAMIRAMIRAMIRAMIRAMIRAMIRAMIRAMIRAMIRAMIRAMIRAMIR
AMIRAMIRAMIR
                %--------------- Pr_BSe_comp compare with mean
neighbor node---------------%
figure
subplot(1,2,1);plot(No_of_quif_set(1:15,10),'ro-',
'MarkerSize',4);hold on;  % naive scheme
subplot(1,2,1);plot(No_of_quif_set(1:15,6),'gd-',
'MarkerSize',4);hold on;   % Wang's scheme randomly select,
tau=1
subplot(1,2,1);plot(No_of_quif_set(1:15,8),'g.-',
'MarkerSize',4);hold on;% Wang's scheme randomly select, tau>1
subplot(1,2,1);plot(No_of_quif_set(1:15,2),'bx-',
'MarkerSize',4);hold on;% proposed scheme - optimized select,
\tau=1
subplot(1,2,1);plot(No_of_quif_set(1:15,4),'b+-',
'MarkerSize',4);
title('(A)');% proposed scheme - optimized select, \tau<1
xlabel('The number of qualified nodes');ylabel('Prob. of BSe
to be compromised');
legend('\fontsize{7} Naive scheme','\fontsize {7} Wang \it{et
al.}  [10],\tau=1',...
    '\fontsize {7} Wang \it{et al.}  [10],\tau<1','\fontsize
{7} Ours, \tau=1','\fontsize {7} Ours, \tau<1','Location',
'NorthEast');


                %--------------- Pr_reliab compare with mean
neighbor node ---------------%
```

```
subplot(1,2,2);plot(No_of_quif_set(1:15,11),'ro-',
'MarkerSize',4);hold on;% naive scheme
subplot(1,2,2);plot(No_of_quif_set(1:15,7),'gd-',
'MarkerSize',4);hold on;% Wang's Infocom 09 scheme - randomly
select, \tau=1
subplot(1,2,2);plot(No_of_quif_set(1:15,9),'g.-',
'MarkerSize',4);hold on;% Wang's Infocom 09 scheme - randomly
select, \tau<1
subplot(1,2,2);plot(No_of_quif_set(1:15,3),'mx-',
'MarkerSize',4);hold on;% proposed scheme - optimized select,
\tau=1
subplot(1,2,2);plot(No_of_quif_set(1:15,5),'b+-',
'MarkerSize',4);% proposed scheme - optimized select, \tau<1
title('(B)');xlabel('The number of qualified nodes');ylabel
('Prob. of data reliability');
legend('\fontsize{7} Naive scheme','\fontsize {7} Wang \it{et
al.}  [10],\tau=1',...
    '\fontsize {7} Wang \it{et al.}  [10],\tau<1','\fontsize
{7} Ours, \tau=1','\fontsize {7} Ours, \tau<1','Location',
'NorthEast');

if adv_TR==30
    title({['n=',num2str(node_no),'   itr=',num2str(nTime),'
spd=',num2str(adv_vol),'   rang=',num2str(adv_TR)]})
elseif adv_TR==60
    title({['n=',num2str(node_no),'   itr=',num2str(nTime),'
spd=',num2str(adv_vol),'   rang=',num2str(adv_TR)]})
elseif adv_TR==90
    title({['n=',num2str(node_no),'   itr=',num2str(nTime),'
spd=',num2str(adv_vol),'   rang=',num2str(adv_TR)]})
else
    title({['n=',num2str(node_no),'   itr=',num2str(nTime),'
spd=',num2str(adv_vol),'   rang=',num2str(adv_TR)]})
end


figure
plot(0:healing,average_net_healing)


if adv_TR==30
    title({['n=',num2str(node_no),'   itr=',num2str(nTime),'
spd=',num2str(adv_vol),'   rang=',num2str(adv_TR)]})
elseif adv_TR==60
    title({['n=',num2str(node_no),'   itr=',num2str(nTime),'
spd=',num2str(adv_vol),'   rang=',num2str(adv_TR)]})
elseif adv_TR==90
    title({['n=',num2str(node_no),'   itr=',num2str(nTime),'
spd=',num2str(adv_vol),'   rang=',num2str(adv_TR)]})
else
```

```
      title({['n=',num2str(node_no),'    itr=',num2str(nTime),'
spd=',num2str(adv_vol),'    rang=',num2str(adv_TR)]})
end
%AMIRAMIRAMIRAMIRAMIRAMIRAMIRAMIRAMIRAMIRAMIRAMIRAMIRAMIR
AMIRAMIRAMIR
```

```
load adv_TR90
figure

              %----------------- Pr_BSe_comp,  tau =1
----------------%
subplot(2,3,1);plot(Iter_node_set(:,6),'ro-','MarkerSize',4);
hold on;  % naive scheme
subplot(2,3,1);plot(Iter_node_set(:,7),'gd-','MarkerSize',4);
hold on;  % Wang's scheme   randomly select
subplot(2,3,1);plot(Iter_node_set(:,8),'bx-','MarkerSize',4);
% Ours - optimized select
title('(A) \tau=1');xlabel('Index of nodes');ylabel('Prob. of
BSe to be compromised');
legend('\fontsize{7} Naive scheme','\fontsize {7} Wang \it{et
al.} [10]','\fontsize {7} Ours','Location','NorthEast');

              %----------------- Pr_BSe_comp,  tau >1
----------------%
subplot(2,3,4);plot(Iter_node_set(:,6),'ro-','MarkerSize',4);
hold on;    % naive scheme
subplot(2,3,4);plot(Iter_node_set(:,14),'gd-','MarkerSize',4);
hold on;  % Wang's scheme - randomly select
subplot(2,3,4);plot(Iter_node_set(:,9),'bx-','MarkerSize',4);
% proposed scheme - optimized select
title('(D) \tau<1');xlabel('Index of nodes');ylabel('Prob. of
BSe to be compromised');
legend('\fontsize{7} Naive scheme','\fontsize {7} Wang \it{et
al.} [10]','\fontsize {7} Ours','Location','NorthEast');

              %----------------- Pr_reliab, tau = 1
----------------%
subplot(2,3,2);plot(Iter_node_set(:,10),'ro-','MarkerSize',4);
hold on;    % naive scheme
subplot(2,3,2);plot(Iter_node_set(:,11),'gd-',
'MarkerSize',4);hold on;    % Wang's scheme - randomly select
```

```
subplot(2,3,2);plot(Iter_node_set(:,12),'bx-','MarkerSize',4);
% proposed scheme - optimized select
title('(B) \tau=1');xlabel('Index of nodes');ylabel('Prob. of
data reliability');
legend('\fontsize{7} Naive scheme','\fontsize {7} Wang \it{et
al.}  [10]','\fontsize {7} Ours','Location','NorthEast');

                %----------------- Pr_reliab, tau > 1
----------------%
subplot(2,3,5);plot(Iter_node_set(:,10),'ro-','MarkerSize',4);
hold on;   % naive scheme
subplot(2,3,5);plot(Iter_node_set(:,15),'gd-','MarkerSize',4);
hold on;   % Wang's scheme - randomly select
subplot(2,3,5);plot(Iter_node_set(:,13),'bx-','MarkerSize',4);
% proposed scheme - optimized select
title('(E) \tau<1');xlabel('Index of nodes');ylabel('Prob. of
data reliability');
legend('\fontsize{7} Naive scheme','\fontsize {7} Wang \it{et
al.}  [10]','\fontsize {7} Ours','Location','NorthEast');

                %--------------- Pr_BSe_comp compare with mean
neighbor node--------------%
subplot(2,3,3);plot(nb_no_comp_reliab_set(1:15,10),'ro-',
'MarkerSize',4);hold on;   % naive scheme
subplot(2,3,3);plot(nb_no_comp_reliab_set(1:15,6),'gd-',
'MarkerSize',4);hold on;   % Wang's scheme randomly select,
tau=1
subplot(2,3,3);plot(nb_no_comp_reliab_set(1:15,8),'g.-',
'MarkerSize',4);hold on;% Wang's scheme randomly select, tau>1
subplot(2,3,3);plot(nb_no_comp_reliab_set(1:15,2),'bx-',
'MarkerSize',4);hold on;% proposed scheme - optimized select,
\tau=1
subplot(2,3,3);plot(nb_no_comp_reliab_set(1:15,4),'b+-',
'MarkerSize',4);
title('(C)');% proposed scheme - optimized select, \tau<1
xlabel('The number of neighbor nodes');ylabel('Prob. of BSe to
be compromised');
legend('\fontsize{7} Naive scheme','\fontsize {7} Wang \it{et
al.}  [10],\tau=1',...
    '\fontsize {7} Wang \it{et al.}  [10],\tau<1','\fontsize
{7} Ours, \tau=1','\fontsize {7} Ours, \tau<1','Location',
'NorthEast');

                %--------------- Pr_reliab compare with mean
neighbor node --------------%

subplot(2,3,6);plot(nb_no_comp_reliab_set(1:15,11),'ro-',
'MarkerSize',4);hold on;% naive scheme
```

```
subplot(2,3,6);plot(nb_no_comp_reliab_set(1:15,7),'gd-',
'MarkerSize',4);hold on;% Wang's Infocom 09 scheme - randomly
select, \tau=1
subplot(2,3,6);plot(nb_no_comp_reliab_set(1:15,9),'g.-',
'MarkerSize',4);hold on;% Wang's Infocom 09 scheme - randomly
select, \tau<1
subplot(2,3,6);plot(nb_no_comp_reliab_set(1:15,3),'bx-',
'MarkerSize',4);hold on;% proposed scheme - optimized select,
\tau=1
subplot(2,3,6);plot(nb_no_comp_reliab_set(1:15,5),'b+-',
'MarkerSize',4);% proposed scheme - optimized select, \tau<1
title('(F)');xlabel('The number of neighbor
nodes');ylabel('Prob. of data reliability');
legend('\fontsize{7} Naive scheme','\fontsize {7} Wang \it{et
al.}  [10],\tau=1',...
    '\fontsize {7} Wang \it{et al.}  [10],\tau<1','\fontsize
{7} Ours, \tau=1','\fontsize {7} Ours, \
tau<1','Location','NorthEast');

if adv_TR==30
    title({['(F)'],['n=',num2str(node_no),'
itr=',num2str(nTime),'    spd=',num2str(adv_vol),'
rang=',num2str(adv_TR)]})
elseif adv_TR==60
    title({['(F)'],['n=',num2str(node_no),'
itr=',num2str(nTime),'    spd=',num2str(adv_vol),'
rang=',num2str(adv_TR)]})
elseif adv_TR==90
    title({['(F)'],['n=',num2str(node_no),'
itr=',num2str(nTime),'    spd=',num2str(adv_vol),'
rang=',num2str(adv_TR)]})
else
    title({['(F)'],['n=',num2str(node_no),'
itr=',num2str(nTime),'    spd=',num2str(adv_vol),'
rang=',num2str(adv_TR)]})
end



%AMIRAMIRAMIRAMIRAMIRAMIRAMIRAMIRAMIRAMIRAMIRAMIRAMIRAMIRAMIR
AMIRAMIRAMIR
%AMIRAMIRAMIRAMIRAMIRAMIRAMIRAMIRAMIRAMIRAMIRAMIRAMIRAMIRAMIR
AMIRAMIRAMIR
%AMIRAMIRAMIRAMIRAMIRAMIRAMIRAMIRAMIRAMIRAMIRAMIRAMIRAMIRAMIR
AMIRAMIRAMIR
                %--------------- Pr_BSe_comp compare with mean
neighbor node--------------%
figure
subplot(1,2,1);plot(No_of_quif_set(1:15,10),'ro-',
'MarkerSize',4);hold on;   % naive scheme
```

```
subplot(1,2,1);plot(No_of_quif_set(1:15,6),'gd-',
'MarkerSize',4);hold on;  % Wang's scheme randomly select,
tau=1
subplot(1,2,1);plot(No_of_quif_set(1:15,8),'g.-',
'MarkerSize',4);hold on;% Wang's scheme randomly select, tau>1
subplot(1,2,1);plot(No_of_quif_set(1:15,2),'bx-',
'MarkerSize',4);hold on;% proposed scheme - optimized select,
\tau=1
subplot(1,2,1);plot(No_of_quif_set(1:15,4),'b+-',
'MarkerSize',4);
title('(A)');% proposed scheme - optimized select, \tau<1
xlabel('The number of qualified nodes');ylabel('Prob. of BSe
to be compromised');
legend('\fontsize{7} Naive scheme','\fontsize {7} Wang \it{et
al.}  [10],\tau=1',...
    '\fontsize {7} Wang \it{et al.}  [10],\tau<1','\fontsize
{7} Ours, \tau=1','\fontsize {7} Ours, \
tau<1','Location','NorthEast');

                %--------------- Pr_reliab compare with mean
neighbor node --------------%

subplot(1,2,2);plot(No_of_quif_set(1:15,11),'ro-',
'MarkerSize',4);hold on;% naive scheme
subplot(1,2,2);plot(No_of_quif_set(1:15,7),'gd-',
'MarkerSize',4);hold on;% Wang's Infocom 09 scheme - randomly
select, \tau=1
subplot(1,2,2);plot(No_of_quif_set(1:15,9),'g.-',
'MarkerSize',4);hold on;% Wang's Infocom 09 scheme - randomly
select, \tau<1
subplot(1,2,2);plot(No_of_quif_set(1:15,3),'mx-',
'MarkerSize',4);hold on;% proposed scheme - optimized select,
\tau=1
subplot(1,2,2);plot(No_of_quif_set(1:15,5),'b+-',
'MarkerSize',4);% proposed scheme - optimized select, \tau<1
title('(B)');xlabel('The number of qualified
nodes');ylabel('Prob. of data reliability');
legend('\fontsize{7} Naive scheme','\fontsize {7} Wang \it{et
al.}  [10],\tau=1',...
    '\fontsize {7} Wang \it{et al.}  [10],\tau<1','\fontsize
{7} Ours, \tau=1','\fontsize {7} Ours, \tau<1','Location',
'NorthEast');

if adv_TR==30
    title({['n=',num2str(node_no),'   itr=',num2str(nTime),'
spd=',num2str(adv_vol),'   rang=',num2str(adv_TR)]})
elseif adv_TR==60
    title({['n=',num2str(node_no),'   itr=',num2str(nTime),'
spd=',num2str(adv_vol),'   rang=',num2str(adv_TR)]})
elseif adv_TR==90
```

```
      title({['n=',num2str(node_no),'    itr=',num2str(nTime),'
spd=',num2str(adv_vol),'    rang=',num2str(adv_TR)]})
else
      title({['n=',num2str(node_no),'    itr=',num2str(nTime),'
spd=',num2str(adv_vol),'    rang=',num2str(adv_TR)]})
end


figure
plot(0:healing,average_net_healing)


if adv_TR==30
      title({['n=',num2str(node_no),'    itr=',num2str(nTime),'
spd=',num2str(adv_vol),'    rang=',num2str(adv_TR)]})
elseif adv_TR==60
      title({['n=',num2str(node_no),'    itr=',num2str(nTime),'
spd=',num2str(adv_vol),'    rang=',num2str(adv_TR)]})
elseif adv_TR==90
      title({['n=',num2str(node_no),'    itr=',num2str(nTime),'
spd=',num2str(adv_vol),'    rang=',num2str(adv_TR)]})
else
      title({['n=',num2str(node_no),'    itr=',num2str(nTime),'
spd-',num2str(adv_vol),'    rang-',num2str(adv_TR)]})
end
%AMIRAMIRAMIRAMIRAMIRAMIRAMIRAMIRAMIRAMIRAMIRAMIRAMIR
AMIRAMIRAMIR




load adv_TR120
figure

              %----------------- Pr_BSe_comp,  tau =1
----------------%
subplot(2,3,1);plot(Iter_node_set(:,6),'ro-','MarkerSize',4);
hold on;  % naive scheme
subplot(2,3,1);plot(Iter_node_set(:,7),'gd-','MarkerSize',4);
hold on;  % Wang's scheme - randomly select
subplot(2,3,1);plot(Iter_node_set(:,8),'bx-','MarkerSize',4);
% Ours - optimized select
title('(A) \tau=1');xlabel('Index of nodes');ylabel('Prob. of
BSe to be compromised');
```

```
legend('\fontsize{7} Naive scheme','\fontsize {7} Wang \it{et
al.}  [10]','\fontsize {7} Ours','Location','NorthEast');

                %----------------- Pr_BSe_comp,  tau
>1----------------%
subplot(2,3,4);plot(Iter_node_set(:,6),'ro-','MarkerSize',4);
hold on;    % naive scheme
subplot(2,3,4);plot(Iter_node_set(:,14),'gd-','MarkerSize',4);
hold on;  % Wang's scheme - randomly select
subplot(2,3,4);plot(Iter_node_set(:,9),'bx-','MarkerSize',4);
% proposed scheme - optimized select
title('(D) \tau<1');xlabel('Index of nodes');ylabel('Prob. of
BSe to be compromised');
legend('\fontsize{7} Naive scheme','\fontsize {7} Wang \it{et
al.}  [10]','\fontsize {7} Ours','Location','NorthEast');

                %----------------- Pr_reliab, tau =
1----------------%
subplot(2,3,2);plot(Iter_node_set(:,10),'ro-','MarkerSize',4);
hold on;    % naive scheme
subplot(2,3,2);plot(Iter_node_set(:,11),'gd-','MarkerSize',4);
hold on;    % Wang's scheme - randomly select
subplot(2,3,2);plot(Iter_node_set(:,12),'bx-','MarkerSize',4);
% proposed scheme - optimized select
title('(B) \tau=1');xlabel('Index of nodes');ylabel('Prob. of
data reliability');
legend('\fontsize{7} Naive scheme','\fontsize {7} Wang \it{et
al.}  [10]','\fontsize {7} Ours','Location','NorthEast');

                %----------------- Pr_reliab, tau >
1----------------%
subplot(2,3,5);plot(Iter_node_set(:,10),'ro-','MarkerSize',4);
hold on;    % naive scheme
subplot(2,3,5);plot(Iter_node_set(:,15),'gd-','MarkerSize',4);
hold on;    % Wang's scheme - randomly select
subplot(2,3,5);plot(Iter_node_set(:,13),'bx-','MarkerSize',4);
% proposed scheme - optimized select
title('(E) \tau<1');xlabel('Index of nodes');ylabel('Prob. of
data reliability');
legend('\fontsize{7} Naive scheme','\fontsize {7} Wang \it{et
al.}  [10]','\fontsize {7} Ours','Location','NorthEast');

                %-------------- Pr_BSe_comp compare with mean
neighbor node--------------%
subplot(2,3,3);plot(nb_no_comp_reliab_set(1:15,10),'ro-',
'MarkerSize',4);hold on;  % naive scheme
subplot(2,3,3);plot(nb_no_comp_reliab_set(1:15,6),'gd-',
'MarkerSize',4);hold on;  % Wang's scheme randomly select,
tau=1
```

```
subplot(2,3,3);plot(nb_no_comp_reliab_set(1:15,8),'g.-',
'MarkerSize',4);hold on;% Wang's scheme randomly select, tau>1
subplot(2,3,3);plot(nb_no_comp_reliab_set(1:15,2),'bx-',
'MarkerSize',4);hold on;% proposed scheme - optimized select,
\tau=1
subplot(2,3,3);plot(nb_no_comp_reliab_set(1:15,4),'b+-',
'MarkerSize',4);
title('(C)');% proposed scheme - optimized select, \tau<1
xlabel('The number of neighbor nodes');ylabel('Prob. of BSe to
be compromised');
legend('\fontsize{7} Naive scheme','\fontsize {7} Wang \it{et
al.}  [10],\tau=1',...
    '\fontsize {7} Wang \it{et al.}  [10],\tau<1','\fontsize
{7} Ours, \tau=1','\fontsize {7} Ours, \
tau<1','Location','NorthEast');

            %--------------- Pr_reliab compare with mean
neighbor node --------------%

subplot(2,3,6);plot(nb_no_comp_reliab_set(1:15,11),'ro-',
'MarkerSize',4);hold on;% naive scheme
subplot(2,3,6);plot(nb_no_comp_reliab_set(1:15,7),'gd-',
'MarkerSize',4);hold on;% Wang's Infocom 09 scheme - randomly
select, \tau=1
subplot(2,3,6);plot(nb_no_comp_reliab_set(1:15,9),'g.-',
'MarkerSize',4);hold on;% Wang's Infocom 09 scheme   randomly
select, \tau<1
subplot(2,3,6);plot(nb_no_comp_reliab_set(1:15,3),'bx-',
'MarkerSize',4);hold on;% proposed scheme - optimized select,
\tau=1
subplot(2,3,6);plot(nb_no_comp_reliab_set(1:15,5),'b+-',
'MarkerSize',4);% proposed scheme - optimized select, \tau<1
title('(F)');xlabel('The number of neighbor nodes');ylabel
('Prob. of data reliability');
legend('\fontsize{7} Naive scheme','\fontsize {7} Wang \it{et
al.}  [10],\tau=1',...
    '\fontsize {7} Wang \it{et al.}  [10],\tau<1','\fontsize
{7} Ours, \tau=1','\fontsize {7} Ours, \tau<1','Location',
'NorthEast');

if adv_TR==30
    title({['(F)'],['n=',num2str(node_no),'
itr=',num2str(nTime),'   spd=',num2str(adv_vol),'
rang=',num2str(adv_TR)]})
elseif adv_TR==60
    title({['(F)'],['n=',num2str(node_no),'
itr=',num2str(nTime),'   spd=',num2str(adv_vol),'
rang=',num2str(adv_TR)]})
elseif adv_TR==90
```

```
     title({['(F)'],['n=',num2str(node_no),'
itr=',num2str(nTime),'    spd=',num2str(adv_vol),'
rang=',num2str(adv_TR)]})
else
     title({['(F)'],['n=',num2str(node_no),'
itr=',num2str(nTime),'    spd=',num2str(adv_vol),'
rang=',num2str(adv_TR)]})
end


%AMIRAMIRAMIRAMIRAMIRAMIRAMIRAMIRAMIRAMIRAMIRAMIRAMIR
AMIRAMIRAMIR
%AMIRAMIRAMIRAMIRAMIRAMIRAMIRAMIRAMIRAMIRAMIRAMIRAMIR
AMIRAMIRAMIR
%AMIRAMIRAMIRAMIRAMIRAMIRAMIRAMIRAMIRAMIRAMIRAMIRAMIR
AMIRAMIRAMIR
                %--------------- Pr_BSe_comp compare with mean
neighbor node--------------%
figure
subplot(1,2,1);plot(No_of_quif_set(1:15,10),'ro-',
'MarkerSize',4);hold on;   % naive scheme
subplot(1,2,1);plot(No_of_quif_set(1:15,6),'gd-',
'MarkerSize',4);hold on;   % Wang's scheme randomly select,
tau=1
subplot(1,2,1);plot(No_of_quif_set(1:15,8),'g.-',
'MarkerSize',4);hold on;% Wang's scheme randomly select, tau>1
subplot(1,2,1);plot(No_of_quif_set(1:15,2),'bx-',
'MarkerSize',4);hold on;% proposed scheme - optimized select,
\tau=1
subplot(1,2,1);plot(No_of_quif_set(1:15,4),'b+-',
'MarkerSize',4);
title('(A)');% proposed scheme - optimized select, \tau<1
xlabel('The number of qualified nodes');ylabel('Prob. of BSe
to be compromised');
legend('\fontsize{7} Naive scheme','\fontsize {7} Wang \it{et
al.}   [10],\tau=1',...
    '\fontsize {7} Wang \it{et al.}   [10],\tau<1','\fontsize
{7} Ours, \tau=1','\fontsize {7} Ours, \
tau<1','Location','NorthEast');


                %--------------- Pr_reliab compare with mean
neighbor node --------------%


subplot(1,2,2);plot(No_of_quif_set(1:15,11),'ro-',
'MarkerSize',4);hold on;% naive scheme
subplot(1,2,2);plot(No_of_quif_set(1:15,7),'gd-',
'MarkerSize',4);hold on;% Wang's Infocom 09 scheme - randomly
select, \tau=1
```

```
subplot(1,2,2);plot(No_of_quif_set(1:15,9),'g.-',
'MarkerSize',4);hold on;% Wang's Infocom 09 scheme - randomly
select, \tau<1
subplot(1,2,2);plot(No_of_quif_set(1:15,3),'mx-',
'MarkerSize',4);hold on;% proposed scheme - optimized select,
\tau=1
subplot(1,2,2);plot(No_of_quif_set(1:15,5),'b+-',
'MarkerSize',4);% proposed scheme - optimized select, \tau<1
title('(B)');xlabel('The number of qualified nodes');ylabel
('Prob. of data reliability');
legend('\fontsize{7} Naive scheme','\fontsize {7} Wang \it{et
al.}  [10],\tau=1',...
    '\fontsize {7} Wang \it{et al.}  [10],\tau<1','\fontsize
{7} Ours, \tau=1','\fontsize {7} Ours, \tau<1','Location',
'NorthEast');

if adv_TR==30
    title({['n=',num2str(node_no),'   itr=',num2str(nTime),'
spd=',num2str(adv_vol),'   rang=',num2str(adv_TR)]})
elseif adv_TR==60
    title({['n=',num2str(node_no),'   itr=',num2str(nTime),'
spd=',num2str(adv_vol),'   rang=',num2str(adv_TR)]})
elseif adv_TR==90
    title({['n-',num2str(node_no),'   itr-',num2str(nTime),'
spd=',num2str(adv_vol),'   rang=',num2str(adv_TR)]})
clsc
    title({['n=',num2str(node_no),'   itr=',num2str(nTime),'
spd=',num2str(adv_vol),'   rang=',num2str(adv_TR)]})
end


figure
plot(0:healing,average_net_healing)


if adv_TR==30
    title({['n=',num2str(node_no),'   itr=',num2str(nTime),'
spd=',num2str(adv_vol),'   rang=',num2str(adv_TR)]})
elseif adv_TR==60
    title({['n=',num2str(node_no),'   itr=',num2str(nTime),'
spd=',num2str(adv_vol),'   rang=',num2str(adv_TR)]})
elseif adv_TR==90
    title({['n=',num2str(node_no),'   itr=',num2str(nTime),'
spd=',num2str(adv_vol),'   rang=',num2str(adv_TR)]})
else
    title({['n=',num2str(node_no),'   itr=',num2str(nTime),'
spd=',num2str(adv_vol),'   rang=',num2str(adv_TR)]})
end
%AMIRAMIRAMIRAMIRAMIRAMIRAMIRAMIRAMIRAMIRAMIRAMIRAMIRAMIRAMIR
AMIRAMIRAMIR
```

```
%###############################################################
###############
%###############################################################
###############
%  "topo" this function used to generate network topology and
compute the following
% 1- the X coordinate of the sensor
% 2- the Y coordinate of the sensor
% 3- the compromised probability of the sensor which is
generated once during
%    the complete one simulation and generated n_Time during
the run of the program
% 1st modification instatde of using a predefine compromised
probability, i
% select it totaly random
function [node] = topo_AS(node_numbr, maxx, maxy)
clf;
node = rand(node_numbr,2);   %AS
% In this line generate two columns contains random values
between 0 and 1
node(:,1) = node(:,1)*maxx;   % Here it generates the X
coordinate of the node
node(:,2) = node(:,2)*maxy;   % Here it generates the Y
coordinate of the node
return;
```

```
%###############################################################
###############
%###############################################################
###############
% This function used to return the following output
    % 1- d: The distance between any two sensors i,j
%----------------------------------------
% This function take the following inputs
%----------------------------------------
    % 1- i:          the index of the sensor Si
    % 2- J:          the index of any other sensor
    % 3- node_numbr: the total number of sensor nodes
    % 4- node:       the matrix "node" of size (500*20)

function [d] = topo_distance_AS(i, j,node_numbr,node)
% return distance  between two nodes i and j
d = 0;
%if (i<=0 || i>node_numbr),return;end % this condition to
check that i lay between 1:node_numbr
%if (j<=0 || j>node_numbr),return;end % this condition to
check that j lay between 1:node_numbr
```

```
diff_x=node(i, 1) - node(j, 1); % "diff_x": The difference
between i,j in X direction
diff_y=node(i, 2) - node(j, 2); % "diff_y": The difference
between i,j in Y direction
d = sqrt((diff_x)^2 + (diff_y)^2);% "d": the absolute distance
between i,j
return;
%######################################################################
##############
%######################################################################
##############
```

Appendix D: MATLAB® Simulation Codes for Chapter 8

```
%*******************************************************************
% ------- Theoretical Results -----------
%*******************************************************************
% without mobility
% this program to plot 2D outputs... this program is the
corrected version
% of the program "push_pull_proposed_2D_old".
% this program will be used to compare the results with the
second paper
% the correction done in both:
%     CORRECTION OF BOTH ERRORS
% 1- the counter begin with 1 instead of 0
% 2- we add the following lines in the Reactive scheme
%           e=combnk_as(n*psr(r-1,k-1)+k-1,t2);      %25-11-2013
%           summ=summ+e/c;                           %25-11-2013
%           e=combnk_as(n*psr_pro(r-1,k-1)+k-1,t2_pro);
%25-11-2013
%           summ=summ+e/c;
%25-11-2013
clc
clear all
n=500;
p=0.2;    %Adversary probability of compromising
rounds=25;   % max no. of rounds
copromise=100;    % max no. of compromised sensors
t1=6;             %proactive peers alone
t2=6;             %reactive peers alone
t1_pro=3;      %proactive peers proposed
t2_pro=3;       %reactive peers proposed
```

365

```
%############# PROACTIVE  ALONE PROACTIVE  ALONE #############
psp(1,1)=0;pgp(1,1)=1;
for r=2:rounds
    for k=2:copromise
        %  disp(['r= ',num2str(r),'  k=  ',num2str(k) ' t=
',num2str(t) ' i=   ',num2str(i),' a= ',num2str(a),' b=
',num2str(b)]);
        pgp(r,k)=1-k/n-psp(r-1,k-1)*(1-t1*(1-p)/
(n-1))^(n*pgp(r-1,k-1)-k);
        psp(r,k)=1-pgp(r,k);
    end
end
pgp(1,:)=1;pgp(:,1)=1;
pgp=[ones(1,k);pgp];
pgp=[ones(1,r+1)',pgp];

psp=[zeros(1,k);psp];
psp=[zeros(1,r+1)',psp];
save proactive


%############## REACTIVE ALONE REACTIVE ALONE #############3
psr(1,1)=0;
for r=2:rounds
    for k=2:copromise
        summ=0;
        for i=1:t2
            a=combnk_as(n*(1-psr(r-1,k-1))-k,i);
            b=combnk_as(n*psr(r-1,k-1)+k-1,t2-i);
            c=combnk_as(n-1,t2);
            %disp(['r= ',num2str(r),'  k=  ',num2str(k) '
t2=  ',num2str(t2) ' i=   ',num2str(i),' a= ',num2str(a),'
b= ',num2str(b)]);
            xx=p^i*a*b/c;
            summ=summ+xx;
        end
        e=combnk_as(n*psr(r-1,k-1)+k-1,t2);    %25-11-2013
        summ=summ+e/c;                         %25-11-2013

        pgr(r,k)=1-k/n-psr(r-1,k-1)*summ;
        psr(r,k)=1-pgr(r,k);
    end
end
pgr(1,:)=1;
pgr(:,1)=1;
pgr=[ones(1,k);pgr];
pgr=[ones(1,r+1)',pgr];
psr=[zeros(1,k);psr];
```

```
psr=[zeros(1,r+1)',psr];
save reactive




%############### PROPOSED PROPOSED PROPOSED ###############
%############### PROPOSED PROPOSED PROPOSED ###############
%############### PROPOSED PROPOSED PROPOSED ###############
%################## 1- PROPOSED PROACTIVE ##################
psp_pro(1,1)=0;
pgp_pro(1,1)=1;
for r=2:rounds
    for k=2:copromise
        %  disp(['r= ',num2str(r),'   k=   ',num2str(k) ' t=
',num2str(t) ' i=   ',num2str(i),'  a= ',num2str(a),' b=
',num2str(b)]);
        pgp_pro(r,k)=1-k/n-psp_pro(r-1,k-1)*(1-t1_pro*(1-p)/
(n-1))^(n*pgp_pro(r-1,k-1)-k);
        psp_pro(r,k)=1-pgp_pro(r,k);
    end
end
pgp_pro(1,:)=1;
pgp_pro(:,1)=1;
pgp_pro=[ones(1,k);pgp_pro];
pgp_pro=[ones(1,r+1)',pgp_pro];
psp_pro=[zeros(1,k);psp_pro];
psp_pro=[zeros(1,r+1)',psp_pro];
%################################################################
#######
%################################################################
#######




%################################################################
#######
%#################### 2-PROPOSED REACTIVE####################3
psr_pro(1,1)=0;
for r=2:rounds
    for k=2:copromise
        summ=0;
        for i=1:t2_pro
            a=combnk_as(n*(1-psr_pro(r-1,k-1))-k,i);
            b=combnk_as(n*psr_pro(r-1,k-1)+k-1,t2_pro-i);
            c=combnk_as(n-1,t2_pro);
            %disp(['r= ',num2str(r),'   k=   ',num2str(k) '
t2= ',num2str(t2) ' i=   ',num2str(i),'  a=  ',num2str(a),'
b=  ',num2str(b)]);
```

```
            xx=p^i*a*b/c;
            summ=summ+xx;
        end
        e=combnk_as(n*psr_pro(r-1,k-1)+k-1,t2_pro);
%25-11-2013
        summ=summ+e/c;
%25-11-2013
        pgr_pro(r,k)=1-k/n-psr_pro(r-1,k-1)*summ;
        psr_pro(r,k)=1-pgr_pro(r,k);
    end
end



pgr_pro(1,:)=1;pgr_pro(:,1)=1;
pgr_pro=[ones(1,k);pgr_pro];
pgr_pro=[ones(1,r+1)',pgr_pro];
psr_pro=[zeros(1,k);psr_pro];
psr_pro=[zeros(1,r+1)',psr_pro];
%##########################################################
#######
%##########################################################
#######
%##########################################################
#######



pstotal=psr_pro.*psp_pro;
pgtotal=1-pstotal;
save proactive_reactive_proposed


figure
box on
hold
plot(0:r,pgp(:,101),'-r*', 'LineWidth',2,'MarkerEdgeColor','k',
'MarkerFaceColor','b','MarkerSize',2)
plot(0:r,pgr(:,101),':gd', 'LineWidth',2,'MarkerEdgeColor','k',
'MarkerFaceColor','b','MarkerSize',2)
plot(0:r,pgtotal(:,101),'--bv', 'LineWidth',2,'MarkerEdgeColor',
'k','MarkerFaceColor','b','MarkerSize',2)
title('CHSHRD both proactive and reactive')
xlabel('Rounds (r)','FontSize',7)
ylabel('Normalized Healthy Sensors','FontSize',7)
legend('Proactive','Reactive','CHSHRD','FontSize',7)
```

```
figure
box on
hold
plot(0:r,psp(:,101),'-r*', 'LineWidth',2,'MarkerEdgeColor','k',
'MarkerFaceColor','b','MarkerSize',2)
plot(0:r,psr(:,101),':gd', 'LineWidth',2,'MarkerEdgeColor','k',
'MarkerFaceColor','b','MarkerSize',2)
plot(0:r,pstotal(:,101),'--bv', 'LineWidth',2,'MarkerEdgeColor',
'k','MarkerFaceColor','b','MarkerSize',2)
title('CHSHRD both proactive and reactive')
xlabel('Rounds (r)','FontSize',7)
ylabel('Normalized Compromised Sensors','FontSize',7)
legend('Proactive','Reactive','CHSHRD','FontSize',7)



figure
box on
hold
plot(0:k,pgp(20,:),'-r*', 'LineWidth',2,'MarkerEdgeColor','k',
'MarkerFaceColor','b','MarkerSize',2)
plot(0:k,pgr(20,:),':gd', 'LineWidth',2,'MarkerEdgeColor','k',
'MarkerFaceColor','b','MarkerSize',2)
plot(0:k,pgtotal(20,:),'--bv', 'LineWidth',2,'MarkerEdgeColor',
'k','MarkerFaceColor','b','MarkerSize',2)
title('CHSHRD both proactive and reactive')
xlabel('Adversary Capapility (k)','FontSize',7)
ylabel('Normalized Healthy Sensors','FontSize',7)
legend('Proactive','Reactive','CHSHRD','FontSize',7)



figure
box on
hold
plot(0:k,psp(20,:),'-r*', 'LineWidth',2,'MarkerEdgeColor','k',
'MarkerFaceColor','b','MarkerSize',2)
plot(0:k,psr(20,:),':gd', 'LineWidth',2,'MarkerEdgeColor','k',
'MarkerFaceColor','b','MarkerSize',2)
plot(0:k,pstotal(20,:),'--bv', 'LineWidth',2,'MarkerEdgeColor',
'k','MarkerFaceColor','b','MarkerSize',2)
title('CHSHRD both proactive and reactive')
xlabel('Adversary Capapility (k)','FontSize',7)
ylabel('Normalized Compromised Sensors','FontSize',7)
legend('Proactive','Reactive','CHSHRD','FontSize',7)
```

```
figure
box on
hold
plot(0:r,pgtotal(:,41),'-r', 'LineWidth',2,'MarkerEdgeColor',
'k','MarkerFaceColor','b','MarkerSize',2)
plot(0:r,pgtotal(:,61),':gd', 'LineWidth',2,'MarkerEdgeColor',
'k','MarkerFaceColor','b','MarkerSize',2)
plot(0:r,pgtotal(:,81),'--bv', 'LineWidth',2,'MarkerEdgeColor',
'k','MarkerFaceColor','b','MarkerSize',2)
plot(0:r,pgtotal(:,101),'-.kp', 'LineWidth',2,'MarkerEdgeColor',
'k','MarkerFaceColor','b','MarkerSize',2)
title('CHSHRD both proactive and reactive')
xlabel('Rounds','FontSize',7)
ylabel('Probability of healthy Sensors','FontSize',7)
legend('k=40','k=60','k=80','k=100','FontSize',7)




figure
box on
hold
plot(0:r,pstotal(:,41),'-r', 'LineWidth',2,'MarkerEdgeColor',
'k','MarkerFaceColor','b','MarkerSize',2)
plot(0:r,pstotal(:,61),':gd', 'LineWidth',2,'MarkerEdgeColor',
'k','MarkerFaceColor','b','MarkerSize',2)
plot(0:r,pstotal(:,81),'--bv', 'LineWidth',2,'MarkerEdgeColor',
'k','MarkerFaceColor','b','MarkerSize',2)
plot(0:r,pstotal(:,101),'-.kp', 'LineWidth',2,'MarkerEdgeColor',
'k','MarkerFaceColor','b','MarkerSize',2)
title('CHSHRD both proactive and reactive')
xlabel('Rounds','FontSize',7)
ylabel('Probability of comprosied Sensors','FontSize',7)
legend('k=40','k=60','k=80','k=100','FontSize',7)




%****************************************************************

function res=combnk_as(n,r)
summm=1;
w=n;
for i=1:r
    summm=summm*w;
    w=w-1;
end
res=summm/factorial(r);
```

```
%****************************************************************



%****************************************************************
% ------- Theoretical Results -----------
%****************************************************************
% with mobility
% this program will be used after the mobility is done, this
mean that, from the
% end of the second round, the process of selfhealing will be
within the cluster it
% Here, in this program the proposed scheme depends on the
proactive peers
% only
% these are the results used in the paper 2, it use the
proactive peers
% only
clc
clear all
n=500;
p-0.2;       %Adversary  probability of compromising
rounds-25;   % max no. of rounds
copromise=100;   % max no. of compromised sensors
t=6;             %proactive peers alone
%t2=6;            %reactive peers alone
t1_pro=6;    %proactive peers proposed
%t2_pro=1;    %reactive peers proposed
An= 500*500;  % An network area
TR=60;        % transmission range
% Ac: cluster area
% Ntc: total number of sensors within the cluster
% Nhc: total number of healed sensors within the cluster
% As_tr: sensor area calculated from the transmission range
% As_cv: sensor area calculated from the coverage
% Rc: cluster radius
% Ah: the area within the cluster which contain the healed
sensors


%###############################################################
#######
%###############################################################
#######
%#################### PROACTIVE  ALONE ####################
psp(1,1)=0;pgp(1,1)=1;
for r=2:rounds
    for k=2:copromise
```

```
        %    disp(['r= ',num2str(r),'    k=   ',num2str(k) '  t=
',num2str(t) '  i=   ',num2str(i),'  a=  ',num2str(a),'  b=
',num2str(b)]);
        % Ac/An=k/n  so Ac=
        Ac=(k/n)*An;               % cluster area
        Rc=sqrt(Ac/pi);            % cluster radius
        As_cv=Ac/k;                % sensor area = "Ac/k" = "An/n"
        As_tr=pi*TR^2;             % sensor area calculated from
the transmission range
        Ntc=k;
        if Rc<TR
            Nhc=Ntc;
        else
            Ah=Ac-(Rc-TR)^2*pi;        % healed area "the area
within the cluster contains the healed sensors "Nhc""
            Nhc=Ah/As_cv;
        end
        %Nhc=ceil(0.4*Ntc);

        p_occupation=(k/n)*(Ac/An) ;
        AAA=psp(r-1,k-1);
        BBB=Ntc*pgp(r-1,k-1);
        FFF=t*(1-p)/(Nhc);
        CCC=1-FFF;
        DDD=(CCC)^(BBB);
        p_sick=AAA*abs(DDD);
        r;k;
        pgp(r,k)=1-p_occupation-p_sick;
         psp(r,k)=1-pgp(r,k);
    end
end
pgp(1,:)=1;
pgp(:,1)=1;
pgp=[ones(1,k);pgp];
pgp=[ones(1,r+1)',pgp];

psp=[zeros(1,k);psp];
psp=[zeros(1,r+1)',psp];
 save proactive


%##############################################################
##################
%##############################################################
##################
%############### PROPOSED PROPOSED PROPOSED ###############
%###################### 1-PROACTIVE ######################
```

```
psp_pro(1,1)=0;
pgp_pro(1,1)=1;
for r=2:rounds
    for k=2:copromise
        %   disp(['r= ',num2str(r),'   k=   ',num2str(k) '   t=
',num2str(t) '  i=   ',num2str(i),'  a=  ',num2str(a),'  b=
',num2str(b)]);
        Ac=(k/n)*An;                % cluster area
        Rc=sqrt(Ac/pi);            % cluster radius
        As_cv=Ac/k;                % sensor area = "Ac/k" = "An/n"
        As_tr=pi*TR^2;             % sensor area calculated from
the transmission range
        Ntc=k;
        if Rc<TR
            Nhc=Ntc
        else
            Ah=Ac-(Rc-TR)^2*pi;      % healed area "the area
within the cluster contains the healed sensors "Nhc""
            Nhc=Ah/As_cv
        end
        %Nhc=ceil(0.4*Ntc);

        p_occupation=(k/n)*(Ac/An)  ;
        aaaa=psp_pro(r-1,k-1);
        bbbb=(1-t1_pro*(1-p)/(Nhc));
        cccc=(Ntc*pgp_pro(r-1,k-1));
        %%%%disp(['aaaa  ',num2str(aaaa),'   bbbb
',num2str(bbbb),'   cccc  ',num2str(cccc),'    aaaa*(bbbb^cccc)=
', num2str(aaaa*(bbbb^cccc))])
        p_sick=abs(aaaa)*(abs(bbbb))^abs(cccc));
        %%%%disp(['p_sick  ',num2str(p_sick),'  r
',num2str(r),'   k  ',num2str(k),' Nhc ',num2str(Nhc)])

        pgp_pro(r,k)=1-p_occupation-p_sick;  ·
        psp_pro(r,k)=1-pgp_pro(r,k);
        %%%%disp(['psp_pro(r,k)    ',num2str(psp_pro(r,k))])
        %%%%disp('*********************************************
*********')
        %pgp_pro(r,k)=1-k/n-psp_pro(r-1,k-1)*(1-t1_pro*(1-p)/
(n-1))^(n*pgp_pro(r-1,k-1)-k);
        %psp_pro(r,k)=1-pgp_pro(r,k);
    end
end
pgp_pro(1,:)=1;
pgp_pro(:,1)=1;
pgp_pro=[ones(1,k);pgp_pro];
pgp_pro=[ones(1,r+1)',pgp_pro];
psp_pro=[zeros(1,k);psp_pro];
psp_pro=[zeros(1,r+1)',psp_pro];
```

```
%pgr_pro(1,:)=1;
%pgr_pro(:,1)=1;
%pgr_pro=[ones(1,k);pgr_pro];
%pgr_pro=[ones(1,r+1)',pgr_pro];
%psr_pro=[zeros(1,k);psr_pro];
%psr_pro=[zeros(1,r+1)',psr_pro];

%pstotal=psr_pro.*psp_pro;     % "pstotal" for the proposed
scheme, depends on
                               %  both proactive and reactive
peers
pstotal=psp_pro;               % "pstotal" for the proposed
scheme, depends only on
                               % the proactive peers only
%pstotal=psr_pro;               % "pstotal" for the proposed
scheme, depends only on
                               % the reactive peers only
pgtotal=1-pstotal;

save proactive_reactive_proposed

figure
box on
hold
plot(0:r,pgp(:,101),'-r*', 'LineWidth',2,'MarkerEdgeColor','k',
'MarkerFaceColor','b','MarkerSize',2)
   %plot(0:r,pgr(:,101),':gd', 'LineWidth',2,'MarkerEdgeColor',
'k','MarkerFaceColor','b','MarkerSize',2)
plot(0:r,pgtotal(:,101),'--bv', 'LineWidth',2,'MarkerEdgeColor',
'k','MarkerFaceColor','b','MarkerSize',2)
xlabel('Rounds (r)','FontSize',7)
ylabel('Normalized Healthy Sensors','FontSize',7)
legend('Proactive only','SHCCM only PROACTIVE')

figure
box on
hold
plot(0:r,psp(:,101),'-r*', 'LineWidth',2,'MarkerEdgeColor','k',
'MarkerFaceColor','b','MarkerSize',2)
   %plot(0:r,psr(:,101),':gd', 'LineWidth',2,'MarkerEdgeColor',
'k','MarkerFaceColor','b','MarkerSize',2)
plot(0:r,pstotal(:,101),'--bv', 'LineWidth',2,'MarkerEdgeColor',
'k','MarkerFaceColor','b','MarkerSize',2)
xlabel('Rounds (r)','FontSize',7)
ylabel('Normalized Compromised Sensors','FontSize',7)
legend('Proactive only','SHCCM only PROACTIVE')
```

```
figure
box on
hold
plot(0:k,pgp(20,:),'-r*', 'LineWidth',2,'MarkerEdgeColor','k',
'MarkerFaceColor','b','MarkerSize',2)
%plot(0:k,pgr(20,:),':gd', 'LineWidth',2,'MarkerEdgeColor','k',
'MarkerFaceColor','b','MarkerSize',2)
plot(0:k,pgtotal(20,:),'--bv', 'LineWidth',2,'MarkerEdgeColor',
'k','MarkerFaceColor','b','MarkerSize',2)
title('SHCCM proactive only')
xlabel('Adversary Capapility (k)','FontSize',7)
ylabel('Normalized Healthy Sensors','FontSize',7)
legend('Proactive only','SHCCM only PROACTIVE')


figure
box on
hold
plot(0:k,psp(20,:),'-r*', 'LineWidth',2,'MarkerEdgeColor','k',
'MarkerFaceColor','b','MarkerSize',2)
%plot(0:k,psr(20,:),':gd', 'LineWidth',2,'MarkerEdgeColor','k',
'MarkerFaceColor','b','MarkerSize',2)
plot(0:k,pstotal(20,:),'--bv', 'LineWidth',2,'MarkerEdgeColor',
'k','MarkerFaceColor','b','MarkerSize',2)
title('SHCCM proactive only')
xlabel('Adversary Capapility (k)','FontSize',7)
ylabel('Normalized Compromised Sensors','FontSize',7)
legend('Proactive only','SHCCM only PROACTIVE')


figure
box on
hold
plot(0:r,pgtotal(:,41),'-r', 'LineWidth',2,'MarkerEdgeColor',
'k','MarkerFaceColor','b','MarkerSize',2)
plot(0:r,pgtotal(:,61),':gd', 'LineWidth',2,'MarkerEdgeColor',
'k','MarkerFaceColor','b','MarkerSize',2)
plot(0:r,pgtotal(:,81),'--bv', 'LineWidth',2,'MarkerEdgeColor',
'k','MarkerFaceColor','b','MarkerSize',2)
plot(0:r,pgtotal(:,101),'-.kp', 'LineWidth',2,'MarkerEdgeColor',
'k','MarkerFaceColor','b','MarkerSize',2)
title('SHCCM proactive only')
xlabel('Rounds','FontSize',7)
ylabel('Probability of healthy Sensors','FontSize',7)
legend('k=40','k=60','k=80','k=100')
```

```
figure
box on
hold
plot(0:r,pstotal(:,41),'-r', 'LineWidth',2,'MarkerEdgeColor',
'k','MarkerFaceColor','b','MarkerSize',2)
plot(0:r,pstotal(:,61),':gd', 'LineWidth',2,'MarkerEdgeColor',
'k','MarkerFaceColor','b','MarkerSize',2)
plot(0:r,pstotal(:,81),'--bv', 'LineWidth',2,'MarkerEdgeColor',
'k','MarkerFaceColor','b','MarkerSize',2)
plot(0:r,pstotal(:,101),'-.kp', 'LineWidth',2,'MarkerEdgeColor',
'k','MarkerFaceColor','b','MarkerSize',2)
title('SHCCM proactive only')
xlabel('Rounds','FontSize',7)
ylabel('Probability of comprosied Sensors','FontSize',7)
legend('k=40','k=60','k=80','k=100')




%****************************************************************
% ------- Simulation Results -----------
%****************************************************************
% no_mobility_single_attack
% "DF:" refer to a definition
% 1st modification instead of using a predefine compromised
probability, i
% select it totally random
clc;
clear all;
rand('state',0);
start=clock;
tic
%##############################################################
##########################
%################### PARAMETERS DEFINITION ##################


figure
nTime=1000; % number of iteration to reduce randomness
heal_rnds=20;

node_no=500;% node number
maxx=500;   % width of the network
maxy=500;   % length of the network

sensor_TR=60;       % Transmission range
sensor_vol=10;
```

```
adv_TR=60;
adv_vol=10; % ADV speed

update_t=1;  % Updating time
 Pr_t=0.3; % Threshold probability of compromising
Pr_up=0.001;% Upbound threshold probability

Iter_node_set=zeros(node_no,27);  % 27 refer to the number of
columns in which data measured stored
Iter_node_dataset=[];  % this matrix changes every nTime
iteration, it contains the data of the matrix (node),
                          % and it is updated every
(nTime) iteration
self_healing=[];
Eo=12424; % intial energy of each sensor

%##########################################################
###########################
%####################### THE BEGINNING #######################

for iTime=1:nTime
    clear node;    % To clear the previous data stored in
matrix "node"
    node(:,28)=Eo;  % intial energy of each sensor
    sick_node_befor_heal=0;
    sick_node_after_hcal=0;
    node = topo_AS(node_no, maxx, maxy);
    % DF "topo_AS()": is a function used to generate two
columns in the node
    % matrix the first and second columns are the X and Y
coordinates of
    % a sensor... this function run only one time each iteration
    % to distribute the nodes at their coordinates (X,Y)
    node =[node,zeros(node_no,25)];

    ro=1;
    %     node=advr_fixed_AS_1(maxx,maxy,node,node_no,
adv_TR,adv_vol,update_t,ro);
    % The first appearance of the Adversary

%node=advr_move_fold_AS_1(maxx,maxy,node,node_no,adv_TR,adv_
vol,update_t,ro);
% do not be used in case single adv attack
%node=advr_move_AS_1(maxx,maxy,node,node_no,adv_TR,adv_
vol,update_t,ro);
% do not be used in case single adv attack
node=adv_rand_loc_AS_1(maxx,maxy,node,node_no,adv_TR,ro);
% it can be used in case single adv attack
```

```
    %################## NODES ARE NOT COMPROMISED   NOT
HEALED#########################
    not_sick_node=find(node(:,3)<0.3);
    current_sick_node=find(node(:,3)>=0.3);

    % clustr([1:length(current_sick_node)],:)=node
(current_sick_node,[1:8]);
    % cluster(:,9)=current_sick_node;




%%length(not_sick_node);length(current_sick_node);length
(not_sick_node)+length(current_sick_node);
    % not_sick_node=not_sick_node';
    %sick_node_befor_heal(healing)=length(sick_node);
    %for  i= not_sick_node
            %disp(['  lower than 0.3  '  num2str(node(i,3))])
%[nb_set,nb_no,qlf_candd_set,qlf_candd_no,not_qlf_candd_
set,not_qlf_candd_no] =...
            %nb_qlf_node_AS(i,Pr_t,sensor_TR,node,node_no);
    %end


    %################## Plotting network  ##################
%    figure(1)
%    plot(node(:,1),node(:,2),'o');

    %############### HEALING PROCESS  ###############
    for healing=1:heal_rnds
        disp(['iteration number  ', num2str(iTime), '
Healing round nunber  ', num2str(healing)]);
        if healing==1
            ONE_HEALING_BEFORE_MOBILITY
        else

            % node(:,8): Pr_BSe_comp_proposed_eq1, tau=1, the
BSe of s_i to be compromised
            %%%%%if healing<430
                ro=healing;



% node=advr_move_fold_AS_1(maxx,maxy,node,node_no,adv_TR,adv_
vol,update_t,ro);
% do not be used in case single adv attack
%node=advr_move_AS_1(maxx,maxy,node,node_no,adv_TR,adv_
vol,update_t,ro);
% do not be used in case single adv attack
```

```
%node=adv_rand_loc_AS_1(maxx,maxy,node,node_no,adv_TR,ro);
% it can be used in case single adv attack




        sick_node=find(node(:,3)>=0.3);    % "sick_node" is
column
        sick_node=sick_node';               % % "sick_node" is
row
        sick_node_befor_heal(healing)=length(sick_node);
        disp(['sick_node before   ' num2str(length(sick_node))
'   Healing round   ' num2str(healing)])

        %  disp('beforeBBBBBBBBBBBBBBBBBBBBBBBBBB')
        node=health_neighbors_before(node);
        % "health_neighbors_before" this function used to find
the number
        % of health neighbours before mobility




        sensor_will_move=[sick_node   heeeleeed  ];
        if find(sick_node~=0)
            %   node = node_motion_varied_cluster(sensor_
will_move, maxx, maxy,sensor_vol,update_t,node,50,ro);
        end
        % "node_motion_fixed" this function is used to perform
mobility

        node=health_neighbors_after(node);
        % "health_neighbors_after" this function used to find
the number
        % of health neighbours after mobility

                    %disp('All    neighbors ')
                    %disp('Sick ID    Before     after      the
difference (before motion-after motion)')
        for iiiii=sick_node;
                %disp([num2str(iiiii),'
',num2str(node(iiiii,1)),'    ',num2str(node(iiiii,2)),'
',num2str(node(iiiii,3)),'          ', num2str(node(ii
```

```
iii,24)),'                 ',num2str(node(iiiii,26)),'
',num2str(node(iiiii,24)-node(iiiii,26))])
         end
                 %disp('Healthe    neighbors ')
                 %disp('Sick ID    Before    after    the
difference (before motion-after motion)')
         for iiiii=sick_node;
                 %disp([ num2str(iiiii),'
',num2str(node(iiiii,1)),'   ',num2str(node(iiiii,2)),'
',num2str(node(iiiii,3)),'         ', num2str(node(ii
iii,25)),'          ',num2str(node(iiiii,27)),'
',num2str(node(iiiii,25)-node(iiiii,27))])
         end
         % the above function will move all nodes or sick nodes
only many times
         % within the network based on the principals that
mobility can enhance
         % the security by moving the sick sensors to another
position in the
         % network with a more healthy neighbors.

         %for i =sick_node
         heeeleeed=[];
         for i =sick_node


             % node=advr_AS_2(maxx,maxy,node,node_no,adv_TR,i,adv_
vol,update_t);
             % node=advr_AS_rand_1(maxx,maxy,node,node_no,adv_TR,i);
             % node=advr_AS_rand_2(maxx,maxy,node,node_no,adv_TR,i);
             % repeat the function mean that there are another
adversary
             % "advr_AS_1" "advr_AS_2" these functions runs for
each si to compute the following :
             % 1-location of  ADV (x,y), which depend on prevous
location, spread, the update time and direction of ADV
             % 2-the distance between the ADV and all sensors in
the area
             % 3-the compromised probability of all sensors
according to their positions from the ADV "node(:,3)"

             %%%disp(['adversay   11  ', num2str(sum(node(:,3)))])
             %%%disp(['adversay   22  ', num2str(sum(node(:,18)))])

             %for jjj=1:node_no,
               %   if node(jjj,3)>node(jjj,18)
                 %else
                 %node(jjj,3)=node(jjj,18);
                 %end
             %end
```

        % the above lines guarantee that if si is compromised
by adv1 and adv2 , then the final compromising probability is
the greater
        % one; for example if si compromising probability is
.5 due to adv1 and it is .6 due to the adv2, then the
effective compromising
        % probability is the greater one (0.6)

[nb_set,nb_no,qlf_candd_set,qlf_candd_no,not_qlf_candd_
set,not_qlf_candd_no] =...
        nb_qlf_node_AS(i,Pr_t,sensor_TR,node,node_no);
        % DF "nb_qlf_node_AS": it is a function used to
compute many parameters such as:
        % 1- Copmute neighbor node set "nb_set" and neighbor
node number "nb_no" which
                % are in the transmission range of si
        % 2- Copmute the qualified candidate set "qlf_candd_
set" and the qualified
                % candidate number "qlf_candd_no", this is a part
of the neighbor set but
                % there compromised probability not exceed the
threshold probability

        node(i,4)=nb_no;   % total number of neighbors for
each sensor i
        node(i,5)=qlf_candd_no;% number of qualified neighbor
nodes for each sensor i

%##
################################
%##
#########################
        %################### Naive scheme Naive scheme Naive
scheme ####################
        %---
-------------------%
        node(i,6)=node(i,3);   % Pr_BSe_comp_naive, Naive
scheme, the BSe of s_i
        %disp(['1- node(i,3)=   ',num2str(node(i,3)),'
node(i,6)=   ',num2str(node(i,6)),   '   ' num2str(i)])
        node(i,10)=1-node(i,3); % Pr_reliab_naive, Naive
scheme, data reliability
%disp(['before Sk_N ID and Comp ' num2str(i) ' '
num2str(node(i,3)) ' Sk_N nb_no ' num2str(nb_no) ' Sk_N
nb_no qlf_candd_no '   num2str(qlf_candd_no)])
        %@@
@@

        %@@
        %@@@@@@ Proposed Scheme Proposed Scheme Proposed
Scheme @@@@@@@@@@

```
        %---------------------------------------------------------
--------------%
        % data cannot be recovered, if one data part is lost.
        % If there is enough qualified candidates, data is
distributed stored. Otherwise, data is stored locally.
        if (qlf_candd_no > 0)&&(prod(qlf_candd_set)< Pr_up)
            % tau=1, that is m = n = t, then Pr_BSe_comp
=prod_{i=1}^{m}P_{i,j}*P_i
            sort_qlf_candd_set=sort(qlf_candd_set);
            % "sort_qlf_candd_set" is an ordered version of
"qlf_candd_set"
            randperm_qlf_candd_set=
qlf_candd_set(randperm(qlf_candd_no));
            % "randperm_qlf_candd_set" is a different order
version of "qlf_candd_set"
            randperm_nb_set = nb_set(randperm(nb_no));
            % "randperm_nb_set" is a different order version
of "nb_set"
            % in other words the same values exist in both
matrices but
            % with different order


            sort_qlf_candd_set_reliab = 1-sort_qlf_candd_
set;       % AMIRAMIRAMIR   ADDED



            randperm_qlf_candd_set_reliab =
1-randperm_qlf_candd_set;
            randperm_nb_set_reliab = 1-randperm_nb_set;
            % note that (qlf_candd_set)  &  (nb_set) contain
the compromised probability
            % and (randperm_qlf_candd_set_reliab) &
(randperm_nb_set_reliab) contain the reliable probability
which
            % equal to (1-compromised probability)

            for  temp_n = 1:length(sort_qlf_candd_set)
                % here we select the first t peers  from
"sort_qlf_candd_set"
                % so that the product of the compromising
probability of these
                % t peers is lower than the upperbound
threshold probability
                if  prod(sort_qlf_candd_set(1:temp_n))> Pr_up
                    %disp(['temp_n 1    ',num2str(temp_n),'
the product    ', num2str(prod(sort_qlf_candd_set(1:temp_n)))])
                else
```

```
                      % "else" mean that  the
"prod(sort_qlf_candd_set(1:temp_n))" is lower than "Pr_up"
                      % disp(['temp_n 2    ',num2str(temp_n),'
the product      ', num2str(prod(sort_qlf_candd_set(1:temp_n)))])
                break;  % this break will terminate the for
loop
                end
                %disp(['temp_n 3    ',num2str(temp_n),'      the
product       ', num2str(prod(sort_qlf_candd_set(1:temp_n)))])
                end
                %disp(['temp_n 4    ',num2str(temp_n),'      the
product       ', num2str(prod(sort_qlf_candd_set(1:temp_n)))])

                node(i,8)=prod(sort_qlf_candd_
set(1:temp_n))*node(i,3);    % tau=1. comp. prob.
                % chose the first "temp_n" nodes from the
"sort_qlf_candd_set"
                %which have the lower compromising probability
and multiply
                %them with each other and with the compromising
probability
                %of the sick node itself..............WE MUST
USE THE SAME
                %NODES WHICH USED TO ENHANCE THE COMPROMOISING
PROBABILITY
                %OF THE SICK NODE .....  TO CACULATE THE DATA
RELIABILITY  OF
                %THIS SICK NODE       SO THAT THE SUM OF COLUMNS 8
AND 12 =1
                % THIS IS THE SAME DONE BY YIREN FOT TAU>1

                %node(i,12)=prod(randperm_qlf_candd_set_
reliab(1:temp_n));  %YIREN tau=1. reliability prob.
                %node(i,12)=prod(sort_qlf_candd_set_
reliab(1:temp_n));  %AMIRAMIR tau=1. reliability prob.
                node(i,12)=1-prod(sort_qlf_candd_set(1:temp_n));
%AMIRAMIR tau=1. reliability prob.

                % Proposed scheme, m>(nt)/(t+2), n=t
                m=ceil((temp_n*temp_n)/(temp_n+2));    % this is
law exist in [58]
                node(i,9)=prod(sort_qlf_candd_
set(1:m))*node(i,3);       % tau>1. comp. prob.
                node(i,13)=1-prod(sort_qlf_candd_set(1:m));
% tau>1. reliability prob.

%&&&&&&&&&&&&&&&&&&&&&&&&&&&&&&&&&&&&&&&&&&&&&&&&&&&&&&&&&&&&&&
&&&&&&&&&&&&&&&&&&&&&&&&&&&
    %&&&&&&&&&&&&&&&&&&&&&&&&&&&&&&&&&&&&&&&&&&&&&&&&&&&&&&&&&&&&&&
&&&&&&&&&&&&&&&
```

```
        %&&&&&&&&&& Wang's Infocom 09 scheme   Wang's Infocom
09 scheme &&&&&&&&&&&&&&&
        %----------------------------------------------------------
--------------%




            node(i,7)=prod(nb_set(1:temp_n))*node(i,3);
% tau=1   YIREN comp. prob
            %node(i,7)=prod(randperm_nb_
set(1:temp_n))*node(i,3);      % tau=1    AMIRAMIR comp. prob

            node(i,11)=prod(randperm_nb_set_reliab(1:temp_n));
% tau=1 reliability prob.
            %  Randomly select m nodes
            node(i,14)=prod(randperm_nb_set(1:m))*node(i,3);
% tau>1 comp. prob
            node(i,15)=1-prod(randperm_nb_set(1:m));
% tau>1 reliability prob.




            %node(i,3)=node(i,8);  % 12/12/2013
            % this line make that if the sensor i is healed
it will help
            % the others during the same round so we can say
in the same
            % round the sick sensor healed first will help
the others
            % sick sensors in the same round. I think this
not accepted
            % because that all sick sensors start the
healing process in
            % the same round.

                    heeeleeed=[heeeleeed  i];

    %          disp([' hhheeellleeeddd222222  ' num2str(qlf_
candd_no)  '  ' num2str(qlf_candd_set')  '  '
num2str(prod(sort_qlf_candd_set(1:temp_n)))  '  '
num2str(prod(sort_qlf_candd_set(1:temp_n))*node(i,3))])

        else
            % this mean that the (qlf_candd_no = 0) which
mean that there is no qualified neighbor nodes
            % and hence there is no self-healing
(cooperation) and any compromised probability will be
```

```
                % the defined compromised probability defined in
columns 3. And any reliable probability will be
                % (1-compromised probability)
                node(i,8)=node(i,3);node(i,7)=node(i,3);
                node(i,9)=node(i,3);node(i,14)=node(i,3);

                node(i,12)=1-node(i,3);node(i,11)=1-node(i,3);
                node(i,13)=1-node(i,3);node(i,15)=1-node(i,3);
          end                          %  end of ( if qlf_
candd_no > 0)
%disp([' after Sk_N ID and Comp  '  num2str(i)  '  '
num2str(node(i,3))  '  Sk_N nb_no  '  num2str(nb_no)  '  Sk_N
nb_no qlf_candd_no  '   num2str(qlf_candd_no)])

%disp(['final- node(i,3)=   ',num2str(node(i,3))  '    '
num2str(i)])
%disp('***********************************************************
********')

    end                             %  end of (for i =
1:node_no)

        disp([' in round    ' num2str(healing) '  is  '
num2str(length(hcccleeed)) '    healed'])

                node(hcccleeed,3)=node(heeeleeed,8);
%12/12/2013       need correction

    sick_node=find(node(:,3)>=0.3);
    sick_node_after_heal(healing)=length(sick_node);
    disp(['sick_node after    ' num2str(length(sick_node))  '
Healing round   ' num2str(healing)])
    disp('***********************************************************
******')

  %    grid on
  %      ang=0:0.01:2*pi;      xp=adv_TR*cos(ang);
yp=adv_TR*sin(ang);
  %      figure(2)
  %      hold   on
  %      plot(node(:,1),node(:,2),'o')
  %       title(['sick node ' num2str(length(sick_node))  '
Healing round   ' num2str(healing)])
  %      ro=1;
  %     plot(node(ro,16)+xp,node(ro,17)+yp)
  %      plot(node(ro,16)+2*xp,node(ro,17)+2*yp)
  %      plot(node(ro,16)+3*xp,node(ro,17)+3*yp)
  %      plot(node(ro,16),node(ro,17),'s-','LineWidth',4,'Mark
erEdgeColor','k','MarkerFaceColor','m','MarkerSize',8);
  %      text(node(ro,16)+5,node(ro,17)+2,'AD')
```

```
%          axis([0  500  0    500])
%
%       hold off
%
%       figure(2)
%       plot(node(:,1),node(:,2),'o');
%              for iiiiii=1:500
%            char a;
%            a=num2str(iiiiii);
%            if (node(iiiiii,3)>=0.3)
%            text(node(iiiiii,1)+2,node(iiiiii,2),'S','color'
,'r')
%            else
%            text(node(iiiiii,1)+2,node(iiiiii,2),'h','color'
,'g')
%            end
%      end

        end
    end




    %sum(node(:,3))  % the sum of all compromised probability
of all nodes
    %length(find(node(:,4)==node(:,5)))

    Iter_node_set=Iter_node_set+node;                    % sum all
the node matrices nTime

%    node(not_sick_node,:)=[];    % Amir

    Iter_node_dataset=[Iter_node_dataset;node];     % store
all the node matrices
    %size(Iter_node_dataset)

self_healing=[self_healing;sick_node_befor_heal,0,sick_node_
after_heal];
end                             %  end of (for
iTime=1:nTime)
net_healing=[self_healing(:,1) self_healing(:,(2+healing):(2*h
ealing+1))];
average_net_healing=mean(net_healing);
%***************************************************************
*******************************
```

```
    %*****************************************************************
************************************
                %*********************** PLOTTING  PLOTTING
PLOTTING ************************
    %---------------------------------------------------------------------
% mean of Iter_node_set
Iter_node_set=Iter_node_set/nTime;

nb_no_comp_reliab_set=unique(Iter_node_dataset(:,4));

No_of_quif_set=unique(Iter_node_dataset(:,5));    %  AMIEAMIR


%size(nb_no_comp_reliab_set)
% nb_no_comp_reliab_set(:,1): the number of neighbor nodes
% column number 4 contains the number of neighbor for each
sensor i while
% the (nb_no_comp_reliab_set) contains the same values as in
(Iter_node_dataset(:,4)) but
% with no repetitions (unique). it will also be sorted
nb_no_comp_reliab_set=[nb_no_comp_reliab_set,zeros(length(nb_
no_comp_reliab_set),10)];


No_of_quif_set=[No_of_quif_set,zeros(length(No_of_quif_
set),10)];    %  AMIEAMIR

% the above line to make "nb_no_comp_reliab_set" has a 11
columns


for temp_nb_no = 1:length(nb_no_comp_reliab_set)
    %if nb_no_comp_reliab_set(1,1)==0
    %    temp_nb_no=temp_nb_no-1;
     %  iiiii=1;
    %else
        iiiii=0;
    %end
    nb_no_comp_reliab_set(temp_nb_no+iiiii,2)=
mean(Iter_node_dataset(find(Iter_node_dataset(:,4)==temp_nb_
no),8));
    % nb_no_comp_reliab_set(:,2): mean_Pr_BSe_comp.
Proposed scheme, tau=1
    % 8th  Col, node(:,8): Pr_BSe_comp_proposed_eq1, tau=1,
the BSe of s_i to be compromised

    nb_no_comp_reliab_set(temp_nb_no+iiiii,3)=
mean(Iter_node_dataset(find(Iter_node_dataset(:,4)==temp_nb_
no),12));
```

```
      % nb_no_comp_reliab_set(:,3): mean_Pr_reliab.
Proposed scheme, tau=1
      % 12th Col, node(:,12): Pr_reliab_proposed_eq1, tau=1,
data reliability

      nb_no_comp_reliab_set(temp_nb_no+iiiii,4)=
mean(Iter_node_dataset(find(Iter_node_dataset(:,4)==temp_nb_
no),9));
      % nb_no_comp_reliab_set(:,4): mean_Pr_BSe_comp.
Proposed scheme, tau>1
      % 9th  Col, node(:,9): Pr_BSe_comp_proposed_less1, tau>1,
the BSe of s_i to be compromised

      nb_no_comp_reliab_set(temp_nb_no+iiiii,5)=
mean(Iter_node_dataset(find(Iter_node_dataset(:,4)==temp_nb_
no),13));
      % nb_no_comp_reliab_set(:,5): mean_Pr_reliab.
Proposed scheme, tau>1
      % 13th Col, node(:,13): Pr_reliab_proposed_less1, tau>1,
data reliability

      nb_no_comp_reliab_set(temp_nb_no+iiiii,6)=
mean(Iter_node_dataset(find(Iter_node_dataset(:,4)==temp_nb_
no),7));
      % nb_no_comp_reliab_set(:,6): mean_Pr_BSe_comp.
wang, tau=1
      % 7th  Col, node(:,7):   Pr_BSe_comp_wang_eq1, tau=1, the
BSe of si

      nb_no_comp_reliab_set(temp_nb_no+iiiii,7)=
mean(Iter_node_dataset(find(Iter_node_dataset(:,4)==temp_nb_
no),11));
      % nb_no_comp_reliab_set(:,7): mean_Pr_reliab.
wang, tau=1
      % 11th Col, node(:,11): Pr_reliab_wang_eq1, tau=1, data
reliability

      nb_no_comp_reliab_set(temp_nb_no+iiiii,8)=
mean(Iter_node_dataset(find(Iter_node_dataset(:,4)==temp_nb_
no),14));
      % nb_no_comp_reliab_set(:,8): mean_Pr_BSe_comp.
wang, tau>1
      % 14th Col, node(:,14): Pr_BSe_comp_wang_less1, tau>1,
the BSe of si

      nb_no_comp_reliab_set(temp_nb_no+iiiii,9)=
mean(Iter_node_dataset(find(Iter_node_dataset(:,4)==temp_nb_
no),15));
```

```
      % nb_no_comp_reliab_set(:,9): mean_Pr_reliab.
wang, tau>1
      % 15th Col, node(:,15): Pr_reliab_wang_less1, tau>1, data
reliability

      nb_no_comp_reliab_set(temp_nb_no+iiiii,10)=
mean(Iter_node_dataset(find(Iter_node_dataset(:,4)==temp_nb_
no),6));
      % nb_no_comp_reliab_set(:,10):mean_Pr_BSe_comp.
naive
      % 6th  Col,  node(:,6):  Pr_BSe_comp_naive, the same as
node(:,3)

      nb_no_comp_reliab_set(temp_nb_no+iiiii,11)=
mean(Iter_node_dataset(find(Iter_node_dataset(:,4)==temp_nb_
no),10));
      % nb_no_comp_reliab_set(:,11):mean_Pr_reliab.
naive
      % 10th Col, node(:,10): Pr_reliab_naive, data reliability

end




%AMIRAMIRAMIRAMIRAMIRAMIRAMIRAMIRAMIRAMIRAMIRAMIRAMIRAMIRAMIR
AMIRAMIRAMIR
%AMIRAMIRAMIRAMIRAMIRAMIRAMIRAMIRAMIRAMIRAMIRAMIRAMIRAMIRAMIR
AMIRAMIRAMIR
%AMIRAMIRAMIRAMIRAMIRAMIRAMIRAMIRAMIRAMIRAMIRAMIRAMIRAMIRAMIR
AMIRAMIRAMIR
for temp_qf_no = 1:length(No_of_quif_set)
   % if No_of_quif_set(1,1)==0
    %     temp_qf_no=temp_qf_no-1;
    %    iii=1;
   % else
        iii=0;
    %end
        temp_qf_no;
        temp_qf_no+iii;
    No_of_quif_set(temp_qf_no+iii,2)=
mean(Iter_node_dataset(find(Iter_node_dataset(:,5)==temp_qf_
no),8));
      % No_of_quif_set(:,2): mean_Pr_BSe_comp.        Proposed
scheme, tau=1
      % 8th  Col, node(:,8): Pr_BSe_comp_proposed_eq1, tau=1,
the BSe of s_i to be compromised
```

```
     No_of_quif_set(temp_qf_no+iii,3)=
mean(Iter_node_dataset(find(Iter_node_dataset(:,5)==temp_qf_
no),12));
     % No_of_quif_set(:,3): mean_Pr_reliab.
Proposed scheme, tau=1
     % 12th Col, node(:,12): Pr_reliab_proposed_eq1, tau=1,
data reliability

     No_of_quif_set(temp_qf_no+iii,4)=
mean(Iter_node_dataset(find(Iter_node_dataset(:,5)==temp_qf_
no),9));
     % No_of_quif_set(:,4): mean_Pr_BSe_comp.         Proposed
scheme, tau>1
     % 9th  Col, node(:,9): Pr_BSe_comp_proposed_less1, tau>1,
the BSe of s_i to be compromised

     No_of_quif_set(temp_qf_no+iii,5)=
mean(Iter_node_dataset(find(Iter_node_dataset(:,5)==temp_qf_
no),13));
     % No_of_quif_set(:,5): mean_Pr_reliab.
Proposed scheme, tau>1
     % 13th Col, node(:,13): Pr_reliab_proposed_less1, tau>1,
data reliability

     No_of_quif_set(temp_qf_no+iii,6)=
mean(Iter_node_dataset(find(Iter_node_dataset(:,5)==temp_qf_
no),7));
     % No_of_quif_set(:,6): mean_Pr_BSe_comp.           wang,
tau=1
     % 7th  Col, node(:,7):   Pr_BSe_comp_wang_eq1, tau=1, the
BSe of si

     No_of_quif_set(temp_qf_no+iii,7)=
mean(Iter_node_dataset(find(Iter_node_dataset(:,5)==temp_qf_
no),11));
     % No_of_quif_set(:,7): mean_Pr_reliab.             wang,
tau=1
     % 11th Col, node(:,11): Pr_reliab_wang_eq1, tau=1, data
reliability

     No_of_quif_set(temp_qf_no+iii,8)=
mean(Iter_node_dataset(find(Iter_node_dataset(:,5)==temp_qf_
no),14));
     % No_of_quif_set(:,8): mean_Pr_BSe_comp.           wang,
tau>1
     % 14th Col, node(:,14): Pr_BSe_comp_wang_less1, tau>1,
the BSe of si
```

```
    No_of_quif_set(temp_qf_no+iii,9)=
mean(Iter_node_dataset(find(Iter_node_dataset(:,5)==temp_qf_
no),15));
    % No_of_quif_set(:,9): mean_Pr_reliab.
wang, tau>1
    % 15th Col, node(:,15): Pr_reliab_wang_less1, tau>1, data
reliability

    No_of_quif_set(temp_qf_no+iii,10)=
mean(Iter_node_dataset(find(Iter_node_dataset(:,5)==temp_qf_
no),6));
    % No_of_quif_set(:,10):mean_Pr_BSe_comp.         naive
    % 6th  Col, node(:,6): Pr_BSe_comp_naive, the same as
node(:,3)

    No_of_quif_set(temp_qf_no+iii,11)=
mean(Iter_node_dataset(find(Iter_node_dataset(:,5)==temp_qf_
no),10));
    % No_of_quif_set(:,11):mean_Pr_reliab.             naive
    % 10th Col, node(:,10): Pr_reliab_naive, data reliability

end
%AMIRAMIRAMIRAMIRAMIRAMIRAMIRAMIRAMIRAMIRAMIRAMIRAMIRAMIR
AMIRAMIRAMIR
%AMIRAMIRAMIRAMIRAMIRAMIRAMIRAMIRAMIRAMIRAMIRAMIRAMIRAMIR
AMIRAMIRAMIR
%AMIRAMIRAMIRAMIRAMIRAMIRAMIRAMIRAMIRAMIRAMIRAMIRAMIRAMIR
AMIRAMIRAMIR




%$$$$$$$$$$$$$$$$$$$$$$$$$$$$$$$$$$$$$$$$$$$$$$$$$$$$$$$$$$$$$
$$$$$$$$$$$$$$$$$$$$$$$$$$$$$$$$$$$$
        %$$$$$$$$$$$$$$$$$$$$$$$$$$$$$$$$$$$$$$$$$ THE PLOTTING
$$$$$$$$$$$$$$$$$$$$$$$$$$$$$$$$$$$$

        %-------------------------------------------------
-------------------%

        %----------------- Pr_BSe_comp,  tau
=1----------------%
```

```
subplot(2,3,1);plot(Iter_node_set(:,6),'ro-
','MarkerSize',4);hold on;   % naive scheme
subplot(2,3,1);plot(Iter_node_set(:,7),'gd-
','MarkerSize',4);hold on;   % Wang's scheme - randomly select
subplot(2,3,1);plot(Iter_node_set(:,8),'bx-','MarkerSize',4);
% Ours - optimized select
title('(A) \tau=1');xlabel('Index of nodes');ylabel('Prob. of
BSe to be compromised');
legend('\fontsize{7} Naive scheme','\fontsize {7} Wang \it{et al.}
[10]','\fontsize {7} Ours','Location','NorthEast');

                %----------------- Pr_BSe_comp,  tau
>1---------------%
subplot(2,3,4);plot(Iter_node_set(:,6),'ro-
','MarkerSize',4);hold on;      % naive scheme
subplot(2,3,4);plot(Iter_node_set(:,14),'gd-
','MarkerSize',4);hold on;   % Wang's scheme - randomly select
subplot(2,3,4);plot(Iter_node_set(:,9),'bx-','MarkerSize',4);
% proposed scheme - optimized select
title('(D) \tau<1');xlabel('Index of nodes');ylabel('Prob. of
BSe to be compromised');
legend('\fontsize{7} Naive scheme','\fontsize {7} Wang \it{et
al.}  [10]','\fontsize {7} Ours','Location','NorthEast');

                %----------------- Pr_reliab, tau =
1---------------%
subplot(2,3,2);plot(Iter_node_set(:,10),'ro-
','MarkerSize',4);hold on;   % naive scheme
subplot(2,3,2);plot(Iter_node_set(:,11),'gd-
','MarkerSize',4);hold on;     % Wang's scheme - randomly
select
subplot(2,3,2);plot(Iter_node_set(:,12),'bx-','MarkerSize',4);
% proposed scheme - optimized select
title('(B) \tau=1');xlabel('Index of nodes');ylabel('Prob. of
data reliability');
legend('\fontsize{7} Naive scheme','\fontsize {7} Wang \it{et al.}
[10]','\fontsize {7} Ours','Location','NorthEast');

                %----------------- Pr_reliab, tau >
1---------------%
subplot(2,3,5);plot(Iter_node_set(:,10),'ro-',
'MarkerSize',4);hold on;     % naive scheme
subplot(2,3,5);plot(Iter_node_set(:,15),'gd-',
'MarkerSize',4);hold on;     % Wang's scheme - randomly select
subplot(2,3,5);plot(Iter_node_set(:,13),'bx-','MarkerSize',4);
% proposed scheme - optimized select
title('(E) \tau<1');xlabel('Index of nodes');ylabel('Prob. of
data reliability');
legend('\fontsize{7} Naive scheme','\fontsize {7} Wang \it{et al.}
[10]','\fontsize {7} Ours','Location','NorthEast');
```

```
            %-------------- Pr_BSe_comp compared with mean
neighbor node--------------%
subplot(2,3,3);plot(nb_no_comp_reliab_set(1:15,10),'ro-',
'MarkerSize',4);hold on;  % naive scheme
subplot(2,3,3);plot(nb_no_comp_reliab_set(1:15,6),'gd-
','MarkerSize',4);hold on;  % Wang's scheme randomly select,
tau=1
subplot(2,3,3);plot(nb_no_comp_reliab_set(1:15,8),'g.-
','MarkerSize',4);hold on;% Wang's scheme randomly select,
tau>1
subplot(2,3,3);plot(nb_no_comp_reliab_set(1:15,2),'bx-
','MarkerSize',4);hold on;% proposed scheme - optimized
select, \tau=1
subplot(2,3,3);plot(nb_no_comp_reliab_set(1:15,4),'b+-
','MarkerSize',4);
title('(C)');% proposed scheme - optimized select, \tau<1
xlabel('The number of neighbor nodes');ylabel('Prob. of BSe to
be compromised');
legend('\fontsize{7} Naive scheme','\fontsize {7} Wang \it{et al.}
[10],\tau=1',...
     '\fontsize {7} Wang \it{et al.}  [10],\tau<1','\fontsize
{7} Ours, \tau=1','\fontsize {7} Ours, \
tau<1','Location','NorthEast');


            %-------------- Pr_reliab compared with mean
neighbor node --------------%

subplot(2,3,6);plot(nb_no_comp_reliab_set(1:15,11),'ro-',
'MarkerSize',4);hold on;% naive scheme
subplot(2,3,6);plot(nb_no_comp_reliab_set(1:15,7),'gd-',
'MarkerSize',4);hold on;% Wang's Infocom 09 scheme - randomly
select, \tau=1
subplot(2,3,6);plot(nb_no_comp_reliab_set(1:15,9),'g.-',
'MarkerSize',4);hold on;% Wang's Infocom 09 scheme - randomly
select, \tau<1
subplot(2,3,6);plot(nb_no_comp_reliab_set(1:15,3),'bx-',
'MarkerSize',4);hold on;% proposed scheme - optimized select,
\tau=1
subplot(2,3,6);plot(nb_no_comp_reliab_set(1:15,5),'b+-',
'MarkerSize',4);% proposed scheme - optimized select, \tau<1
title('(F)');xlabel('The number of neighbor
nodes');ylabel('Prob. of data reliability');
legend('\fontsize{7} Naive scheme','\fontsize {7} Wang \it{et al.}
[10],\tau=1',...
     '\fontsize {7} Wang \it{et al.}  [10],\tau<1','\fontsize
{7} Ours, \tau=1','\fontsize {7} Ours, \
tau<1','Location','NorthEast');


if adv_TR==30
```

```
     title({['(F)'],['n=',num2str(node_no),'
itr=',num2str(nTime),'    spd=',num2str(adv_vol),'
rang=',num2str(adv_TR)]})
elseif adv_TR==60
     title({['(F)'],['n=',num2str(node_no),'
itr=',num2str(nTime),'    spd=',num2str(adv_vol),'
rang=',num2str(adv_TR)]})
elseif adv_TR==90
     title({['(F)'],['n=',num2str(node_no),'
itr=',num2str(nTime),'    spd=',num2str(adv_vol),'
rang=',num2str(adv_TR)]})
else
     title({['(F)'],['n=',num2str(node_no),'
itr=',num2str(nTime),'    spd=',num2str(adv_vol),'
rang=',num2str(adv_TR)]})
end




%AMIRAMIRAMIRAMIRAMIRAMIRAMIRAMIRAMIRAMIRAMIRAMIRAMIRAMIR
AMIRAMIRAMIR
%AMIRAMIRAMIRAMIRAMIRAMIRAMIRAMIRAMIRAMIRAMIRAMIRAMIRAMIR
AMIRAMIRAMIR
%AMIRAMIRAMIRAMIRAMIRAMIRAMIRAMIRAMIRAMIRAMIRAMIRAMIRAMIR
AMIRAMIRAMIR
               %--------------- Pr_BSe_comp compared with mean
neighbor node--------------%
figure
subplot(1,2,1);plot(No_of_quif_set(1:length(No_of_quif_
set(:,10)),10),'ro-','MarkerSize',4);hold on;   % naive scheme
subplot(1,2,1);plot(No_of_quif_set(1:length(No_of_quif_
set(:,6)),6),'gd-','MarkerSize',4);hold on;   % Wang's scheme
randomly select, tau=1
subplot(1,2,1);plot(No_of_quif_set(1:length(No_of_quif_
set(:,8)),8),'g.-','MarkerSize',4);hold on;% Wang's scheme
randomly select, tau>1
subplot(1,2,1);plot(No_of_quif_set(1:length(No_of_quif_
set(:,2)),2),'bx-','MarkerSize',4);hold on;% proposed scheme
- optimized select, \tau=1
subplot(1,2,1);plot(No_of_quif_set(1:length(No_of_quif_
set(:,4)),4),'b+-','MarkerSize',4);
```

```
title('(A)');% proposed scheme - optimized select, \tau<1
xlabel('The number of qualified nodes');ylabel('Prob. of BSe
to be compromised');
legend('\fontsize{7} Naive scheme','\fontsize {7} Wang \it{et al.}
[10],\tau=1',...
    '\fontsize {7} Wang \it{et al.}  [10],\tau<1','\fontsize
{7} Ours, \tau=1','\fontsize {7} Ours, \
tau<1','Location','NorthEast');


                %--------------- Pr_reliab compared with mean
neighbor node ---------------%

subplot(1,2,2);plot(No_of_quif_set(1:length(No_of_quif_
set(:,11)),11),'ro-','MarkerSize',4);hold on;% naive scheme
subplot(1,2,2);plot(No_of_quif_set(1:length(No_of_quif_
set(:,7)),7),'gd-','MarkerSize',4);hold on;% Wang's Infocom 09
scheme - randomly select, \tau=1
subplot(1,2,2);plot(No_of_quif_set(1:length(No_of_quif_
set(:,9)),9),'g.-','MarkerSize',4);hold on;% Wang's Infocom 09
scheme - randomly select, \tau<1
subplot(1,2,2);plot(No_of_quif_set(1:length(No_of_quif_
set(:,3)),3),'mx-','MarkerSize',4);hold on;% proposed scheme
- optimized select, \tau=1
subplot(1,2,2);plot(No_of_quif_set(1:length(No_of_quif_
set(:,5)),5),'b+-','MarkerSize',4);% proposed scheme -
optimized select, \tau<1
title('(B)');xlabel('The number of qualified nodes');
ylabel('Prob. of data reliability');
legend('\fontsize{7} Naive scheme','\fontsize {7} Wang \it{et al.}
[10],\tau=1',...
    '\fontsize {7} Wang \it{et al.}  [10],\tau<1','\fontsize
{7} Ours, \tau=1','\fontsize {7} Ours, \
tau<1','Location','NorthEast');

figure
plot(1:healing+1,average_net_healing/node_no)

%AMIRAMIRAMIRAMIRAMIRAMIRAMIRAMIRAMIRAMIRAMIRAMIRAMIRAMIR
AMIRAMIRAMIR
%AMIRAMIRAMIRAMIRAMIRAMIRAMIRAMIRAMIRAMIRAMIRAMIRAMIRAMIR
AMIRAMIRAMIR
%AMIRAMIRAMIRAMIRAMIRAMIRAMIRAMIRAMIRAMIRAMIRAMIRAMIRAMIR
AMIRAMIRAMIR
```

```
if adv_TR==30
    save adv_tr30
    title({['(F)'],['n=',num2str(node_no),'
itr=',num2str(nTime),'   rounds   ',num2str(heal_rnds),'   ADV
spd=',num2str(adv_vol),'   ADV range=',num2str(adv_TR),'
sensor rang=',num2str(sensor_TR),'   sensor vol
',num2str(sensor_vol)]})
elseif adv_TR==60
    save
adv_TR60_100it_50healing_round_3step_random_motion_for_each_
sensor_increment_&_decrement_in_location
    title({['(F)'],['n=',num2str(node_no),'
itr=',num2str(nTime),'   rounds   ',num2str(heal_rnds),'   ADV
spd=',num2str(adv_vol),'   ADV range=',num2str(adv_TR),'
sensor rang=',num2str(sensor_TR),'   sensor vol
',num2str(sensor_vol)]})
elseif adv_TR==90
    save adv_TR90
    title({['(F)'],['n=',num2str(node_no),'
itr=',num2str(nTime),'   rounds   ',num2str(heal_rnds),'   ADV
spd=',num2str(adv_vol),'   ADV range=',num2str(adv_TR),'
sensor rang=',num2str(sensor_TR),'   sensor vol
',num2str(sensor_vol)]})
else
    save adv_TR120
    title({['(F)'],['n=',num2str(node_no),'
itr=',num2str(nTime),'   rounds   ',num2str(heal_rnds),'   ADV
spd=',num2str(adv_vol),'   ADV range=',num2str(adv_TR),'
sensor rang=',num2str(sensor_TR),'   sensor vol
',num2str(sensor_vol)]})
end
toc

endtime=clock;
disp(['start time  =  '   num2str(start)])
disp(['end time   =  '   num2str(endtime)])
beep,beep,beep,beep,beep,beep,beep,beep,beep,beep,beep,beep,beep,
beep;




ssss=0;
for dddd=1:1000
    ssss=ssss+length(find(Iter_node_dataset(:,4)==dddd));
end
disp(['    ' num2str(ssss)])
%length(find(Iter_node_dataset(:,4)==1))+length(find(Iter_
node_dataset(:,4)==2))+length(find(Iter_node_dataset(:,4)==3))
+length(find(Iter_node_dataset(:,4)==4))+length(find(Iter_
node_dataset(:,4)==5))+length(find(Iter_node_dataset(:,4)==6))
```

```
+length(find(Iter_node_dataset(:,4)==7))+length(find(Iter_
node_dataset(:,4)==8))+length(find(Iter_node_dataset(:,4)==9))
+length(find(Iter_node_dataset(:,4)==10))+length(find(Iter_
node_dataset(:,4)==11))+length(find(Iter_node_dataset(:,4)==12
))+length(find(Iter_node_dataset(:,4)==13))+length(find(I
ter_node_dataset(:,4)==14))+length(find(Iter_node_dataset(:,4)
==15))+length(find(Iter_node_dataset(:,4)==16))+length(find(I
ter_node_dataset(:,4)==17))+length(find(Iter_node_dataset(:,4)
==18))+length(find(Iter_node_dataset(:,4)==19))+length(find(I
ter_node_dataset(:,4)==20))+length(find(Iter_node_dataset(:,4)
==21))+length(find(Iter_node_dataset(:,4)==22))+length(find(I
ter_node_dataset(:,4)==23))+length(find(Iter_node_dataset(:,4)
==24))+length(find(Iter_node_dataset(:,4)==25))+length(find(I
ter_node_dataset(:,4)==26))+length(find(Iter_node_dataset(:,4)
==27))+length(find(Iter_node_dataset(:,4)==28))+length(find(I
ter_node_dataset(:,4)==29))+length(find(Iter_node_dataset(:,4)
==30))+length(find(Iter_node_dataset(:,4)==31))+length(find(I
ter_node_dataset(:,4)==32))+length(find(Iter_node_dataset(:,4)
==33))+length(find(Iter_node_dataset(:,4)==34))+length(find(I
ter_node_dataset(:,4)==35))+length(find(Iter_node_dataset(:,4)
==36))+length(find(Iter_node_dataset(:,4)==37))+length(find(I
ter_node_dataset(:,4)==38))+length(find(Iter_node_dataset(:,4)
==39))+length(find(Iter_node_dataset(:,4)==40))+length(find(I
ter_node_dataset(:,4)==41))+length(find(Iter_node_dataset(:,4)
==42))+length(find(Iter_node_dataset(:,4)==43))+length(find(I
ter_node_dataset(:,4)==44))+length(find(Iter_node_dataset(:,4)
==45))+length(find(Iter_node_dataset(:,4)==46))+length(find(I
ter_node_dataset(:,4)==47))+length(find(Iter_node_dataset(:,4)
==48))+length(find(Iter_node_dataset(:,4)==49))+length(find(I
ter_node_dataset(:,4)==50))+length(find(Iter_node_dataset(:,4)
==51))+length(find(Iter_node_dataset(:,4)==52))+length(find(I
ter_node_dataset(:,4)==53))+length(find(Iter_node_dataset(:,4)
==54))+length(find(Iter_node_dataset(:,4)==55))+length(find(I
ter_node_dataset(:,4)==56))+length(find(Iter_node_dataset(:,4)
==57))+length(find(Iter_node_dataset(:,4)==58))+length(find(I
ter_node_dataset(:,4)==59))+length(find(Iter_node_dataset(:,4)
==60))

beep;for i=1:1000000,i;end;beep;for i=1:1000000,i;end;beep;for
i=1:1000000,i;end;beep;for i=1:1000000,i;end;
beep;for i=1:1000000,i;end;beep;for i=1:1000000,i;end;beep;for
i=1:1000000,i;end;beep;for i=1:1000000,i;end;
beep;for i=1:1000000,i;end;beep;for i=1:1000000,i;end;beep;for
i=1:1000000,i;end;beep;for i=1:1000000,i;end;
beep;for i=1:1000000,i;end;beep;for i=1:1000000,i;end;beep;for
i=1:1000000,i;end;beep;for i=1:1000000,i;end;
beep;for i=1:1000000,i;end;beep;for i=1:1000000,i;end;beep;for
i=1:1000000,i;end;beep;for i=1:1000000,i;end;
```

```
%*****************************************************************
%*****************************************************************
% This function has the following assumption
% 2- the adversary is assumed to be fixed
%-------------------------------------------
% 1- the adversary can compromise the nodes in its
transmission range
%---------------------------------------------------
% This function used to give the following outputs
%---------------------------------------------------
% 1- the "x_loc" of the adversary located in column 16
% 2- the "y_loc" of the adversary located in column 17
% 3- the compromising probability of sensors according to
their positions
%     with respect to the adversary location
%-----------------------------------------
% This function take the following inputs
%-----------------------------------------
% 1- maxmumx: the maxmum distance in x direction
% 2- maxmumy: the maxmum distance in y direction
% 3- node_number: the total number of node
% 4- node: the "node" matrix which hold sensor data
% 3- adv_range: the adversary range
% 4- node_index_i: the index of the current sensor Si
% 5- adv_vol: the adversary velocity
% 6- update_t: the update time of adversary motion
function
[node]=adv_rand_loc_AS_1(maxmumx,maxmumy,node,node_number,adv_
range,round)
    node(round,16)=rand*maxmumx;  % this is the X position of
the adversary
    node(round,17)=rand*maxmumy;  % this is the y position of
the adversary

    % the direction of motion of adversary range from 0:2*pi

    for jj=1:node_number
        % this "for-loop" used to fill the third column of
matrix "node(:,3)"
        % with the compromising probability of each sensor
according to
        % its position with respect to the adversary position

adv_dist=sqrt((node(jj,1)-node(round,16))^2+(node(jj,2)-
node(round,17))^2);
        node(jj,21)=adv_dist;
        % adv_dist: is the distance between the adv location
to the location
        % of the current sensor node Si
        if  adv_dist<=adv_range
```

```
            % node(jj,3)=0.6*rand;
            node(jj,3)=max(0.6+0.4*rand,node(jj,3));      %
0.6:1
            %node(jj,3)=0.6*rand+0.4;plot(node(jj, 1),
node(jj, 2), 'k.', 'MarkerSize', 7);node(i, 1), node(i, 2)
refer to X and Y coordinate of node i  %text(node(jj,
1),node(jj, 2),num2str(jj));
        elseif (adv_dist>adv_range && adv_dist<=2*adv_range)
            node(jj,3)=max(0.4+0.2*rand,node(jj,3));      %
0.4:0.6
            % node(jj,3)=0.2+0.2*rand;plot(node(jj, 1),
node(jj, 2), 'b.', 'MarkerSize', 7);text(node(jj, 1),node(jj,
2),num2str(jj));
        elseif (adv_dist>1.5*adv_range &&
adv_dist<=3*adv_range)
            node(jj,3)=max(0.2+0.2*rand,node(jj,3));      %
0.2:0.4
            %node(jj,3)=0.2*rand;plot(node(jj, 1), node(jj,
2), 'r.', 'MarkerSize', 17);text(node(jj, 1),node(jj,
2),num2str(jj));
        else
            node(jj,3)=max(0.2*rand,node(jj,3));      % 0 : 0.2
            %node(jj,3)=0.001;plot(node(jj, 1), node(jj, 2),
'g.', 'MarkerSize', 7);text(node(jj, 1),node(jj,
2),num2str(jj));
        end
%disp('&&&&&&&&&&&&&&&&&&&&&&&&&&&&&&&&&&&&&&&&&&&&&&&&&&&
&&&&&&&');disp(['adversary- node(jj,3)=
',num2str(node(jj,3))  '    '    num2str(jj)]);d
isp('&&&&&&&&&&&&&&&&&&&&&&&&&&&&&&&&&&&&&&&&&&&&&&&&&&&&&&&
&&&&&&&')
    end

    %title({['Network Topology: ',num2str(node_number), '
sensor nodes '];[num2str(black_node), '% black nodes, '   ,
num2str(blue_node),'% blue nodes, ',num2str(red_node), '% red
nodes, ',num2str(green_node),'% green nodes.'];['with
different compromise probability'];['black='
,num2str(black_node/node_number),'%,blue=',num2str(blue_node/
node_number),'%,red=',num2str(red_node/node_number),'%,green='
,num2str(green_node/node_number),'%.']});
    %xlabel('The Horizontal X');
    %ylabel('The Vertical Y');
    %axis([0, maxmumx, 0, maxmumy]);
    %set(gca, 'XTick', [0;maxmumx]);
    %set(gca, 'YTick', [0;maxmumy]);

    % DF "gca":  Get handle to Current Axis
    % DF "'XTick', [0; maxx]": make X axis begin with 0 and
ends with maxx
```

```
    % if we write "set(gca, 'XTick', 0:20:maxx)" thus will
divide the X
    % axis as 0,20,40,60,80.........,maxx

return;



%**************************************************************
**
%**************************************************************
**
% This function has the following assumption
%---------------------------------------------
% 1- the adversary can compromise the nodes in its
transmission range
% 2- We have only one adversary, it is assumed to be moving in
a random
%    direction in each round.
%    If the adv is out at one border it will be reflected at
the same border
%----------------------------------------------------
% This function used to give the following outputs
%----------------------------------------------------
% 1- the "x_loc" of the adversary
% 2- the "y_loc" of the adversary
%---------------------------------------
% This function take the following inputs
%---------------------------------------
% 1- maxmumx: the maxmum distance in x direction
% 2- maxmumy: the maxmum distance in y direction
% 3- node_number: the total number of node
% 4- node: the "node" matrix which hold sensor data
% 3- adv_range: the transmission range of normal sensor node
% 4- round: the index of the current sensor Si
% 5- adv_vol: the adversary velocity
% 6- update_t: the update time of adversary motion
function
[node]=advr_move_AS_1(maxmumx,maxmumy,node,node_number,adv_
range,adv_vol,update_t,round)
%size(node)
%node(:,1) = node(:,1)*maxmumx;  % Here it generates the X
coordinate of the node
%node(:,2) = node(:,2)*maxmumy;  % Here it generates the Y
coordinate of the node
if   round==1
    node(round,16)=rand*maxmumx;
    node(round,17)=rand*maxmumy;
```

```
    % the direction of motion of adversary range from 0:2*pi
else
    % "else" refer to round>1
    direction=2*pi*rand;              % random direction
    d_travel=adv_vol*update_t;        % fixed amount of motion

    node(round,16)=node((round-1),16)+d_travel*cos(direction);
% updata X direction of ADV
    node(round,17)=node((round-1),17)+d_travel*sin(direction);
% updata Y direction of ADV



    if (node(round,16)<0 ||node(round,16)>maxmumx)
        %disp('#################### adv out of area   x
direction ###############################')
        %disp(['old direction  ' num2str(direction)  '
before  ' num2str(node(round,16)) ])
        %disp([' previous location '
num2str(node(round-1,16))  ' d_travel ' num2str(d_travel)])
        direction=direction+pi;

node(round,16)=node((round-1),16)+d_travel*cos(direction);
        %disp(['new direction  ' num2str(direction) '
distance  ' num2str(d_travel*cos(direction))  ' after '
num2str(node(round,16))])
        beep
        %disp('###########################################
######################################')
    end
    if (node(round,17)<0 ||node(round,17)>maxmumy)
        %disp('#################### adv out of area   y
direction ###############################')
        %disp(['old direction  ' num2str(direction)  '
before  ' num2str(node(round,17)) ])
        %disp([' previous location '
num2str(node(round-1,17))  ' d_travel ' num2str(d_travel) ])
        direction=direction+pi;

node(round,17)=node((round-1),17)+d_travel*sin(direction);
        %disp(['new direction  ' num2str(direction)   '
distance  ' num2str(d_travel*sin(direction))  ' after '
num2str(node(round,17))])
        beep
        %disp('###########################################
######################################')
    end
end
```

```
      for jj=1:node_number

adv_dist=sqrt((node(jj,1)-node(round,16))^2+(node(jj,2)-
node(round,17))^2);
          node(jj,21)=adv_dist;

          % adv_dist: is the distance between the adv location
to the location
          % of the current sensor node Si
          if  adv_dist<=adv_range
              node(jj,3)=max(0.6+0.4*rand,node(jj,3));     %    0.6:1
              %plot(node(jj, 1), node(jj, 2), 'k.',
'MarkerSize', 7);
              % node(i, 1), node(i, 2) refer to X and Y
coordinate of node i
              %text(node(jj, 1),node(jj, 2),num2str(jj));
          elseif (adv_dist>adv_range && adv_dist<=2*adv_range)
              node(jj,3)=max(0.4+0.2*rand,node(jj,3));     %
0.4:0.6
              %plot(node(jj, 1), node(jj, 2), 'b.',
'MarkerSize', 7);
              %text(node(jj, 1),node(jj, 2),num2str(jj));
          elseif (adv_dist>2*adv_range &&
adv_dist<=3*adv_range)
              node(jj,3)=max(0.2+0.2*rand,node(jj,3));     %
0.2:0.4
              %plot(node(jj, 1), node(jj, 2), 'r.',
'MarkerSize', 20);
              %text(node(jj, 1),node(jj, 2),num2str(jj));
          else
              node(jj,3)=max(0.2*rand,node(jj,3));     % 0 :
0.2
              %plot(node(jj, 1), node(jj, 2), 'g.',
'MarkerSize', 7);
              %text(node(jj, 1),node(jj, 2),num2str(jj));
          end
      end
      %black_node=length(find(node(:,3)>=0.6));
      %blue_node=length(find((node(:,3)>=0.4 &
node(:,3)<0.6)));
      %red_node=length(find((node(:,3)>=0.2 & node(:,3)<0.4)));
      %green_node=length(find(node(:,3)<0.2));

      %title({['Network Topology: ',num2str(node_number), '
sensor nodes '];[num2str(black_node), '% black nodes, '
,num2str(blue_node),'% blue nodes, ',num2str(red_node), '% red
nodes, ',num2str(green_node),'% green nodes.'];['with
different compromise probability'];['black='
,num2str(black_node/node_number),'%,blue=',num2str(blue_node/
```

```
node_number),'%,red=',num2str(red_node/node_number),'%,green='
,num2str(green_node/node_number),'%.']});
      %xlabel('The Horizontal X');
      %ylabel('The Vertical Y');
      %axis([0, maxmumx, 0, maxmumy]);
      %set(gca, 'XTick', [0;maxmumx]);
      %set(gca, 'YTick', [0;maxmumy]);

      % DF "gca":  Get handle to Current Axis
      % DF "'XTick', [0; maxx]": make X axis begin with 0 and
ends with maxx
      % if we write "set(gca, 'XTick', 0:20:maxx)" thus will
divide the X
      % axis as 0,20,40,60,80........,maxx

return;




%****************************************************************
**
%AAAAAAAAAAAAAAAAAAAAAAAAAAAAAAAAAAAAAAAAAAAAAAAAAAAAAAAAAAAAAAAA
**
% This function has the following assumption
%-------------------------------------------
% 1- the adversary can compromise the nodes in its
transmission range
% 2- We have only adversary, it is assumed to be moving in a
random
%    direction in each round.
%    If the adv is out at one border it will be folded at the
other border side
%--------------------------------------------------
% This function used to give the following outputs
%--------------------------------------------------
% 1- the "x_loc" of the adversary
% 2- the "y_loc" of the adversary
%-----------------------------------------
% This function take the following inputs
%-----------------------------------------
% 1- maxmumx: the maxmum distance in x direction
% 2- maxmumy: the maxmum distance in y direction
% 3- node_number: the total number of node
% 4- node: the "node" matrix which hold sensor data
% 3- adv_range: the transmission range of normal sensor node
% 4- round: the index of the current sensor Si
% 5- adv_vol: the adversary velocity
% 6- update_t: the update time of adversary motion
```

```
function
[node]=advr_move_fold_AS_1(maxmumx,maxmumy,node,node_
number,adv_range,adv_vol,update_t,round)
%size(node)
%node(:,1) = node(:,1)*maxmumx;  % Here it generates the X
coordinate of the node
%node(:,2) = node(:,2)*maxmumy;  % Here it generates the Y
coordinate of the node
if  round==1
    node(round,16)=rand*maxmumx
    node(round,17)=rand*maxmumy
    % the direction of motion of adversary range from 0:2*pi
else
    % "else" refer to round>1
    direction=2*pi*rand;          % random direction
    d_travel=adv_vol*update_t;    % fixed amount of motion

    node(round,16)=node((round-1),16)+d_travel*cos(direction);
% updata X direction of ADV
    node(round,17)=node((round-1),17)+d_travel*sin(direction);
% updata Y direction of ADV



    if (n3ode(round,16)<0)
        %disp('#################### adv out of area   x
direction ###############################')
        %disp(['old direction  ' num2str(direction)  '
before  ' num2str(node(round,16)) ])
        %disp([' previous location '
num2str(node(round-1,16))  ' d_travel ' num2str(d_travel)])
        node(round,16)=node(round,16)+maxmumx;
        %disp(['new direction  ' num2str(direction) '
distance  ' num2str(d_travel*cos(direction))  ' after '
num2str(node(round,16))])
        beep
        %disp('##########################################
#####################################')
    elseif (node(round,16)>maxmumx)
        node(round,16)=node(round,16)-maxmumx;
    end
    if (node(round,17)<0)
        %disp('#################### adv out of area   y
direction ###############################')
        %disp(['old direction  ' num2str(direction)  '
before  ' num2str(node(round,17)) ])
        %disp([' previous location '
num2str(node(round-1,17))  ' d_travel ' num2str(d_travel) ])
        node(round,17)=node(round,17)+maxmumy;
```

```
        %disp(['new direction  ' num2str(direction)   '
distance  ' num2str(d_travel*sin(direction))  ' after '
num2str(node(round,17))])
        beep
        %disp('###########################################
######################################')
    elseif (node(round,17)>maxmumy)
        node(round,17)=node(round,17)-maxmumy;
    end
end



    for jj=1:node_number

adv_dist=sqrt((node(jj,1)-node(round,16))^2+(node(jj,2)-
node(round,17))^2);
        node(jj,21)=adv_dist;

        % adv_dist: is the distance between the adv location
to the location
        % of the current sensor node Si
        if  adv_dist<=adv_range
            node(jj,3)=max(0.6+0.4*rand,node(jj,3));      %   0.6:1
            %plot(node(jj, 1), node(jj, 2), 'k.',
'MarkerSize', 7);
            % node(i, 1), node(i, 2) refer to X and Y
coordinate of node i
            %text(node(jj, 1),node(jj, 2),num2str(jj));
        elseif (adv_dist>adv_range && adv_dist<=2*adv_range)
            node(jj,3)=max(0.4+0.2*rand,node(jj,3));      %
0.4:0.6
            %plot(node(jj, 1), node(jj, 2), 'b.',
'MarkerSize', 7);
            %text(node(jj, 1),node(jj, 2),num2str(jj));
        elseif (adv_dist>2*adv_range &&
adv_dist<=3*adv_range)
            node(jj,3)=max(0.2+0.2*rand,node(jj,3));      %
0.2:0.4
            %plot(node(jj, 1), node(jj, 2), 'r.',
'MarkerSize', 20);
            %text(node(jj, 1),node(jj, 2),num2str(jj));
        else
            node(jj,3)=max(0.2*rand,node(jj,3));       % 0 : 0.2
            %plot(node(jj, 1), node(jj, 2), 'g.',
'MarkerSize', 7);
            %text(node(jj, 1),node(jj, 2),num2str(jj));
        end
    end
```

```
    %black_node=length(find(node(:,3)>=0.6));
    %blue_node=length(find((node(:,3)>=0.4 &
node(:,3)<0.6)));
    %red_node=length(find((node(:,3)>=0.2 & node(:,3)<0.4)));
    %green_node=length(find(node(:,3)<0.2));

    %title({['Network Topology: ',num2str(node_number), '
sensor nodes '];[num2str(black_node), '% black nodes, '
,num2str(blue_node),'% blue nodes, ',num2str(red_node), '% red
nodes, ',num2str(green_node),'% green nodes.'];['with
different compromise probability'];['black='
,num2str(black_node/node_number),'%,blue=',num2str(blue_node/
node_number),'%,red=',num2str(red_node/node_number),'%,green='
,num2str(green_node/node_number),'%.']});
    %xlabel('The Horizontal X');
    %ylabel('The Vertical Y');
    %axis([0, maxmumx, 0, maxmumy]);
    %set(gca, 'XTick', [0;maxmumx]);
    %set(gca, 'YTick', [0;maxmumy]);

    % DF "gca":  Get handle to Current Axis
    % DF "'XTick', [0; maxx]": make X axis begin with 0 and
ends with maxx
    % if we write "set(gca, 'XTick', 0:20:maxx)" thus will
divide the X
    % axis as 0,20,40,60,80.........,maxx




return;

%****************************************************************
**
%****************************************************************
**
% THIS FUNCTION USED TO GIVE THE FOLLOWING After MOTION:
% this function is identical to the function
"health_neighbors_before"
% but they differ in the output columns
%@@@@@@@@@@@@@@@@@@@@@@@@@@@@@@@@@@@@@@@@@@@@@@@@@@@@@@@@@@@@@@@@
% 1- the total numbers of the neighbours lies in the range of
each sick
%     sensor "neighbors_in_range" and store it in the matrix
node
%     in column 26 "node(:,26)"
%@@@@@@@@@@@@@@@@@@@@@@@@@@@@@@@@@@@@@@@@@@@@@@@@@@@@@@@@@@@@@@@@@@@@@@
@@@@@@@@@@@@@@
% 2- the total numbers of the healthy neighbours lies in the
range of each
```

```
%      sick sensor "healthy_neighbors", the healthy neighbors
are selected
%      from the above computed total numbers of neighbors.
%      The healthy neighbors stored in the matrix node in column
27 "node(:,27)"
%
%
%
%
function [node]=health_neighbors_after(node)
distances=zeros(500,500);
        sick_nodes=find(node(:,3)>=0.3);
        for ii=sick_nodes'
            ii;

            xx=(node(ii,1)-node(:,1));
            yy=(node(ii,2)-node(:,2));
            distances(ii,:)=sqrt(xx.^2+yy.^2);
            neighbors_in_range=(find(distances(ii,:)<=60&dista
nces(ii,:)>0));
            node(ii,26)=length(neighbors_in_range);
            %"neighbors_in_range" refer to the total number of
the neighbors in the
            %range of the sick sensor ii
            healthy_neighbors=length(find(node(neighb
ors_in_range,3)<.3));
            node(ii,27)=healthy_neighbors;
            %    disp(['ii ', num2str(ii),'   neighbors_in_
range  ', num2str(length(neighbors_in_range)), ' healthy_
neighbors  ', num2str(healthy_neighbors),'   ',
num2str(node(ii,26))])
            %"healthy_neighbors" refer to the healthy
neighbors exist in
            %the range of the sick sensor.
            %"healthy_neighbors" are selected from
"neighbors_in_range"

        end

end


%*****************************************************************
**
%*****************************************************************
**
% THIS FUNCTION USED TO GIVE THE FOLLOWING BEFORE MOTION:
% this function is identical to the function
"health_neighbors_after"
```

```
% but they differ in the output columns
%@@@@@@@@@@@@@@@@@@@@@@@@@@@@@@@@@@@@@@@@@@@@@@@@@@@@@@@@@@@@@@@
% 1- the total numbers of the neighbours lies in the range of
each sick
%      sensor "neighbors_in_range" and store it in the matrix
node
%      in column 24 "node(:,24)"
%@@@@@@@@@@@@@@@@@@@@@@@@@@@@@@@@@@@@@@@@@@@@@@@@@@@@@@@@@@@@@@@@@@
@@@@@@@@@@@@@@
% 2- the total numbers of the healthy neighbours lies in the
range of each
%      sick sensor "healthy_neighbors", the healthy neighbors
are selected
%      from the above computed total numbers of neighbors.
%      The healthy neighbors stored in the matrix node in column
25 "node(:,25)"
%
%
%
%


function [node]=health_neighbors_before(node)
%distances=zeros(500,500);

        sick_nodes=find(node(:,3)>=0.3);
        for ii=sick_nodes'
             ii;

             xx=(node(ii,1)-node(:,1));
             % xx is the distance in X direction between sick
node ii and all
             % sensor in the network

             yy=(node(ii,2)-node(:,2));
             % yy is the distance in Y direction between sick
node ii and all
             % sensor in the network

             distances(ii,:)=sqrt(xx.^2+yy.^2);
             ii;
             neighbors_in_range=(find(distances(ii,:)<=60&dista
nces(ii,:)>0));
             % give the index "ID" of all neighbours of sick
sensor ii

             neiiiii=length(neighbors_in_range);
             % give the number of all neighbours of sick sensor ii
```

```
            node(ii,24)=length(neighbors_in_range);
            % put the number of neighbours of sick sensor ii
in the matrix
            % node in column 24

            heeeeee=(find(node(neighbors_in_range,3)<.3));
            % give the index of nodes with respect to the
matrix "neighbors_in_range"
            % with compromising probability <0.3

            heeepeee=neighbors_in_range(heeeeee);
            % give the index of the health nodes

            asasasas=node(heeepeee,3);
            % give the compromising probability of health nodes
            healthy_neighbors=length(find(node(neighb
ors_in_range,3)<.3));
            node(ii,25)=healthy_neighbors;
            %   disp(['ii ', num2str(ii),'   neighbors_in_
range ', num2str(length(neighbors_in_range)), ' healthy_
neighbors ', num2str(healthy_neighbors),'   ',
num2str(node(ii,24))])
            %"healthy_neighbors" refer to the healthy
neighbors exist in
            %the range of the sick sensor.
            %"healthy_neighbors" are selected from
"neighbors_in_range"

      end

end


%*****************************************************************
%*****************************************************************
%   This function ""node_motion_fixed"" used to move sensors
within network
%   in different direction by fixed amount.
%   the motion will be:
    % 1- in the same direction or each sensor has its own
direction
    % 2- by the same amount
%   the function compute the following:
    % 1- the new X coordinate of the sensor after the motion
    % 2- the new Y coordinate of the sensor after the motion

% THIS FUNCTION TAKE THE FOLLOWING INPUTS
% 1- "node_will_move":    The indices of the sensors the will
move
```

```
% 2- "maxx":              The width of the sensing area in X
direction
% 3- "maxy":              The length of the sensing area in Y
direction
% 4- "velocity":         The velocity of sensor motion
% 5- "dt":               The updating time
% 6- "node":             The matrix node
% 7- "Number_of_motion_steps": The number of steps (motions)
the sensor will moves

% NOTE THAT: The compromised probability of the sensor which
is generated
            % once during the complete one Iteration and
generated
            % n_Time during the run of the program

function [node] = node_motion_fixed(node_will_move, maxx,
maxy,velocity,dt,node,Number_of_motion_steps)
distance_of_motion=velocity*dt;
% the distance of motion is the same for each moving sensor
direction_of_motion=2*pi*rand(length(node_will_move),1);
% "direction_of_motion" is different for each moving sensor

% direction_of_motion=rand*2*pi;
  % "direction_of_motion" is constant for all moving  sensors

for countr=1:1:Number_of_motion_steps

    % update the X-location of the sensor in column 1 of
matrix "node" and
    % save the moving distance in X-direction in column 22 of
matrix "node"
motion_in_Xdirc=distance_of_motion*cos(direction_of_motion)
    % "motion_in_Ydirc" this value added to all moved sensors
in X direction
node(node_will_move,22)=node(node_will_move,22)+motion_in_
Xdirc;
    node(node_will_move,1) = node(node_will_move,1)+motion_
in_Xdirc;
    out_region_in_X=find(node(:,1)>maxx)   ;
    % Check for out of network region due to moving (greater
than the maxx)
    node(out_region_in_X,1)=node(out_region_in_X,1)-maxx;
    % If it is out of region the sensor will be folded (enter
from the other side)
    out_region_in_X=find(node(:,1)<0)   ;
    % Check for out of network region due to moving (lower
than zero)
    node(out_region_in_X,1)=node(out_region_in_X,1)+maxx;
```

```
    % If it is out of region the sensor will be folded (enter
from the other side)

    % update the Y-location of the sensor in column 2 of
matrix "node" and
    % save the moving distance in X-direction in column 23 of
matrix "node"
motion_in_Ydirc=distance_of_motion*sin(direction_of_motion)
    % "motion_in_Ydirc" this value added to all moved sensors
in Y direction
node(node_will_move,23)=node(node_will_move,23)+motion_in_Ydirc;
    node(node_will_move,2) =
node(node_will_move,2)+motion_in_Ydirc;
    % Here the Y coordinate of the all sensors is updated by
the same amount
    out_region_in_Y=find(node(:,2)>maxy);
    % Check for out of network region due to moving (greater
than the maxy)
    node(out_region_in_Y,2)=node(out_region_in_Y,2)-maxy;
    % If it is out of region the sensor will be folded (enter
from the other side)
    out_region_in_Y=find(node(:,2)<0);
    % Check for out of network region due to moving (lower
than zero)
    node(out_region_in_Y,2)=node(out_region_in_Y,2)+maxy;
    % If it is out of region the sensor will be folded (enter
from the other side)
end

return;

%*****************************************************************
%*****************************************************************

%   This function ""node_motion_varied"" used to move sensors
within network
%   in different direction by a different amount.

%   the motion will be:
    % 1- in the same direction or each sensor has its own
direction
    % 2- by the same amount
%   the function compute the following:
    % 1- the new X coordinate of the sensor after the motion
    % 2- the new Y coordinate of the sensor after the motion

% THIS FUNCTION TAKE THE FOLLOWING INPUTS
% 1- "node_will_move":   The indices of the sensors the will
move
```

```
% 2- "maxx":              The width of the sensing area in X
direction
% 3- "maxy":              The length of the sensing area in Y
direction
% 4- "velocity":         The velocity of sensor motion
% 5- "dt":               The updating time
% 6- "node":             The matrix node
% 7- "Number_of_motion_steps": The number of steps (motions)
the sensor will moves

% NOTE THAT: The compromised probability of the sensor which
is generated
            % once during the complete one Iteration and
generated
            % n_Time during the run of the program

function [node] = node_motion_varied(node_will_move, maxx,
maxy,velocity,dt,node,Number_of_motion_steps)
distance_of_motion=velocity*dt*rand(length(node_will_move),1);
% the distance of motion is varied for each moving sensor
direction_of_motion=2*pi*rand(length(node_will_move),1);
% "direction_of_motion" is different for each moving sensor

% direction_of_motion=rand*2*pi;
  % "direction_of_motion" is constant for all moving  sensors

for countr=1:1:Number_of_motion_steps

    % update the X-location of the sensor in column 1 of
matrix "node" and
    % save the moving distance in X-direction in column 22 of
matrix "node"
motion_in_Xdirc=distance_of_motion.*cos(direction_of_motion);
    % "motion_in_Ydirc" this value added to all moved sensors
in X direction
node(node_will_move,22)=node(node_will_move,22)+motion_in_Xdirc;
    node(node_will_move,1) =
node(node_will_move,1)+motion_in_Xdirc;
    out_region_in_X=find(node(:,1)>maxx)   ;
    % Check for out of network region due to moving (greater
than the maxx)
    node(out_region_in_X,1)=node(out_region_in_X,1)-maxx;
    % If it is out of region the sensor will be folded (enter
from the other side)
    out_region_in_X=find(node(:,1)<0)   ;
    % Check for out of network region due to moving (lower
than zero)
    node(out_region_in_X,1)=node(out_region_in_X,1)+maxx;
    % If it is out of region the sensor will be folded (enter
from the other side)
```

```
    % update the Y-location of the sensor in column 2 of
matrix "node" and
    % save the moving distance in X-direction in column 23 of
matrix "node"
motion_in_Ydirc=distance_of_motion.*sin(direction_of_motion);
    % "motion_in_Ydirc" this value added to all moved sensors
in Y direction
node(node_will_move,23)=node(node_will_move,23)+motion_in_
Ydirc;
    node(node_will_move,2) =
node(node_will_move,2)+motion_in_Ydirc;
    % Here the Y coordinate of the all sensors is updated by
the same amount
    out_region_in_Y=find(node(:,2)>maxy);
    % Check for out of network region due to moving (greater
than the maxy)
    node(out_region_in_Y,2)=node(out_region_in_Y,2)-maxy;
    % If it is out of region the sensor will be folded (enter
from the other side)
    out_region_in_Y=find(node(:,2)<0);
    % Check for out of network region due to moving (lower
than zero)
    node(out_region_in_Y,2)=node(out_region_in_Y,2)+maxy;
    % If it is out of region the sensor will be folded (enter
from the other side)
end

return;


%**************************************************************
**
%**************************************************************
**
% This function ""node_motion_varied_cluster"" used to move
sensors within
% the cluster of revoked sensors in different direction by a
different amount

% The motion will be within the cluster

% In round one and after the adv leaving the healing process
is performed
% for one time within the network as in paper 1, so that the
sick sensors
% at the cluster border will be self healed

% After that we perform motion within the cluster NOT outside
the cluster,
% the direction of motion will be one of two
```

```
    % 1- motion of some of healed sensors from the cluster
border to the
        % cluster center
    % 2- motion of some of sick sensors from inside the cluster
to near
        % the cluster border this is reverse to the first
motion
    % instead of these two type of motions, we can make random
motion within
    % the cluster after the first healing process


%   the motion will be:
    % 1- in the same direction or each sensor has its own
direction
    % 2- by the same amount
%   the function compute the following:
    % 1- the new X coordinate of the sensor after the motion
    % 2- the new Y coordinate of the sensor after the motion

% THIS FUNCTION TAKE THE FOLLOWING INPUTS
% 1- "node_will_move":    The indices of the sensors the will
move
% 2- "maxx":              The width of the sensing area in X
direction
% 3- "maxy":              The length of the sensing area in Y
direction
% 4- "velocity":          The velocity of sensor motion
% 5- "dt":                The updating time
% 6- "node":              The matrix node
% 7- "Number_of_motion_steps": The nnmber of steps (motions)
the sensor will moves

% NOTE THAT: The compromised probability of the sensor which
is generated
                % once during the complete one Iteration and
generated
                % n_Time during the run of the program

function [node] = node_motion_varied_cluster(node_will_move,
maxx, maxy,velocity,dt,node,Number_of_motion_steps,round)

%node_will_move;
%Xc=node(round,16);
%Yc=node(round,17);
%xcmax=max(node(node_will_move,1));
%xcmin=min(node(node_will_move,1));
%ycmax=max(node(node_will_move,2));
%ycmin=min(node(node_will_move,2));
%Xc=(xcmax-xcmin)/2   ;     % center of the cluster
```

```
%Yc=(ycmax-ycmin)/2   ;    % center of the cluster
%Rc=((xcmax-Xc)+(ycmax-Yc))/2;




distance_of_motion=velocity*dt*rand(length(node_will_move),1);
distance_of_motion=Number_of_motion_steps*distance_of_motion
% the distance of motion is varied for each moving sensor
direction_of_motion=2*pi*rand(length(node_will_move),1);
% "direction_of_motion" is different for each moving sensor

% direction_of_motion=rand*2*pi;
  % "direction_of_motion" is constant for all moving  sensors

    % update the X-location of the sensor in column 1 of
matrix "node" and
    % save the moving distance in X-direction in column 22 of
matrix "node"
    motion_in_Xdirc=distance_of_motion.*cos(direction_of_
motion);
    % "motion_in_Ydirc" this value added to all moved sensors
in X direction
    node(node_will_move,22)=node(node_will_move,22)+motion_
in_Xdirc;
    node(node_will_move,1) =
node(node_will_move,1)+motion_in_Xdirc;
    out_region_in_X=find(node(:,1)>maxx)   ;
    % Check for out of network region due to moving (greater
than the maxx)
    node(out_region_in_X,1)=node(out_region_in_X,1)-maxx;
    % If it is out of region the sensor will be folded (enter
from the other side)
    out_region_in_X=find(node(:,1)<0)   ;
    % Check for out of network region due to moving (lower
than zero)
    node(out_region_in_X,1)=node(out_region_in_X,1)+maxx;
    % If it is out of region the sensor will be folded (enter
from the other side)

    % update the Y-location of the sensor in column 2 of
matrix "node" and
    % save the moving distance in X-direction in column 23 of
matrix "node"
    motion_in_Ydirc=distance_of_motion.*sin(direction_of_
motion);
    % "motion_in_Ydirc" this value added to all moved sensors
in Y direction
```

```
    node(node_will_move,23)=node(node_will_move,23)+motion_
in_Ydirc;
    node(node_will_move,2) =
node(node_will_move,2)+motion_in_Ydirc;
    % Here the Y coordinate of the all sensors is updated by
the same amount
    out_region_in_Y=find(node(:,2)>maxy);
    % Check for out of network region due to moving (greater
than the maxy)
    node(out_region_in_Y,2)=node(out_region_in_Y,2)-maxy;
    % If it is out of region the sensor will be folded (enter
from the other side)
    out_region_in_Y=find(node(:,2)<0);
    % Check for out of network region due to moving (lower
than zero)
    node(out_region_in_Y,2)=node(out_region_in_Y,2)+maxy;
    % If it is out of region the sensor will be folded (enter
from the other side)




return;
%******************************************************************
%******************************************************************

        sick_node=find(node(:,3)>=0.3) ;  % "sick_node" is
column
        aaaa=sick_node;
        sick_node=sick_node';               % % "sick_node" is
row
        sick_node_befor_heal(healing)=length(sick_node);
        disp(['sick_node before   ' num2str(length(sick_node))
'   Healing round   ' num2str(healing)])




%        figure(2)
%        plot(node(:,1),node(:,2),'o');
%        for iiiiii=1:500
%            char a;
%            a=num2str(iiiiii);
%            if (node(iiiiii,3)>=0.3);text(node(iiiiii,1),node
(iiiiii,2),'S','color','r')
%            else
```

```
%              text(node(iiiiii,1),node(iiiiii,2),'H','color'
,'g')
%              end
%          end



    heeeleeed=[];

    for i =sick_node

            beforheal=node(i,3);
        [nb_set,nb_no,qlf_candd_set,qlf_candd_no,not_qlf_
candd_set,not_qlf_candd_no] =...
            nb_qlf_node_AS(i,Pr_t,sensor_TR,node,node_no);
            node(i,4)=nb_no;         % total number of neighbors for
each sensor i
            node(i,5)=qlf_candd_no;% number of qualified neighbor
nodes for each sensor i

            %################### Naive scheme Naive scheme Naive
scheme ###################
            node(i,6)=node(i,3);     % Pr_BSe_comp_naive, Naive
scheme, the BSe of s_i
            node(i,10)=1-node(i,3); % Pr_reliab_naive, Naive
scheme, data reliability

            %@@@@@@ Proposed Scheme Proposed Scheme  Proposed
Scheme  @@@@@@@@@@
            if (qlf_candd_no > 0)&&(prod(qlf_candd_set)< Pr_up)
                % tau=1, that is m = n = t, then Pr_BSe_comp
=prod_{i=1}^{m}P_{i,j}*P_i
                sort_qlf_candd_set=sort(qlf_candd_set);
                % "sort_qlf_candd_set" is an ordered version of
"qlf_candd_set"
                randperm_qlf_candd_set= qlf_candd_set
(randperm(qlf_candd_no));
                % "randperm_qlf_candd_set" is a different order
version of "qlf_candd_set"
                randperm_nb_set = nb_set(randperm(nb_no));
                sort_qlf_candd_set_reliab = 1-sort_qlf_candd_
set;    % AMIRAMIRAMIR    ADDED
```

```
                randperm_qlf_candd_set_reliab =
1-randperm_qlf_candd_set;
                randperm_nb_set_reliab = 1-randperm_nb_set;

                for  temp_n = 1:length(sort_qlf_candd_set)
                    if  prod(sort_qlf_candd_set(1:temp_n))>
Pr_up
                    else
                        break;  % this break will terminate
the for loop
                    end
                end
            node(i,8)=prod(sort_qlf_candd_set(1:temp_n))
*node(i,3);  % tau=1. comp. prob.
            node(i,12)=1-prod(sort_qlf_candd_set(1:temp_n));
%AMIRAMIR tau=1. reliability prob.
            % Proposed scheme, m>(nt)/(t+2), n=t
            m=ceil((temp_n*temp_n)/(temp_n+2));    % this is
law exist in [58]
            node(i,9)=prod(sort_qlf_candd_set(1:m))
*node(i,3);     % tau>1. comp. prob.
            node(i,13)=1-prod(sort_qlf_candd_set(1:m));
% tau>1. reliability prob.


            %&&&&&&&&&& Wang's Infocom 09 scheme   Wang's
Infocom 09 scheme &&&&&&&&&&&&&&&&&
            node(i,7)=prod(nb_set(1:temp_n))*node(i,3);
% tau=1   YIREN comp. prob
            %node(i,7)=prod(randperm_nb_set(1:temp_n))
*node(i,3);     % tau=1   AMIRAMIR comp. prob
            node(i,11)=prod(randperm_nb_set_reliab(1:temp_n));
% tau=1 reliability prob.
            node(i,14)=prod(randperm_nb_set(1:m))*node(i,3);
% tau>1 comp. prob
            node(i,15)=1-prod(randperm_nb_set(1:m));
% tau>1 reliability prob.


            heeeleeed=[heeeleeed  i];




%    disp([' hhheeellleeeddd111111  ' num2str(qlf_candd_no)
'  '  num2str(qlf_candd_set')  '  '  num2str(prod(sort_qlf_
candd_set(1:temp_n)))  '  '
num2str(prod(sort_qlf_candd_set(1:temp_n))*node(i,3))])
```

```
        else
            node(i,8)=node(i,3);node(i,7)=node(i,3);
            node(i,9)=node(i,3);node(i,14)=node(i,3);
            node(i,12)=1-node(i,3);node(i,11)=1-node(i,3);
            node(i,13)=1-node(i,3);node(i,15)=1-node(i,3);
        end                                % end of ( if qlf_
candd_no > 0)
        afterheal=node(i,3);
    end                                % end of (for i =
1:node_no)

    disp(['healed in first round   ' num2str(length
(heeeleeed))])


                node(heeeleeed,3)=node(heeeleeed,8);
%12/12/2013       need correction


    sick_node=find(node(:,3)>=0.3);
    bbbb=sick_node;
    sick_node_after_heal(healing)=length(sick_node);
    disp(['sick_node after    ' num2str(length(sick_node))  '
Healing round   ' num2str(healing)])
    disp('*****************************************************
******')


%*********************************************************************
%*********************************************************************
%load adv_TR60


            %----------------- Pr_BSe_comp,  tau =1
----------------%
subplot(2,3,1);plot(Iter_node_set(:,6),'ro-',
'MarkerSize',4);hold on;  % naive scheme
subplot(2,3,1);plot(Iter_node_set(:,7),'gd-',
'MarkerSize',4);hold on;  % Wang's scheme - randomly select
subplot(2,3,1);plot(Iter_node_set(:,8),'bx-','MarkerSize',4);
% Ours - optimized select
title('(A) \tau=1');xlabel('Index of nodes');ylabel('Prob. of
BSe to be compromised');
legend('\fontsize{7} Naive scheme','\fontsize {7} Wang \it{et al.}
[10]','\fontsize {7} Ours','Location','NorthEast');

            %----------------- Pr_BSe_comp,  tau
>1----------------%
subplot(2,3,4);plot(Iter_node_set(:,6),'ro-',
'MarkerSize',4);hold on;    % naive scheme
subplot(2,3,4);plot(Iter_node_set(:,14),'gd-',
'MarkerSize',4);hold on;  % Wang's scheme - randomly select
```

```
subplot(2,3,4);plot(Iter_node_set(:,9),'bx-','MarkerSize',4);
% proposed scheme - optimized select
title('(D) \tau<1');xlabel('Index of nodes');ylabel('Prob. of
BSe to be compromised');
legend('\fontsize{7} Naive scheme','\fontsize {7} Wang \it{et al.}
[10]','\fontsize {7} Ours','Location','NorthEast');


                %----------------- Pr_reliab, tau =
1----------------%
subplot(2,3,2);plot(Iter_node_set(:,10),'ro-',
'MarkerSize',4);hold on;    % naive scheme
subplot(2,3,2);plot(Iter_node_set(:,11),'gd-',
'MarkerSize',4);hold on;    % Wang's scheme - randomly select
subplot(2,3,2);plot(Iter_node_set(:,12),'bx-','MarkerSize',4);
% proposed scheme - optimized select
title('(B) \tau=1');xlabel('Index of nodes');ylabel('Prob. of
data reliability');
legend('\fontsize{7} Naive scheme','\fontsize {7} Wang \it{et al.}
[10]','\fontsize {7} Ours','Location','NorthEast');


                %----------------- Pr_reliab, tau >
1----------------%
subplot(2,3,5);plot(Iter_node_set(:,10),'ro-',
'MarkerSize',4);hold on;    % naive scheme
subplot(2,3,5);plot(Iter_node_set(:,15),'gd-',
'MarkerSize',4);hold on;    % Wang's scheme - randomly select
subplot(2,3,5);plot(Iter_node_set(:,13),'bx-','MarkerSize',4);
% proposed scheme - optimized select
title('(E) \tau<1');xlabel('Index of nodes');ylabel('Prob. of
data reliability');
legend('\fontsize{7} Naive scheme','\fontsize {7} Wang \it{et al.}
[10]','\fontsize {7} Ours','Location','NorthEast');


                %--------------- Pr_BSe_comp compared with mean
neighbor node--------------%
subplot(2,3,3);plot(nb_no_comp_reliab_set(1:15,10),'ro-',
'MarkerSize',4);hold on;  % naive scheme
subplot(2,3,3);plot(nb_no_comp_reliab_set(1:15,6),'gd-',
'MarkerSize',4);hold on;  % Wang's scheme randomly select,
tau=1
subplot(2,3,3);plot(nb_no_comp_reliab_set(1:15,8),'g.-',
'MarkerSize',4);hold on;% Wang's scheme randomly select, tau>1
subplot(2,3,3);plot(nb_no_comp_reliab_set(1:15,2),'bx-',
'MarkerSize',4);hold on;% proposed scheme - optimized select,
\tau=1
subplot(2,3,3);plot(nb_no_comp_reliab_set(1:15,4),'b+-',
'MarkerSize',4);
title('(C)');% proposed scheme - optimized select, \tau<1
```

```
xlabel('The number of neighbor nodes');ylabel('Prob. of BSe to
be compromised');
legend('\fontsize{7} Naive scheme','\fontsize {7} Wang \it{et al.}
[10],\tau=1',...
    '\fontsize {7} Wang \it{et al.}  [10],\tau<1','\fontsize
{7} Ours, \tau=1','\fontsize {7} Ours, \tau<1',
'Location','NorthEast');

            %--------------- Pr_reliab compared with mean
neighbor node --------------%

subplot(2,3,6);plot(nb_no_comp_reliab_set(1:15,11),'ro-',
'MarkerSize',4);hold on;% naive scheme
subplot(2,3,6);plot(nb_no_comp_reliab_set(1:15,7),'gd-',
'MarkerSize',4);hold on;% Wang's Infocom 09 scheme - randomly
select, \tau=1
subplot(2,3,6);plot(nb_no_comp_reliab_set(1:15,9),'g.-',
'MarkerSize',4);hold on;% Wang's Infocom 09 scheme - randomly
select, \tau<1
subplot(2,3,6);plot(nb_no_comp_reliab_set(1:15,3),'bx-',
'MarkerSize',4);hold on;% proposed scheme - optimized select,
\tau=1
subplot(2,3,6);plot(nb_no_comp_reliab_set(1:15,5),'b+-
','MarkerSize',4);% proposed scheme  optimized select, \tau<1
title('(F)');xlabel('The number of neighbor
nodes');ylabel('Prob. of data reliability');
legend('\fontsize{7} Naive scheme','\fontsize {7} Wang \it{et al.}
[10],\tau=1',...
    '\fontsize {7} Wang \it{et al.}  [10],\tau<1','\fontsize
{7} Ours, \tau=1','\fontsize {7} Ours, \tau<1','Location',
'NorthEast');

if adv_TR==30
    title({['(F)'],['n=',num2str(node_no),'
itr=',num2str(nTime),'   spd=',num2str(adv_vol),'
rang=',num2str(adv_TR)]})
elseif adv_TR==60
    title({['(F)'],['n=',num2str(node_no),'
itr=',num2str(nTime),'   spd=',num2str(adv_vol),'
rang=',num2str(adv_TR)]})
elseif adv_TR==90
    title({['(F)'],['n=',num2str(node_no),'
itr=',num2str(nTime),'   spd=',num2str(adv_vol),'
rang=',num2str(adv_TR)]})
else
    title({['(F)'],['n=',num2str(node_no),'
itr=',num2str(nTime),'   spd=',num2str(adv_vol),'
rang=',num2str(adv_TR)]})
end
```

```
%AMIRAMIRAMIRAMIRAMIRAMIRAMIRAMIRAMIRAMIRAMIRAMIRAMIR
AMIRAMIRAMIR
%AMIRAMIRAMIRAMIRAMIRAMIRAMIRAMIRAMIRAMIRAMIRAMIRAMIR
AMIRAMIRAMIR
%AMIRAMIRAMIRAMIRAMIRAMIRAMIRAMIRAMIRAMIRAMIRAMIRAMIR
AMIRAMIRAMIR
                %-------------- Pr_BSe_comp compared with mean
neighbor node--------------%
figure
subplot(1,2,1);plot(No_of_quif_set(1:15,10),'ro-',
'MarkerSize',4);hold on;  % naive scheme
subplot(1,2,1);plot(No_of_quif_set(1:15,6),'gd-',
'MarkerSize',4);hold on;  % Wang's scheme randomly select,
tau=1
subplot(1,2,1);plot(No_of_quif_set(1:15,8),'g.-',
'MarkerSize',4);hold on;% Wang's scheme randomly select, tau>1
subplot(1,2,1);plot(No_of_quif_set(1:15,2),'bx-',
'MarkerSize',4);hold on;% proposed scheme - optimized select,
\tau=1
subplot(1,2,1);plot(No_of_quif_set(1:15,4),'b+-',
'MarkerSize',4);
title('(A)');% proposed scheme - optimized select, \tau<1
xlabel('The number of qualified nodes');ylabel('Prob. of BSe
to be compromised');
legend('\fontsize{7} Naive scheme','\fontsize {7} Wang \it{et al.}
[10],\tau=1',...
    '\fontsize {7} Wang \it{et al.}  [10],\tau<1','\fontsize
{7} Ours, \tau=1','\fontsize {7} Ours, \tau<1','Location',
'NorthEast');

                %-------------- Pr_reliab compared with mean
neighbor node --------------%

subplot(1,2,2);plot(No_of_quif_set(1:15,11),'ro-
','MarkerSize',4);hold on;% naive scheme
subplot(1,2,2);plot(No_of_quif_set(1:15,7),'gd-
','MarkerSize',4);hold on;% Wang's Infocom 09 scheme -
randomly select, \tau=1
subplot(1,2,2);plot(No_of_quif_set(1:15,9),'g.-
','MarkerSize',4);hold on;% Wang's Infocom 09 scheme -
randomly select, \tau<1
subplot(1,2,2);plot(No_of_quif_set(1:15,3),'mx-
','MarkerSize',4);hold on;% proposed scheme - optimized
select, \tau=1
subplot(1,2,2);plot(No_of_quif_set(1:15,5),'b+-
','MarkerSize',4);% proposed scheme - optimized select, \tau<1
title('(B)');xlabel('The number of qualified nodes');ylabel
('Prob. of data reliability');
legend('\fontsize{7} Naive scheme','\fontsize {7} Wang \it{et al.}
[10],\tau=1',...
```

```
        '\fontsize {7} Wang \it{et al.}  [10],\tau<1','\fontsize
{7} Ours, \tau=1','\fontsize {7} Ours, \
tau<1','Location','NorthEast');


if adv_TR==30
        title({['n=',num2str(node_no),'    itr=',num2str(nTime),'
spd=',num2str(adv_vol),'    rang=',num2str(adv_TR)]})
elseif adv_TR==60
        title({['n=',num2str(node_no),'    itr=',num2str(nTime),'
spd=',num2str(adv_vol),'    rang=',num2str(adv_TR)]})
elseif adv_TR==90
        title({['n=',num2str(node_no),'    itr=',num2str(nTime),'
spd=',num2str(adv_vol),'    rang=',num2str(adv_TR)]})
else
        title({['n=',num2str(node_no),'    itr=',num2str(nTime),'
spd=',num2str(adv_vol),'    rang=',num2str(adv_TR)]})
end



figure
plot(0:healing,average_net_healing)

if adv_TR==30
        title({['n=',num2str(node_no),'    itr=',num2str(nTime),'
spd=',num2str(adv_vol),'    rang=',num2str(adv_TR)]})
elseif adv_TR==60
        title({['n=',num2str(node_no),'    itr=',num2str(nTime),'
spd=',num2str(adv_vol),'    rang=',num2str(adv_TR)]})
elseif adv_TR==90
        title({['n=',num2str(node_no),'    itr=',num2str(nTime),'
spd=',num2str(adv_vol),'    rang=',num2str(adv_TR)]})
else
        title({['n=',num2str(node_no),'    itr=',num2str(nTime),'
spd=',num2str(adv_vol),'    rang=',num2str(adv_TR)]})
end

%AMIRAMIRAMIRAMIRAMIRAMIRAMIRAMIRAMIRAMIRAMIRAMIRAMIRAMIRAMIR
AMIRAMIRAMIR


%****************************************************************


%****************************************************************
%****************************************************************
        % ------- Simulation Results -----------
%****************************************************************
        % With mobility_single_attack
```

```
% MAIN DECLERATION
% "DF:" refer to a definition
% "AS" refer to mthat line is added by Amir Salah
% 1st modification instead of using a predefined compromised
probability, i
% select it totally random
clc;
clear all;
rand('state',0);
start=clock;
tic
%############################################################
###########################
%################### PARAMETERS DEFINITION #################
figure
nTime=1000; % number of iteration to reduce randomness
heal_rnds=10;

node_no=500;% node number
maxx=500;    % width of the network
maxy=500;    % length of the network

sensor_TR=60;       % Transmission range
sensor_vol=10;

adv_TR=60;
adv_vol=10; % ADV speed

motion_steps=5;

update_t=1;  % Updating time
 Pr_t=0.3; % Threshold probability of compromising
Pr_up=0.001;% Upbound threshold probability

Iter_node_set=zeros(node_no,27);  % 27 refer to the number of
columns in which data measured stored
Iter_node_dataset=[]; % this matrix changes every nTime
iteration, it contains the data of the matrix (node),
                          % and it is updated every
(nTime) iteration
self_healing=[];
Eo=12424; % intial energy of each sensor
%##############################################################
###########################
%##################### THE BEGINNING ######################

for iTime=1:nTime
```

```
     clear node;    % To clear the previous data stored in
matrix "node"
     node(:,28)=Eo;  % intial energy of each sensor
     sick_node_befor_heal=0;
     sick_node_after_heal=0;
     node =  topo_AS(node_no, maxx, maxy);
     % DF "topo_AS()": is a function used to generate two
columns in the node
     % matrix the first and second columns are the X and Y
coordinates of
     % a sensor... this function run only one time each
iteration

     % to distribute the nodes at their coordinates (X,Y)
     node =[node,zeros(node_no,25)];

     ro=1;
     %     node=advr_fixed_AS_1(maxx,maxy,node,node_no,adv_
TR,adv_vol,update_t,ro);
     % The first appearance of the Adversary

%node=advr_move_fold_AS_1(maxx,maxy,node,node_no,adv_TR,adv_
vol,update_t,ro);
% do not be used in case single adv attack
%node=advr_move_AS_1(maxx,maxy,node,node_no,adv_TR,adv_
vol,updatc_t,ro);
% do not be used in case single adv attack
node=adv_rand_loc_AS_1(maxx,maxy,node,node_no,adv_TR,ro);
% it can be used in case single adv attack




     %######### NODES ARE NOT COMPROMISED  NOT HEALED##########
     not_sick_node=find(node(:,3)<0.3);
     current_sick_node=find(node(:,3)>=0.3);

     % clustr([1:length(current_sick_node)],:)=node(current_
sick_node,[1:8]);
     % cluster(:,9)=current_sick_node;

%%length(not_sick_node);length(current_sick_node);
length(not_sick_node)+length(current_sick_node);
     % not_sick_node=not_sick_node';
     %sick_node_befor_heal(healing)=length(sick_node);
     %for  i= not_sick_node
             %disp([' lower than 0.3  ' num2str(node(i,3))])
```

```
%[nb_set,nb_no,qlf_candd_set,qlf_candd_no,not_qlf_candd_
set,not_qlf_candd_no] =...
             %nb_qlf_node_AS(i,Pr_t,sensor_TR,node,node_no);
     %end

     %############### Plotting network ####################
%      figure(1)
%      plot(node(:,1),node(:,2),'o');


     %############# HEALING PROCESS ##################
     for healing=1:heal_rnds
          disp(['iteration number  ', num2str(iTime), '
Healing round nunber  ', num2str(healing)]);
          if healing==1
              ONE_HEALING_BEFORE_MOBILITY
          else

              % node(:,8): Pr_BSe_comp_proposed_eq1, tau=1, the
BSe of s_i to be compromised
              %%%%%if healing<430
                  ro=healing;

% node=advr_move_fold_AS_1(maxx,maxy,node,node_no,adv_TR,
adv_vol,update_t,ro);
% do not be used in case single adv attack
%node=advr_move_AS_1(maxx,maxy,node,node_no,adv_TR,adv_
vol,update_t,ro);
% do not be used in case single adv attack
%node=adv_rand_loc_AS_1(maxx,maxy,node,node_no,adv_TR,ro);
% it can be used in case single adv attack




          sick_node=find(node(:,3)>=0.3);    % "sick_node" is
column
          sick_node=sick_node';                % % "sick_node" is row
          sick_node_befor_heal(healing)=length(sick_node);
          disp(['sick_node before  ' num2str(length(sick_node))
'   Healing round  ' num2str(healing)])

          %   disp('beforeBBBBBBBBBBBBBBBBBBBBBBBBBB')
          node=health_neighbors_before(node);
```

```
        % "health_neighbors_before" this function used to find
the number
        % of health neighbours before mobility




        sensor_will_move=[sick_node  heeeleeed  ];
        if find(sick_node~=0)
                node = node_motion_varied_cluster(sensor_will_
move, maxx, maxy,sensor_vol,update_t,node,motion_steps,ro);
        end
        % "node_motion_fixed" this function is used to perform
mobility

        node=health_neighbors_after(node);
        % "health_neighbors_after" this function used to find
the number
        % of health neighbours after mobility

                %disp('All     neighbors ')
                %disp('Sick ID    Before      after       the
difference (before motion-after motion)')
        for iiiii=sick_node;
                %disp([num2str(iiiii),'   ',num2str(node
(iiiii,1)),'   ',num2str(node(iiiii,2)),'   ',num2str(node
(iiiii,3)),'           ', num2str(node(iiiii,24)),'           ',
num2str(node(iiiii,26)),'           ',
num2str(node(iiiii,24)-node(iiiii,26))])
        end
                %disp('Healthe    neighbors ')
                %disp('Sick ID    Before      after       the
difference (before motion-after motion)')
        for iiiii=sick_node;
                %disp([ num2str(iiiii),'   ',
num2str(node(iiiii,1)),'   ',num2str(node(iiiii,2)),'   ',num2s
tr(node(iiiii,3)),'           ', num2str(node(iiiii,25)),'        ',
num2str(node(iiiii,27)),'           ',num2str(node(iiiii,25)-node
(iiiii,27))])
        end
        % the above function will move all nodes or sick nodes
only many times
        % within the network based on the pricipals that
mobility can enhance
        % the security by moving the sick sensors to another
position in the
        % network with a more healty neighbors.
```

```
        %for i =sick_node
        heeeleeed=[];
        for i =sick_node


        % node=advr_AS_2(maxx,maxy,node,node_no,adv_TR,i,
adv_vol,update_t);
            % node=advr_AS_rand_1(maxx,maxy,node,node_no,adv_TR,i);
            % node=advr_AS_rand_2(maxx,maxy,node,node_no,adv_TR,i);
        % repeate the function mean that there are another
adversary
        % "advr_AS_1" "advr_AS_2" these functions runs for
each si to compute the following :
        % 1-location of  ADV (x,y), which depend on previous
location, spread, the update time and direction of ADV
        % 2-the distance between the ADV and all sensors in
the area
        % 3-the compromised probability of all sensors
according to their positions from the ADV "node(:,3)"

        %%%disp(['adversay   11  ', num2str(sum(node(:,3)))])
        %%%disp(['adversay   22  ', num2str(sum(node(:,18)))])

        %for jjj=1:node_no,
            %  if node(jjj,3)>node(jjj,18)
            %else
            %node(jjj,3)=node(jjj,18);
            %end
        %end
        % the above lines guarantee that if si is compromised
by adv1 and.adv2 , then the final compromising probability is
the greater
        % one; for example if si compromising probability is
.5 due to adv1 and it is .6 due to the adv2, then the
effective compromising
        % probability is the greater one (0.6)

        [nb_set,nb_no,qlf_candd_set,qlf_candd_no,not_qlf_
candd_set,not_qlf_candd_no] =...
            nb_qlf_node_AS(i,Pr_t,sensor_TR,node,node_no);
        % DF "nb_qlf_node_AS": it is a function used to
compute many parameters such as:
        % 1- Copmute neighbor node set "nb_set" and neighbor
node number "nb_no" which
            % are in the transmission range of si
        % 2- Copmute the qualified candidate set "qlf_candd_
set" and the qualified
            % candidate number "qlf_candd_no", this is a part
of the neighbor set but
            % there compromised probability not exceed the
threshold probability
```

```
        node(i,4)=nb_no;          % total number of neighbors for
each sensor i
        node(i,5)=qlf_candd_no;% number of qualified neighbor
nodes for each sensor i

%#############################################################
################################

%#############################################################
##########################
        %#################### Naive scheme Naive scheme Naive
scheme ####################

%-------------------------------------------------------------
-----------%
        node(i,6)=node(i,3);      % Pr_BSe_comp_naive, Naive
scheme, the BSe of s_i
        %disp(['1- node(i,3)=   ',num2str(node(i,3)),'
node(i,6)=   ',num2str(node(i,6)),  '      ' num2str(i)])
        node(i,10)=1-node(i,3); % Pr_reliab_naive, Naive
scheme, data reliability
%disp(['before Sk_N ID and Comp  ' num2str(i)  '  '
num2str(node(i,3))   ' Sk_N nb_no ' num2str(nb_no)  ' Sk_N
nb_no qlf_candd_no '   num2str(qlf_candd_no)])
        %@@@@@@@@@@@@@@@@@@@@@@@@@@@@@@@@@@@@@@@@@@@@@@@@@@@@@@@@
        %@@@@@@@@@@@@@@@@@@@@@@@@@@@@@@@@@@@@@@@@@@@@@@@@@@@@@@
        %@@@@@@ Proposed Scheme Proposed Scheme  Proposed
Scheme  @@@@@@@@@@

        %-----------------------------------------------------
------------%
        % data cannot be recovered, if one data part is lost.
        % If there is enough qualified candidates, data is
distributed stored. Otherwise, data is stored locally.
        if (qlf_candd_no > 0)&&(prod(qlf_candd_set)< Pr_up)
            % tau=1, that is m = n = t, then Pr_BSe_comp
=prod_{i=1}^{m}P_{i,j}*P_i
            sort_qlf_candd_set=sort(qlf_candd_set);
            % "sort_qlf_candd_set" is an ordered version of
"qlf_candd_set"
            randperm_qlf_candd_set=
qlf_candd_set(randperm(qlf_candd_no));
            % "randperm_qlf_candd_set" is a different order
version of "qlf_candd_set"
            randperm_nb_set = nb_set(randperm(nb_no));
            % "randperm_nb_set" is a different order version
of "nb_set"
            % in other words the same values exist in both
matrices but
            % with different order
```

```
                sort_qlf_candd_set_reliab = 1-sort_qlf_candd_
set;      % AMIRAMIRAMIR   ADDED


            randperm_qlf_candd_set_reliab =
1-randperm_qlf_candd_set;
            randperm_nb_set_reliab = 1-randperm_nb_set;
            % note that (qlf_candd_set) &  (nb_set) contain
the comprised probability
            % and (randperm_qlf_candd_set_reliab) &
(randperm_nb_set_reliab) contain the reliable probability
which
            % equal to (1-comprised probability)

            for  temp_n = 1:length(sort_qlf_candd_set)
                % here we select the first t peers  from
"sort_qlf_candd_set"
                % so that the product of the compromising
probability of these
                % t peers is lower than the upperbound
threshold probability
                if  prod(sort_qlf_candd_set(1:temp_n))>
Pr_up
                    %disp(['temp_n 1   ',num2str(temp_n),'
the product     ',
num2str(prod(sort_qlf_candd_set(1:temp_n)))])
                else
                    % "else" mean that  the "prod(sort_
qlf_candd_set(1:temp_n))" is lower than "Pr_up"
                    % disp(['temp_n 2   ',num2str(temp_n),'
the product     ',
num2str(prod(sort_qlf_candd_set(1:temp_n)))])
                    break;  % this break will terminate the for
loop
                end
                %disp(['temp_n 3    ',num2str(temp_n),'
the product     ',
num2str(prod(sort_qlf_candd_set(1:temp_n)))])
            end
            %disp(['temp_n 4   ',num2str(temp_n),'     the
product     ', num2str(prod(sort_qlf_candd_set(1:temp_n)))])

            node(i,8)=prod(sort_qlf_candd_
set(1:temp_n))*node(i,3);   % tau=1. comp. prob.
            % chose the first "temp_n" nodes from the
"sort_qlf_candd_set"
            %which have the lower compromising probability
and multiplu
            %them with each other and with the compromising
probability
```

```
                %of the sick node itself.............WE MUST
USE THE SAME
                %NODES WHICH USED TO ENHANCE THE COMPROMOISING
PROBABILITY
                %OF THE SICK NODE ..... TO CACULATE THE DATA
RELIABILITY  OF
                %THIS SICK NODE       SO THAT THE SUM OF COLUMNS 8
AND 12 =1
                % THIS IS THE SAME DONE BY YIREN FOT TAU>1


                %node(i,12)=prod(randperm_qlf_candd_set_reliab
(1:temp_n));   %YIREN tau=1. reliability prob.
                %node(i,12)=prod(sort_qlf_candd_set_reliab
(1:temp_n));   %AMIRAMIR tau=1. reliability prob.
                node(i,12)=1-prod(sort_qlf_candd_set(1:temp_n));
%AMIRAMIR tau=1. reliability prob.


                % Proposed scheme, m>(nt)/(t+2), n=t
                m=ceil((temp_n*temp_n)/(temp_n+2));     % this is
law exist in [58]
                node(i,9)=prod(sort_qlf_candd_set(1:m))
*node(i,3);       % tau>1. comp. prob.
                node(i,13)=1-prod(sort_qlf_candd_set(1:m));
% tau>1. reliability prob.


%&&&&&&&&&&&&&&&&&&&&&&&&&&&&&&&&&&&&&&&&&&&&&&&&&&&&&&&&&&&&
&&&&&&&&&&&&&&&&&&&&&&&&&&&
   %&&&&&&&&&&&&&&&&&&&&&&&&&&&&&&&&&&&&&&&&&&&&&&&&&&&&&&&&&&&&&
&&&&&&&&&&&&&&&
        %&&&&&&&&&& Wang's Infocom 09 scheme  Wang's Infocom
09 scheme &&&&&&&&&&&&&&&&&
                %------------------------------------------------
-------------------%


                node(i,7)=prod(nb_set(1:temp_n))*node(i,3);
% tau-1   YIREN comp. prob
                %node(i,7)=prod(randperm_nb_set(1:temp_n))
*node(i,3);     % tau=1   AMIRAMIR comp. prob




                node(i,11)=prod(randperm_nb_set_
reliab(1:temp_n));          % tau=1 reliability prob.
                %   Randomly select m nodes
                node(i,14)=prod(randperm_nb_set(1:m))*node(i,3);
% tau>1 comp. prob
                node(i,15)=1-prod(randperm_nb_set(1:m));
% tau>1 reliability prob.
```

```
            %node(i,3)=node(i,8);  % 12/12/2013
            % this line make that if the sensor i is healed
it will help
            % the others during the same round so we can say
in the same
            % round the sick sensor healed first will help
the others
            % sick sensors in the same round. I think this
not accepted
            % because that all sick sensors start the
healing process in
            % the same round.


                         heeeleeed=[heeeleeed  i];

    %           disp([' hhheeellleeeddd222222 ' num2str(qlf_
candd_no)  '  '   num2str(qlf_candd_set')  '  '
num2str(prod(sort_qlf_candd_set(1:temp_n)))  '  '
num2str(prod(sort_qlf_candd_set(1:temp_n))*node(i,3))])

        else
            % this mean that the (qlf_candd_no = 0) which
mean that there is no qualified neighbor nodes
            % and hence there is no self-healing
(cooperation) and any comporomised probability will be
            % the defined compromised probability defined in
columns 3. And any reliable probability will be
            % (1-comprised probability)
            node(i,8)=node(i,3);node(i,7)=node(i,3);
            node(i,9)=node(i,3);node(i,14)=node(i,3);

            node(i,12)=1-node(i,3);node(i,11)=1-node(i,3);
            node(i,13)=1-node(i,3);node(i,15)=1-node(i,3);
        end                          %  end of ( if qlf_
candd_no > 0)
%disp([' after Sk_N ID and Comp ' num2str(i)  '  '
num2str(node(i,3))   ' Sk_N nb_no ' num2str(nb_no) ' Sk_N
nb_no qlf_candd_no '   num2str(qlf_candd_no)])


        %disp(['final- node(i,3)=   ',num2str(node(i,3))  '
'  num2str(i)])
        %disp('*********************************************
******************')

    end                          %  end of (for i =
1:node_no)
```

```
        disp([' in round   ' num2str(healing) '  is  '
num2str(length(heeeleeed)) '     healed'])

                node(heeeleeed,3)=node(heeeleeed,8);
%12/12/2013      need correction




    sick_node=find(node(:,3)>=0.3);
    sick_node_after_heal(healing)=length(sick_node);
    disp(['sick_node after     ' num2str(length(sick_node))  '
Healing round   ' num2str(healing)])
    disp('**************************************************
******')

  %   grid on
  %      ang=0:0.01:2*pi;     xp=adv_TR*cos(ang);
yp=adv_TR*sin(ang);
  %      figure(2)
  %       hold   on
  %      plot(node(:,1),node(:,2),'o')
  %       title(['sick node ' num2str(length(sick_node))  '
Healing round   ' num2str(healing)])
  %      ro=1;
  %    plot(node(ro,16)+xp,node(ro,17)+yp)
  %     plot(node(ro,16)+2*xp,node(ro,17)+2*yp)
  %     plot(node(ro,16)+3*xp,node(ro,17)+3*yp)
  %      plot(node(ro,16),node(ro,17),'s-','LineWidth',4,'Mark
erEdgeColor','k','MarkerFaceColor','m','MarkerSize',8);
  %      text(node(ro,16)+5,node(ro,17)+2,'AD')
  %       axis([0   500   0    500])
  %
  %      hold off
  %
  %      figure(2)
  %      plot(node(:,1),node(:,2),'o');
  %           for iiiiii=1:500
  %          char a;
  %          a=num2str(iiiiii);
  %          if (node(iiiiii,3)>=0.3)
  %          text(node(iiiiii,1)+2,node(iiiiii,2),'S','color'
,'r')
  %          else
  %          text(node(iiiiii,1)+2,node(iiiiii,2),'h','color'
,'g')
  %          end
  %      end
```

```
        end
    end


    %sum(node(:,3))  % the sum of all compromised probability
of all nodes
    %length(find(node(:,4)==node(:,5)))

    Iter_node_set=Iter_node_set+node;                    % sum all
the node matrices nTime

%    node(not_sick_node,:)=[];     % Amir

    Iter_node_dataset=[Iter_node_dataset;node];    % store
all the node matrices
    %size(Iter_node_dataset)

self_healing=[self_healing;sick_node_befor_heal,0,sick_node_
after_heal];
end                               % end of (for iTime=1:nTime)
net_healing=[self_healing(:,1) self_healing(:,(2+healing):(2*h
ealing+1))];
average_net_healing=mean(net_healing);
%*************************************************************
*******************************
        %*********************************************************
*******************************
                %*********************** PLOTTING   PLOTTING
PLOTTING ***********************
 %----------------------------------------------------------------------
% mean of Iter_node_set
Iter_node_set=Iter_node_set/nTime;

nb_no_comp_reliab_set=unique(Iter_node_dataset(:,4));

No_of_quif_set=unique(Iter_node_dataset(:,5));    %  AMIEAMIR

%size(nb_no_comp_reliab_set)
% nb_no_comp_reliab_set(:,1): the number of neighbor nodes
% column number 4 contains the number of neighbor for each
sensor i while
% the (nb_no_comp_reliab_set) contains the same values as in
(Iter_node_dataset(:,4)) but
% with no repetitions (unique). it will also be sorted
nb_no_comp_reliab_set=[nb_no_comp_reliab_set,zeros(length(nb_
no_comp_reliab_set),10)];

No_of_quif_set=[No_of_quif_set,zeros(length(No_of_quif_
set),10)];    %  AMIEAMIR
```

```
% the above line to make "nb_no_comp_reliab_set" has a 11
columns

for temp_nb_no = 1:length(nb_no_comp_reliab_set)
    %if nb_no_comp_reliab_set(1,1)==0
    %    temp_nb_no=temp_nb_no-1;
    %    iiiii=1;
    %else
        iiiii=0;
    %end
    nb_no_comp_reliab_set(temp_nb_no+iiiii,2)=
mean(Iter_node_dataset(find(Iter_node_dataset(:,4)==temp_nb_
no),8));
    % nb_no_comp_reliab_set(:,2): mean_Pr_BSe_comp.
Proposed scheme, tau=1
    % 8th  Col, node(:,8): Pr_BSe_comp_proposed_eq1, tau=1,
the BSe of s_i to be compromised

    nb_no_comp_reliab_set(temp_nb_no+iiiii,3)=
mean(Iter_node_dataset(find(Iter_node_dataset(:,4)==temp_nb_
no),12));
    % nb_no_comp_reliab_set(:,3): mean_Pr_reliab.
Proposed scheme, tau=1
     % 12th Col, node(:,12): Pr_reliab_proposed_eq1, tau=1,
data reliability

    nb_no_comp_reliab_set(temp_nb_no+iiiii,4)=
mean(Iter_node_dataset(find(Iter_node_dataset(:,4)==temp_nb_
no),9));
    % nb_no_comp_reliab_set(:,4): mean_Pr_BSe_comp.
Proposed scheme, tau>1
    % 9th  Col, node(:,9): Pr_BSe_comp_proposed_less1, tau>1,
the BSe of s_i to be compromised

    nb_no_comp_reliab_set(temp_nb_no+iiiii,5)=
mean(Iter_node_dataset(find(Iter_node_dataset(:,4)==temp_nb_
no),13));
    % nb_no_comp_reliab_set(:,5): mean_Pr_reliab.
Proposed scheme, tau>1
    % 13th Col, node(:,13): Pr_reliab_proposed_less1, tau>1,
data reliability

    nb_no_comp_reliab_set(temp_nb_no+iiiii,6)=
mean(Iter_node_dataset(find(Iter_node_dataset(:,4)==temp_nb_
no),7));
    % nb_no_comp_reliab_set(:,6): mean_Pr_BSe_comp.
wang, tau=1
    % 7th  Col, node(:,7):   Pr_BSe_comp_wang_eq1, tau=1, the
BSe of si
```

```
        nb_no_comp_reliab_set(temp_nb_no+iiiii,7)=
mean(Iter_node_dataset(find(Iter_node_dataset(:,4)==temp_nb_
no),11));
        % nb_no_comp_reliab_set(:,7): mean_Pr_reliab.
wang, tau=1
        % 11th Col, node(:,11): Pr_reliab_wang_eq1, tau=1, data
reliability

        nb_no_comp_reliab_set(temp_nb_no+iiiii,8)=
mean(Iter_node_dataset(find(Iter_node_dataset(:,4)==temp_nb_
no),14));
        % nb_no_comp_reliab_set(:,8): mean_Pr_BSe_comp.
wang, tau>1
        % 14th Col, node(:,14): Pr_BSe_comp_wang_less1, tau>1,
the BSe of si

        nb_no_comp_reliab_set(temp_nb_no+iiiii,9)=
mean(Iter_node_dataset(find(Iter_node_dataset(:,4)==temp_nb_
no),15));
        % nb_no_comp_reliab_set(:,9): mean_Pr_reliab.
wang, tau>1
        % 15th Col, node(:,15): Pr_reliab_wang_less1, tau>1, data
reliability

        nb_no_comp_reliab_set(temp_nb_no+iiiii,10)=
mean(Iter_node_dataset(find(Iter_node_dataset(:,4)==temp_nb_
no),6));
        % nb_no_comp_reliab_set(:,10):mean_Pr_BSe_comp.
naive
        % 6th  Col, node(:,6): Pr_BSe_comp_naive, the same as
node(:,3)

        nb_no_comp_reliab_set(temp_nb_no+iiiii,11)=
mean(Iter_node_dataset(find(Iter_node_dataset(:,4)==temp_nb_
no),10));
        % nb_no_comp_reliab_set(:,11):mean_Pr_reliab.
naive
        % 10th Col, node(:,10): Pr_reliab_naive, data reliability

end




%AMIRAMIRAMIRAMIRAMIRAMIRAMIRAMIRAMIRAMIRAMIRAMIRAMIR
AMIRAMIRAMIR
%AMIRAMIRAMIRAMIRAMIRAMIRAMIRAMIRAMIRAMIRAMIRAMIRAMIR
AMIRAMIRAMIR
```

```
%AMIRAMIRAMIRAMIRAMIRAMIRAMIRAMIRAMIRAMIRAMIRAMIRAMIRAMIR
AMIRAMIRAMIR
for temp_qf_no = 1:length(No_of_quif_set)
    % if No_of_quif_set(1,1)==0
    %       temp_qf_no=temp_qf_no-1;
    %    iii=1;
    % else
        iii=0;
    %end
        temp_qf_no;
        temp_qf_no+iii;
    No_of_quif_set(temp_qf_no+iii,2)=
mean(Iter_node_dataset(find(Iter_node_dataset(:,5)==temp_qf_
no),8));
        % No_of_quif_set(:,2): mean_Pr_BSe_comp.      Proposed
scheme, tau=1
        % 8th  Col, node(:,8): Pr_BSe_comp_proposed_eq1, tau=1,
the BSe of s_i to be compromised

        No_of_quif_set(temp_qf_no+iii,3)=
mean(Iter_node_dataset(find(Iter_node_dataset(:,5)==temp_qf_
no),12));
        % No_of_quif_set(:,3): mean_Pr_reliab.
Proposed scheme, tau=1
        % 12th Col, node(:,12): Pr_reliab_proposed_eq1, tau=1,
data reliability

        No_of_quif_set(temp_qf_no+iii,4)=
mean(Iter_node_dataset(find(Iter_node_dataset(:,5)==temp_qf_
no),9));
        % No_of_quif_set(:,4): mean_Pr_BSe_comp.      Proposed
scheme, tau>1
        % 9th  Col, node(:,9): Pr_BSe_comp_proposed_less1, tau>1,
the BSe of s_i to be compromised

        No_of_quif_set(temp_qf_no+iii,5)=
mean(Iter_node_dataset(find(Iter_node_dataset(:,5)==temp_qf_
no),13));
        % No_of_quif_set(:,5): mean_Pr_reliab.
Proposed scheme, tau>1
        % 13th Col, node(:,13): Pr_reliab_proposed_less1, tau>1,
data reliability

        No_of_quif_set(temp_qf_no+iii,6)=
mean(Iter_node_dataset(find(Iter_node_dataset(:,5)==temp_qf_
no),7));
        % No_of_quif_set(:,6): mean_Pr_BSe_comp.       wang, tau=1
        % 7th  Col, node(:,7):   Pr_BSe_comp_wang_eq1, tau=1, the
BSe of si
```

```
        No_of_quif_set(temp_qf_no+iii,7)=
mean(Iter_node_dataset(find(Iter_node_dataset(:,5)==temp_qf_
no),11));
        % No_of_quif_set(:,7): mean_Pr_reliab.                    wang,
tau=1
        % 11th Col, node(:,11): Pr_reliab_wang_eq1, tau=1, data
reliability

        No_of_quif_set(temp_qf_no+iii,8)=
mean(Iter_node_dataset(find(Iter_node_dataset(:,5)==temp_qf_
no),14));
        % No_of_quif_set(:,8): mean_Pr_BSe_comp.          wang, tau>1
        % 14th Col, node(:,14): Pr_BSe_comp_wang_less1, tau>1,
the BSe of si

        No_of_quif_set(temp_qf_no+iii,9)=
mean(Iter_node_dataset(find(Iter_node_dataset(:,5)==temp_qf_
no),15));
        % No_of_quif_set(:,9): mean_Pr_reliab.
wang, tau>1
        % 15th Col, node(:,15): Pr_reliab_wang_less1, tau>1, data
reliability

        No_of_quif_set(temp_qf_no+iii,10)=
mean(Iter_node_dataset(find(Iter_node_dataset(:,5)==temp_qf_
no),6));
        % No_of_quif_set(:,10):mean_Pr_BSe_comp.        naive
        % 6th  Col, node(:,6): Pr_BSe_comp_naive, the same as
node(:,3)

        No_of_quif_set(temp_qf_no+iii,11)=
mean(Iter_node_dataset(find(Iter_node_dataset(:,5)==temp_qf_
no),10));
        % No_of_quif_set(:,11):mean_Pr_reliab.                    naive
        % 10th Col, node(:,10): Pr_reliab_naive, data reliability

end
%AMIRAMIRAMIRAMIRAMIRAMIRAMIRAMIRAMIRAMIRAMIRAMIRAMIRAMIR
AMIRAMIRAMIR
%AMIRAMIRAMIRAMIRAMIRAMIRAMIRAMIRAMIRAMIRAMIRAMIRAMIRAMIR
AMIRAMIRAMIR
%AMIRAMIRAMIRAMIRAMIRAMIRAMIRAMIRAMIRAMIRAMIRAMIRAMIRAMIR
AMIRAMIRAMIR

%$$$$$$$$$$$$$$$$$$$$$$$$$$$$$$$$$$$$$$$$$$$$$$$$$$$$$$$$$$$$
$$$$$$$$$$$$$$$$$$$$$$$$$$$$$$$$$
        %$$$$$$$$$$$$$$$$$$$$$$$$$$$$$$$$$$$$$$$$ THE PLOTTING
$$$$$$$$$$$$$$$$$$$$$$$$$$$$$$$$$
                %---------------------------------------------
---------------------%
```

```
                %----------------- Pr_BSe_comp,  tau =1
----------------%
subplot(2,3,1);plot(Iter_node_set(:,6),'ro-
','MarkerSize',4);hold on;  % naive scheme
subplot(2,3,1);plot(Iter_node_set(:,7),'gd-
','MarkerSize',4);hold on;  % Wang's scheme - randomly select
subplot(2,3,1);plot(Iter_node_set(:,8),'bx-','MarkerSize',4);
% Ours - optimized select
title('(A) \tau=1');xlabel('Index of nodes');ylabel('Prob. of
BSe to be compromised');
legend('\fontsize{7} Naive scheme','\fontsize {7} Wang \it{et
al.}  [10]','\fontsize {7} Ours','Location','NorthEast');

                %----------------- Pr_BSe_comp,  tau
>1----------------%
subplot(2,3,4);plot(Iter_node_set(:,6),'ro-
','MarkerSize',4);hold on;  % naive scheme
subplot(2,3,4);plot(Iter_node_set(:,14),'gd-
','MarkerSize',4);hold on;  % Wang's scheme - randomly select
subplot(2,3,4);plot(Iter_node_set(:,9),'bx-','MarkerSize',4);
% proposed scheme - optimized select
title('(D) \tau<1');xlabel('Index of nodes');ylabel('Prob. of
BSe to be compromised');
legend('\fontsize{7} Naive scheme','\fontsize {7} Wang \it{et
al.}  [10]','\fontsize {7} Ours','Location','NorthEast');

                %----------------- Pr_reliab, tau =
1----------------%
subplot(2,3,2);plot(Iter_node_set(:,10),'ro-
','MarkerSize',4);hold on;    % naive scheme
subplot(2,3,2);plot(Iter_node_set(:,11),'gd-
','MarkerSize',4);hold on;    % Wang's scheme - randomly
select
subplot(2,3,2);plot(Iter_node_set(:,12),'bx-','MarkerSize',4);
% proposed scheme - optimized select
title('(B) \tau=1');xlabel('Index of nodes');ylabel('Prob. of
data reliability');
legend('\fontsize{7} Naive scheme','\fontsize {7} Wang \it{et
al.}  [10]','\fontsize {7} Ours','Location','NorthEast');

                %----------------- Pr_reliab, tau > 1
----------------%
subplot(2,3,5);plot(Iter_node_set(:,10),'ro-
','MarkerSize',4);hold on;    % naive scheme
subplot(2,3,5);plot(Iter_node_set(:,15),'gd-
','MarkerSize',4);hold on;    % Wang's scheme - randomly select
subplot(2,3,5);plot(Iter_node_set(:,13),'bx-','MarkerSize',4);
% proposed scheme - optimized select
title('(E) \tau<1');xlabel('Index of nodes');ylabel('Prob. of
data reliability');
```

```
legend('\fontsize{7} Naive scheme','\fontsize {7} Wang \it{et
al.}  [10]','\fontsize {7} Ours','Location','NorthEast');

            %--------------- Pr_BSe_comp compared with mean
neighbor node--------------%

subplot(2,3,3);plot(nb_no_comp_reliab_set(1:15,10),'ro-
','MarkerSize',4);hold on;  % naive scheme
subplot(2,3,3);plot(nb_no_comp_reliab_set(1:15,6),'gd-
','MarkerSize',4);hold on;   % Wang's scheme randomly select,
tau=1
subplot(2,3,3);plot(nb_no_comp_reliab_set(1:15,8),'g.-
','MarkerSize',4);hold on;% Wang's scheme randomly select,
tau>1
subplot(2,3,3);plot(nb_no_comp_reliab_set(1:15,2),'bx-
','MarkerSize',4);hold on;% proposed scheme - optimized
select, \tau=1
subplot(2,3,3);plot(nb_no_comp_reliab_set(1:15,4),'b+-
','MarkerSize',4);
title('(C)');% proposed scheme - optimized select, \tau<1
xlabel('The number of neighbor nodes');ylabel('Prob. of BSe to
be compromised');
legend('\fontsize{7} Naive scheme','\fontsize {7} Wang \it{et
al.}  [10],\tau=1',...
   '\fontsize {7} Wang \it{et al.}  [10],\tau<1','\fontsize
{7} Ours, \tau=1','\fontsize {7} Ours, \
tau<1','Location','NorthEast');

            %--------------- Pr_reliab compared with mean
neighbor node --------------%

subplot(2,3,6);plot(nb_no_comp_reliab_set(1:15,11),'ro-
','MarkerSize',4);hold on;% naive scheme
subplot(2,3,6);plot(nb_no_comp_reliab_set(1:15,7),'gd-
','MarkerSize',4);hold on;% Wang's Infocom 09 scheme -
randomly select, \tau=1
subplot(2,3,6);plot(nb_no_comp_reliab_set(1:15,9),'g.-
','MarkerSize',4);hold on;% Wang's Infocom 09 scheme -
randomly select, \tau<1
subplot(2,3,6);plot(nb_no_comp_reliab_set(1:15,3),'bx-
','MarkerSize',4);hold on;% proposed scheme - optimized
select, \tau=1
subplot(2,3,6);plot(nb_no_comp_reliab_set(1:15,5),'b+-
','MarkerSize',4);% proposed scheme - optimized select, \tau<1
title('(F)');xlabel('The number of neighbor
nodes');ylabel('Prob. of data reliability');
legend('\fontsize{7} Naive scheme','\fontsize {7} Wang \it{et
al.}  [10],\tau=1',...
```

```
        '\fontsize {7} Wang \it{et al.}  [10],\tau<1','\fontsize
{7} Ours, \tau=1','\fontsize {7} Ours, \
tau<1','Location','NorthEast');

if adv_TR==30
     title({['(F)'],['n=',num2str(node_no),'
itr=',num2str(nTime),'   spd=',num2str(adv_vol),'
rang=',num2str(adv_TR)]})
elseif adv_TR==60
     title({['(F)'],['n=',num2str(node_no),'
itr=',num2str(nTime),'   spd=',num2str(adv_vol),'
rang=',num2str(adv_TR)]})
elseif adv_TR==90
     title({['(F)'],['n=',num2str(node_no),'
itr=',num2str(nTime),'   spd=',num2str(adv_vol),'
rang=',num2str(adv_TR)]})
else
     title({['(F)'],['n=',num2str(node_no),'
itr=',num2str(nTime),'   spd=',num2str(adv_vol),'
rang=',num2str(adv_TR)]})
end


%AMIRAMIRAMIRAMIRAMIRAMIRAMIRAMIRAMIRAMIRAMIRAMIRAMIR
AMIRAMIRAMIR
%AMIRAMIRAMIRAMIRAMIRAMIRAMIRAMIRAMIRAMIRAMIRAMIRAMIR
AMIRAMIRAMIR
%AMIRAMIRAMIRAMIRAMIRAMIRAMIRAMIRAMIRAMIRAMIRAMIRAMIR
AMIRAMIRAMIR
             %--------------- Pr_BSe_comp compared with mean
neighbor node--------------%
figure
subplot(1,2,1);plot(No_of_quif_set(1:length(No_of_quif_
set(:,10)),10),'ro-','MarkerSize',4);hold on;  % naive scheme
subplot(1,2,1);plot(No_of_quif_set(1:length(No_of_quif_
set(:,6)),6),'gd-','MarkerSize',4);hold on;  % Wang's scheme
randomly select, tau=1
subplot(1,2,1);plot(No_of_quif_set(1:length(No_of_quif_
set(:,8)),8),'g.-','MarkerSize',4);hold on;% Wang's scheme
randomly select, tau>1
subplot(1,2,1);plot(No_of_quif_set(1:length(No_of_quif_
set(:,2)),2),'bx-','MarkerSize',4);hold on;% proposed scheme
- optimized select, \tau=1
subplot(1,2,1);plot(No_of_quif_set(1:length(No_of_quif_
set(:,4)),4),'b+-','MarkerSize',4);
title('(A)');% proposed scheme - optimized select, \tau<1
xlabel('The number of qualified nodes');ylabel('Prob. of BSe
to be compromised');
legend('\fontsize{7} Naive scheme','\fontsize {7} Wang \it{et
al.}  [10],\tau=1',...
```

```
      '\fontsize {7} Wang \it{et al.}  [10],\tau<1','\fontsize
{7} Ours, \tau=1','\fontsize {7} Ours, \
tau<1','Location','NorthEast');

              %-------------- Pr_reliab compared with mean
neighbor node --------------%

subplot(1,2,2);plot(No_of_quif_set(1:length(No_of_quif_
set(:,11)),11),'ro-','MarkerSize',4);hold on;% naive scheme
subplot(1,2,2);plot(No_of_quif_set(1:length(No_of_quif_
set(:,7)),7),'gd-','MarkerSize',4);hold on;% Wang's Infocom 09
scheme - randomly select, \tau=1
subplot(1,2,2);plot(No_of_quif_set(1:length(No_of_quif_
set(:,9)),9),'g.-','MarkerSize',4);hold on;% Wang's Infocom 09
scheme - randomly select, \tau<1
subplot(1,2,2);plot(No_of_quif_set(1:length(No_of_quif_
set(:,3)),3),'mx-','MarkerSize',4);hold on;% proposed scheme
- optimized select, \tau=1
subplot(1,2,2);plot(No_of_quif_set(1:length(No_of_quif_
set(:,5)),5),'b+-','MarkerSize',4);% proposed scheme -
optimized select, \tau<1
title('(B)');xlabel('The number of qualified
nodes');ylabel('Prob. of data reliability');
legend('\fontsize{7} Naive scheme','\fontsize {7} Wang \it{et
al.}  [10],\tau=1',...
      '\fontsize {7} Wang \it{et al.}  [10],\tau<1','\fontsize
{7} Ours, \tau=1','\fontsize {7} Ours, \
tau<1','Location','NorthEast');

figure
plot(1:healing+1,average_net_healing/node_no)

%AMIRAMIRAMIRAMIRAMIRAMIRAMIRAMIRAMIRAMIRAMIRAMIRAMIRAMIRAMIR
AMIRAMIRAMIR
%AMIRAMIRAMIRAMIRAMIRAMIRAMIRAMIRAMIRAMIRAMIRAMIRAMIRAMIRAMIR
AMIRAMIRAMIR
%AMIRAMIRAMIRAMIRAMIRAMIRAMIRAMIRAMIRAMIRAMIRAMIRAMIRAMIRAMIR
AMIRAMIRAMIR


if adv_TR==30
    save adv_tr30
    title({['(F)'],['n=',num2str(node_no),'
itr=',num2str(nTime),'  rounds  ',num2str(heal_rnds),'  ADV
spd=',num2str(adv_vol),'   ADV range=',num2str(adv_TR),'
sensor rang=',num2str(sensor_TR),'   sensor vol
',num2str(sensor_vol)]})
elseif adv_TR==60
```

```
     save
adv_TR60_100it_50healing_round_3step_random_motion_for_each_
sensor_increment_&_decrement_in_location
     title({['(F)'],['n=',num2str(node_no),'
itr=',num2str(nTime),'   rounds   ',num2str(heal_rnds),'   ADV
spd=',num2str(adv_vol),'   ADV range=',num2str(adv_TR),'
sensor rang=',num2str(sensor_TR),'   sensor vol
',num2str(sensor_vol)]})
elseif adv_TR==90
    save adv_TR90
     title({['(F)'],['n=',num2str(node_no),'
itr=',num2str(nTime),'   rounds   ',num2str(heal_rnds),'   ADV
spd=',num2str(adv_vol),'   ADV range=',num2str(adv_TR),'
sensor rang=',num2str(sensor_TR),'   sensor vol
',num2str(sensor_vol)]})
else
    save adv_TR120
     title({['(F)'],['n=',num2str(node_no),'
itr=',num2str(nTime),'   rounds   ',num2str(heal_rnds),'   ADV
spd=',num2str(adv_vol),'   ADV range=',num2str(adv_TR),'
sensor rang=',num2str(sensor_TR),'   sensor vol
',num2str(sensor_vol)]})
end
toc


endtime=clock;
disp(['start time  = '   num2str(start)])
disp(['end time   = '   num2str(endtime)])
beep,beep,beep,beep,beep,beep,beep,beep,beep,beep,beep,beep,be
ep,beep;

ssss=0;
for dddd=1:1000
    ssss=ssss+length(find(Iter_node_dataset(:,4)==dddd));
end
disp(['    ' num2str(ssss)])
%length(find(Iter_node_dataset(:,4)==1))+length(find(Iter_
node_dataset(:,4)==2))+length(find(Iter_node_dataset(:,4)==3))
+length(find(Iter_node_dataset(:,4)==4))+length(find(Iter_
node_dataset(:,4)==5))+length(find(Iter_node_dataset(:,4)==6))
+length(find(Iter_node_dataset(:,4)==7))+length(find(Iter_
node_dataset(:,4)==8))+length(find(Iter_node_dataset(:,4)==9))
+length(find(Iter_node_dataset(:,4)==10))+length(find(Iter_
node_dataset(:,4)==11))+length(find(Iter_node_dataset(:,4)==12
))+length(find(Iter_node_dataset(:,4)==13))+length(find(I
ter_node_dataset(:,4)==14))+length(find(Iter_node_dataset(:,4)
==15))+length(find(Iter_node_dataset(:,4)==16))+length(find(I
ter_node_dataset(:,4)==17))+length(find(Iter_node_dataset(:,4)
==18))+length(find(Iter_node_dataset(:,4)==19))+length(find(I
ter_node_dataset(:,4)==20))+length(find(Iter_node_dataset(:,4)
```

```
==21))+length(find(Iter_node_dataset(:,4)==22))+length(find(I
ter_node_dataset(:,4)==23))+length(find(Iter_node_dataset(:,4)
==24))+length(find(Iter_node_dataset(:,4)==25))+length(find(I
ter_node_dataset(:,4)==26))+length(find(Iter_node_dataset(:,4)
==27))+length(find(Iter_node_dataset(:,4)==28))+length(find(I
ter_node_dataset(:,4)==29))+length(find(Iter_node_dataset(:,4)
==30))+length(find(Iter_node_dataset(:,4)==31))+length(find(I
ter_node_dataset(:,4)==32))+length(find(Iter_node_dataset(:,4)
==33))+length(find(Iter_node_dataset(:,4)==34))+length(find(I
ter_node_dataset(:,4)==35))+length(find(Iter_node_dataset(:,4)
==36))+length(find(Iter_node_dataset(:,4)==37))+length(find(I
ter_node_dataset(:,4)==38))+length(find(Iter_node_dataset(:,4)
==39))+length(find(Iter_node_dataset(:,4)==40))+length(find(I
ter_node_dataset(:,4)==41))+length(find(Iter_node_dataset(:,4)
==42))+length(find(Iter_node_dataset(:,4)==43))+length(find(I
ter_node_dataset(:,4)==44))+length(find(Iter_node_dataset(:,4)
==45))+length(find(Iter_node_dataset(:,4)==46))+length(find(I
ter_node_dataset(:,4)==47))+length(find(Iter_node_dataset(:,4)
==48))+length(find(Iter_node_dataset(:,4)==49))+length(find(I
ter_node_dataset(:,4)==50))+length(find(Iter_node_dataset(:,4)
==51))+length(find(Iter_node_dataset(:,4)==52))+length(find(I
ter_node_dataset(:,4)==53))+length(find(Iter_node_dataset(:,4)
==54))+length(find(Iter_node_dataset(:,4)==55))+length(find(I
ter_node_dataset(:,4)==56))+length(find(Iter_node_dataset(:,4)
==57))+length(find(Iter_node_dataset(:,4)==58))+length(find(I
ter_node_dataset(:,4)==59))+length(find(I
ter_node_dataset(:,4)==60))

beep;for i=1:1000000,i;end;beep;for i=1:1000000,i;end;beep;for
i=1:1000000,i;end;beep;for i=1:1000000,i;end;
beep;for i=1:1000000,i;end;beep;for i=1:1000000,i;end;beep;for
i=1:1000000,i;end;beep;for i=1:1000000,i;end;
beep;for i=1:1000000,i;end;beep;for i=1:1000000,i;end;beep;for
i=1:1000000,i;end;beep;for i=1:1000000,i;end;
beep;for i=1:1000000,i;end;beep;for i=1:1000000,i;end;beep;for
i=1:1000000,i;end;beep;for i=1:1000000,i;end;
beep;for i=1:1000000,i;end;beep;for i=1:1000000,i;end;beep;for
i=1:1000000,i;end;beep;for i=1:1000000,i;end;

%**************************************************************
**
%**************************************************************
**
```

Appendix E: MATLAB®
Simulation Codes
for Chapter 9

```
%*******************************************************************
%      ------- Theoretical Results -----------
%*******************************************************************
**
% MAIN DECLERATION    16-1-2017
% "DF:" refer to a definition
% 1st modification instead of using a predefined compromised
probability, i
% select it totally random
clc;
clear all;
rand('state',0);
start=clock;
tic
%###################################################################
#########################
%################# PARAMETERTS DEFINITION ##################

nTime=200; % number of iteration to reduce randomness
heal_rnds=10;
node_no=500;% node number
maxx=500;    % width of the network
maxy=500;    % length of the network

sensor_TR=50;      % Transmission range
sensor_vol=10;    % Sensor velocity note used now
c_p=0.2;        %compromising probability (the ability of ADV to
compromise network) for example if p=0.2 this mean the ADV has
the ability to compromise 0.2 from the network
adv_TR=sqrt(((c_p*node_no)*(maxx*maxy/node_no))/3.1429);
adv_vol=10; % ADV velocity note used now
```

```
single_attack=1;            % "single_attack=1" refer to single
attack which attacks the network only one time during the
iteration
many_attacks_parallel=0;  % "many_attacks_parallel=1" refer to
MANY attacks those attack the network "No_of_parallel_attacks"
times at the same time during the same round at different
location before the self healing process started
No_of_parallel_attacks=2; % "No_of_parallel_attacks" refers to
the number of parallel attacks
many_attacks_serial=0;     % "many_attacks_serial=1" refer to
MANY attacks those attack the network at a successive times
during the successive rounds at different location before and
during the self healing process

motion_steps=22;      % In case of no mobility (no nodes will
move(CHSHRD scheme)), only we put "motion_steps=0"
                      % In case of mobility (some nodes will
move(SH-CCM scheme)), only we put "motion_step > 0"
moving__type=2;       % If "moving__type=1" this mean that the
sensors allowed to move are (only healed)or(only sick)or(both
healed and sick)
                      % If "moving__type=2" this mean that only
some of the best healed sensors will be allowed to move toward
the cluster center (ADV position)
select_heal_move=1; % This refer to the number of the best
healed sensors will be allowed to move toward the cluster
center (ADV position)

Folded_Network=2;

% in papers 394 and 568 and book 569 we have the following
statements:
% The energy consumed due to mobility will be given by
% Em=m·dm,
% m: is the movement parameter, measured in Joule/meter,
%     is a constant based on the aforementioned factors,
%     and dm is the distance traversed by the mobile sensor in
meter.
% " According to [16], a wheeled vehicle with rubber tires at
one kilogram
%     moving on concrete must overcome 0.1N force of dynamic
friction
%     in other words it expends 0.1J/m (N=J/m)"
% " As mentioned in Section 3,m at 3 J/m is suitable for
wheeled robots
%     at 30-kg moving on flat surface. Depending on the
application,
% this mean that
%             30  kgm sensor will have    m=3     J/m
%  so for   1   Kgm sensor will have    m=0.1   J/m
```

```
%   also for 0.1 Kgm sensor will have    m=0.01 J/m
%   also for 0.2 Kgm sensor will have    m=0.02 J/m
%   also for 0.3 Kgm sensor will have    m=0.03 J/m
%   also for 0.5 Kgm sensor will have    m=0.05 J/m


sensor_mass=0.5;    %this refer to the mass of the sensor by
Kilogram

mobility_coefficiant=0.1;

update_t=1;   % Updating time
Pr_t=0.3; % Threshold probability of compromising
Pr_up=0.001;% Upbound threshold probability


Communication_cost_in_Nb_selct=0;
Communication_cost_in_Nb_qualified_selct=0;

Iter_node_set=zeros(node_no,32);   % 29 refer to the number of
columns in which data measured stored
Iter_node_dataset=[];   % this matrix changes every nTime
iteration, it contains the data of the matrix (node),
                            % and it is updated every
(nTime) iteration
self_healing=[];


alhpa= 4   % pathloss exponent ranges from 2:6 , we use it
equal 4
etxPA=0.000000000001     % the energy consumed by power
amplifier to transmit one bit by joul
ecct= 0.0000001  % the energy consumed by the transmitter
circuit to transmit one bit by joul
eccr= 0.0000001  % the energy consumed by the receiver circuit
to receive one bit  by joul
packet_size=1000 % the packet size in byte
No_of_bits=8*packet_size  % the number of bits per packet
Eo=12424; % initial energy of each sensor by joul

distance_energy_matrix=[];
% This matrix stores both the following:
  % 1- Distance between sick sensor and one of its peers
  % 2- The transmission energy consumption corresponding to
this distance
counter1=1;

coverage_matrix_over_point_all_itt=zeros(node_no,node_no);
      % This matrix stores the coverage number (the number of
sensors cover
    % a specific point) of a number of (500*500) points in the
network
```

```
%#############################################################
############################
%##################### THE BEGINNING #####################

for iTime=1:nTime

    vooice=rem(iTime,25);
    if vooice==0
        beep
        ss=num2str(iTime);
        tts(ss,'3',-5,8000)
    end

    distance_energy_matrix=[];
    % This matrix stores both the following:
    % 1- Distance between sick sensor and one of its peers
    % 2- The transmission energy consumption corresponding to
this distance
    counter1=1;

    clear node;     % To clear the previous data stored in
matrix "node"
    sick_node_befor_heal=0;
    % this parameter stors the number of sick sensors before
each healing
    % round starting from the first round
    sick_node_after_heal=0;
    % this parameter stores the number of sick sensors after
each healing
    % round starting from the first round
    still_sickkkkkk=0;
    % This variable refer to the sick sensors those have no
qualified
    % neighbours "qlf_candd_no=0"

    energy_consumed_all_heal_round=0;
    total_Comm_distance_all_heal_round=0;
    total_sick_node_per_heal_round=0;
    five_power_parts_all_heal_round=[];
    coverage_matrix_all_round=[];
    sum_power_parts_all_heal_round=0;

    %DDDDDDDDDDDDDDDDDDDDDDDDDDDDDDDDDDDDDDDDDDDDDDDDDDDDDDDDDDD
DDDDDDDDDDDDDD
    %DDDDDDDDDDDDDDDDDDDDDDDDDDDDDDDDDDDDDDDDDDDDDDDDDDDDDDDDDDD
DDDDDDDDDDDDDD
    %DDDDDDDDDDDDDDDDDDDDDDDDDDDDDDDDDDDDDDDDDDDDDDDDDDDDDDDDDDD
DDDDDDDDDDDDDD
    % NETWORK DEPLOYMENT AND INITIALAIZATION
```

```
    [X_cordinats , Y_cordinats] =  topo_AS(node_no, maxx,
maxy);
    node(:,1)=X_cordinats;
    node(:,2)=Y_cordinats;

    disp(['1   size of node matrix  ', num2str(size(node))])
    % DF "topo_AS()": is a function used to generate columns 1
and 2 in
    % matrix node as the following:
    % column 1 = node(:,1) contains the X coordinates of all
sensors
    % column 2 = node(:,2) contains the Y coordinates of all
sensors
    % This function run only one time each iteration to
distribute the nodes
    % at their coordinates (X,Y).
    node =[node,zeros(node_no,25)]; % at this line size(node)=
500 * 27
    node(:,28)=Eo; % initial energy of each sensor stored in
column (28)
    node(:,29)=0;  % consumed energy of each sensor due to
transmission (TX_COMM_COST)
    node(:,30)=0;  % consumed energy of each sensor due to
reception (RX_COMM_COST)
    node(:,31)=0;  % consumed energy of each sensor due to
mobility (MOBILITY_COST)
    node(:,32)=0;
    ro=1;
    for jjj=1:node_no


[nb_set,nb_no,nb_set_index,qlf_candd_set,qlf_candd_no,qlf_
candd_index,not_qlf_candd_set,not_qlf_candd_no,not_qlf_candd_
set_index] =...
        nb_qlf_node_AS(jjj,Pr_t,sensor_TR,node,node_no);
        qualified_neighbours(jjj)=qlf_candd_no;
        % "qualified_neighbours" stores the "qlf_candd_no" for
each sensors
    end
    no_of_non_zero_elements=length(find(qualified_neighbours));
    while no_of_non_zero_elements<node_no
        disp('NETWORK HAS AN ISOLATED SENSORs NODE which has
no neighbours')
        %tts('network has an isolated sensors','3',-5,8000)
        [X_cordinats , Y_cordinats] =  topo_AS(node_no, maxx,
maxy);
        node(:,1)=X_cordinats;
        node(:,2)=Y_cordinats;
        for jjj=1:node_no
```

```
[nb_set,nb_no,nb_set_index,qlf_candd_set,qlf_candd_no,qlf_
candd_index,not_qlf_candd_set,not_qlf_candd_no,not_qlf_candd_
set_index] =...
            nb_qlf_node_AS(jjj,Pr_t,sensor_TR,node,node_no);
            qualified_neighbours(jjj)=qlf_candd_no;
      end
    no_of_non_zero_elements=length(find(qualified_neighbours));
    end

%%%%%%%%Plot_Circle_range_Around_Sensor(node,1,adv_TR,sensor_
TR,single_attack,many_attacks_serial,many_attacks_parallel,No_
of_parallel_attacks)
f=1;
    disp('GODE  NETWORK DEPLOYMENT  GODE  NETWORK DEPLOYMENT
GODE  NETWORK DEPLOYMENT')
    % the above "for" and "while"  loop is to guarantee that
the network
    % deployment has not any isolated sensors
    %DDDDDDDDDDDDDDDDDDDDDDDDDDDDDDDDDDDDDDDDDDDDDDDDDDDDDDDDDDDD
DDDDDDDDDDDDDD
    %DDDDDDDDDDDDDDDDDDDDDDDDDDDDDDDDDDDDDDDDDDDDDDDDDDDDDDDDDDDD
DDDDDDDDDDDDDD
    %DDDDDDDDDDDDDDDDDDDDDDDDDDDDDDDDDDDDDDDDDDDDDDDDDDDDDDDDDDDD
DDDDDDDDDDDDDD


    %AAAAAAAAAAAAAAAAAAAAAAAAAAAAAAAAAAAAAAAAAAAAAAAAAAAAAAAAA
AAAAAAAAAAAAAA
    %AAAAAAAAAAAAAAAAAAAAAAAAAAAAAAAAAAAAAAAAAAAAAAAAAAAAAAAAA
AAAAAAAAAAAAAA
    %AAAAAAAAAAAAAAAAAAAAAAAAAAAAAAAAAAAAAAAAAAAAAAAAAAAAAAAAA
AAAAAAAAAAAAAA
    if (single_attack==1||many_attacks_serial==1)
        [node    ,
no_of_sick_sensor_curentrly_sick]=adv_rand_loc_
AS_1(maxx,maxy,node,node_no,adv_TR,ro);
                disp(['  QQQQQQQQQQQQQ  no_of_sick_sensor_
curentrly_sick=
',num2str(no_of_sick_sensor_curentrly_sick)])
    elseif many_attacks_parallel==1
            [node
]=adv_rand_loc_AS_1_parllel(maxx,maxy,node,node_no,adv_TR,No_
of_parallel_attacks);
            % in this case we should change the self healing
process because we
            % have "No_of_parallel_attacks" clusters exist at
the same time
    end
```

```
    %AAAAAAAAAAAAAAAAAAAAAAAAAAAAAAAAAAAAAAAAAAAAAAAAAAAAAAAAA
AAAAAAAAAAAAA
    %AAAAAAAAAAAAAAAAAAAAAAAAAAAAAAAAAAAAAAAAAAAAAAAAAAAAAAAAA
AAAAAAAAAAAAA
    %AAAAAAAAAAAAAAAAAAAAAAAAAAAAAAAAAAAAAAAAAAAAAAAAAAAAAAAAA
AAAAAAAAAAAAA


%%%%%%%Plot_Circle_range_Around_Sensor(node,1,adv_TR,sensor_
TR,single_attack,many_attacks_serial,many_attacks_parallel,No_
of_parallel_attacks)
f=2;


    %##########################################################
#############
    % The start of the healing process
    %-------------------------------%

    for healing=1:heal_rnds

        disp(['iteration number  ', num2str(iTime), '
Healling round nunber  ', num2str(healing)]);
        disp(['node coulm 22  and 23',num2str(node(:,22)')])
        disp(['node coulm 22  and 23',num2str(node(:,23)')])
        disp(['    ',num2str(sum(sqrt((node(:,22)).^2+(node
(:,23)).^2)))])

    % nodes that are sick during the first round and after the
first
    % attack and before healing
        total_power_consumption_in_comm_per_iteration=0;
        total_distance=0;  % "total_distance"  stores the
total distance over which the communication between sick
sensor and its peers are carried and vice verse (between peers
and sick sensor)
        total_neighbours=0;
        total_qualified_neigbours=0;
        total_processs=0;
        part1=0;part2=0;part3=0;part4=0;part5=0;
        node(:,22)=0;
        node(:,23)=0;
        node(:,24)=0;
        node(:,25)=0;
        node(:,26)=0;
        node(:,27)=0;
```

```
        if healing==1
           current_sick_node_before_healing=find(node(:,3)>=0.3);
           % the above line compute the number of
(current_sick_node_before_healing)
           % after the ADV attacks and before the healing started

%%%%%%%%Plot_Circle_range_Around_Sensor(node,healing,adv_
TR,sensor_TR,single_attack,many_attacks_serial,many_attacks_
parallel)
f=3;
                % this function used to plot the circle range of
any sensor

                ONE_HEALLING_BEFORE_MOBILITY
                % the above function used to perform the first
healing process
                % during the first healing round before mobility
occurrence and
                % at the network level
                disp([' the sick sensors healed in first round   '
num2str(length(heeeleeed))])
                node(heeeleeed,3)=node(heeeleeed,8);
                disp ([' the healed sensors BSe=   ',
num2str(node(heeeleeed,3)')])
                sick_node=find(node(:,3)>=0.3);
                sick_node_after_heal(healing)=length(sick_node);
                disp(['the No. of sick_node after Healing round
No.   ' , num2str(healing),  '   is  '   ,num2str(length(sick_
node))  ])
                disp('*********************************************
**************')

                current_sick_node_after_healing=find
(node(:,3)>=0.3);
                % the above line compute the number of
(current_sick_node_after_healing)
                % after the healing started

%%%%%%%%Plot_Circle_range_Around_Sensor(node,healing,adv_
TR,sensor_TR,single_attack,many_attacks_serial,many_attacks_
parallel,No_of_parallel_attacks)
f=4;

        else    %else of (if healing==1)

%%%%%%%%Plot_Circle_range_Around_Sensor(node,healing,adv_
TR,sensor_TR,single_attack,many_attacks_serial,many_attacks_
parallel,No_of_parallel_attacks)
 f=5;
                % this function used to plot the circle range of
any sensor
```

```
        ro=healing;

        node=health_neighbors_before(node,sensor_TR);
        % DF: "health_neighbors_before" this function used
to find the number
        % of health neighbours before mobility


        %MMMMMMMMMMMMMMMMMMMMMMMMMMMMMMMMMMMMMMMMMMMMMMMMM
MMMMMMMMMMMMMMM
        %MMMMMMMMMMMMMMMMMMMMMMMMMMMMMMMMMMMMMMMMMMMMMMMMM
MMMMMMMMMMMMMMM
        %MMMMMMMMMMMMMMMMMMMMMMMMMMMMMMMMMMMMMMMMMMMMMMMMM
MMMMMMMMMMMMMMM
        % MOBILITY APPLIED IN THIS PART after the first
round of first
        % healing and before the second attack appear (in
case of many attacks)
        %-----------------------------%
        if find(sick_node~=0)   % This condition to ensure
that when we have no sick node we have no mobility
            if ro==2&&single_attack==1
                ro
                [node] =
node_motion_varied_cluster_Fold_modify1(sick_node,heeeleeed,
maxx,maxy,sensor_vol,update_t,node,motion_steps,ro,adv_
TR,moving__type,select_heal_move);
                % The above function apply mobility
considering the network is Folded but with the following
modification:
                % 1- the motion is restricted inside the
square region surrounded the ADV location
                % 2- it is expected that we will have not
empty region
            elseif many_attacks_serial==1
                ro
                [node] =
node_motion_varied_cluster_Fold_modify1(sick_node,heeeleeed,
maxx,maxy,sensor_vol,update_t,node,motion_steps,ro,adv_
TR,moving__type,select_heal_move);
                % The above function apply mobility
considering the network is Folded but with the following
modification:
                % 1- the motion is restricted inside the
square region surrounded the ADV location
                % 2- it is expected that we will have not
empty region
            elseif (ro==2&&many_attacks_parallel==1)
                ro
```

```
                    [node]=node_motion_varied_cluster_Fold_
modify1_parallel( node_no , maxx,maxy,sensor_vol,update_t,node,
motion_steps,No_of_parallel_attacks,adv_TR,moving__type,
select_heal_move);
            end
        end
        % the above function will move all nodes or sick
nodes only
        % many times within the network based on the
principals that
        % mobility can enhance the security by moving the
sick sensors
        % to another position in the network with a more
healthy neighbors.

%----------------------------------------------------------------%
        %MMMMMMMMMMMMMMMMMMMMMMMMMMMMMMMMMMMMMMMMMMMMMMMMMMMM
MMMMMMMMMMMMMMM
        %MMMMMMMMMMMMMMMMMMMMMMMMMMMMMMMMMMMMMMMMMMMMMMMMMMMM
MMMMMMMMMMMMMMM

        %viscircles([new_X,  new_Y],sensor_TR,'EdgeColor',
'k','LineWidth',3)


        node=health_neighbors_after(node,sensor_TR);
        % DF: "health_neighbors_after" this function used
to find the number
        % of health neighbours after mobility




        %AAAAAAAAAAAAAAAAAAAAAAAAAAAAAAAAAAAAAAAAAAAAAAAAAA
AAAAAAAAAAAAAA
        %AAAAAAAAAAAAAAAAAAAAAAAAAAAAAAAAAAAAAAAAAAAAAAAAAA
AAAAAAAAAAAAAA

        if many_attacks_serial==1
            %4- it can be used in case single adv attack

node=adv_rand_loc_AS_1(maxx,maxy,node,node_no,adv_TR,ro);
        end

%----------------------------------------------------------%
        %AAAAAAAAAAAAAAAAAAAAAAAAAAAAAAAAAAAAAAAAAAAAAAAAAA
AAAAAAAAAAAAAA
        %AAAAAAAAAAAAAAAAAAAAAAAAAAAAAAAAAAAAAAAAAAAAAAAAAA
AAAAAAAAAAAAAA
```

```
                current_sick_node_before_healing=find
(node(:,3)>=0.3);
                % the above line compute the number of
(current_sick_node_before_healing)
                % after the ADV attacks and before the healing
started


                sick_node=find(node(:,3)>=0.3);    % "sick_node" is
column
                sick_node=sick_node';              %  % "sick_
node" is row
                sick_node_befor_heal(healing)=length(sick_node);
                %%%disp(['sick_node before      '
num2str(length(sick_node))   '   Healing round   '
num2str(healing)])
                %    disp('beforeBBBBBBBBBBBBBBBBBBBBBBBBBB')




        %HHHHHHHHHHHHHHHHHHHHHHHHHHHHHHHHHHHHHHHHHHHHHHHHHHH
HHHHHHHHHHHHHHH
        %HHHHHHHHHHHHHHHHHHHHHHHHHHHHHHHHHHHHHHHHHHHHHHHHHHH
HHHHHHHHHHHHHHH
        %HHHHHHHHHHHHHHHHHHHHHHHHHHHHHHHHHHHHHHHHHHHHHHHHHHH
HHHHHHHHHHHHHHH
                heeeleeed=[];
                for i =sick_node,
                    %%%%%%disp(['selected_held_ID_to_move=
',num2str(selected_held_ID), '    and its BSe is
',num2str(node(selected_held_ID,3)) ])
                        %%%%%%dist_bet_selected_held_ID_and_i=sqrt
((node(i,1)-node(selected_held_ID,1)).^2+(node(i,2)-
node(selected_held_ID,2)).^2)
                    %%%%%%disp([' i= ', num2str(i),'   at distant=
',num2str(dist_bet_selected_held_ID_and_i),'   from the healed
id=   ',num2str(selected_held_ID)])%,'   number of sick node
after for 1=  ', num2str(length(sick_node))])

                    %for jjj=1:node_no,
                        %if node(jjj,3)>node(jjj,18)
                        %else
                            %node(jjj,3)=node(jjj,18);
                        %end
                    %end
```

```
                % the above lines guarantee that if si is
compromised by
                % adv1 and adv2 , then the final compromising
probability
                % is the greater one; for example if si
compromising
                % probability is .5 due to adv1 and it is .6
due to
                % the adv2, then the effective compromising
probability
                % is the greater one (0.6)


[nb_set,nb_no,nb_set_index,qlf_candd_set,qlf_candd_no,qlf
_candd_index,not_qlf_candd_set,not_qlf_candd_no,not_qlf_candd_
set_index] =...
                nb_qlf_node_AS(i,Pr_t,sensor_TR,node,node_no);

                % DF "nb_qlf_node_AS": it is a function used
to compute
                % many parameters such as:
                % 1- Copmute neighbor node set "nb_set" and
neighbor node
                % number "nb_no" which are in the transmission
range of si
                % 2- Copmute the qualified candidate set
"qlf_candd_set"
                % and the qualified candidate number "qlf_
candd_no", this
                % is a part of the neighbor set but there
compromised
                % probability not exceed the threshold
probability.


                node(i,4)=nb_no;        % total number of
neighbors for each sensor i
                node(i,5)=qlf_candd_no;% number of qualified
neighbor nodes for each sensor i
                %%%%%disp(['sick sensor i  is=  ',
num2str(i),'  all neighbours=  ', num2str(nb_no),'  qlf_candd_
no=  ', num2str(qlf_candd_no),'   qlf_candd_index=  ',
num2str(qlf_candd_index)])
                %%%disp(['(prod(qlf_candd_set)=  ',
num2str(prod(qlf_candd_set))])
                %%%disp('@@@@@@@@@@@@@@@@@@@@@@@@@@@@@@@@@@@@')


                total_neighbours=total_neighbours+nb_no;
```

```
total_qualified_neigbours=total_qualified_neigbours+qlf_candd_
no;

                %CCCCCCCCCCCCCCCCCCCCCCCCCCCCCCCCCCCCCCCCCCCCC
CCCCCCCCCCCCCC

                %CCCCCCCCCCCCCCCCCCCCCCCCCCCCCCCCCCCCCCCCCCCCC
CCCCCCCCCCCCCC
                % COMMUNICATION COST
    %-------------------------------------------------------------
                % each sick sensor i will transmit a packet to
all surrounding sensors within its range
                % only neighbors will receive this packet
                % "nb_set" will transmit a response packet to
declare about it self
                % sick sensor will receive these response
packet and for a table contains all its neighbors
                if (nb_no>0&&Communication_cost_in_Nb_selct==1)
                    for sens_index=nb_set_index;
                        distat= topo_distance_AS(i,sens_index,
node);
                        ETX=((distat^alhpa)*etxPA+ecct)*
No_of_bits;
                        % the energy consumed in the
transmitter
                        ERX=eccr*No_of_bits;
                        % the energy consumed in the receiver
                        part1=part1+2*(ETX+ERX);

total_power_consumption_in_comm_per_iteration=total_power_
consumption_in_comm_per_iteration+2*(ETX+ERX);
                        node(i,29)=node(i,29)+ETX;
                        node(sens_index,30)=node
(sens_index,30)+ERX;
                        node(sens_index,29)=node
(sens_index,29)+ETX;
                        node(i,30)=node(i,30)+ERX;
                        total_distance=total_distance+2*
distat;
                        total_processs=total_processs+4;
                        distance_energy_matrix(counter1,:)=
[distat   ETX ];
                        counter1=counter1+1;
                    end
                end
                %%%disp(['nb_no
',num2str(total_power_consumption_in_comm_per_iteration)])
                % sick sensor i will transmit a request packet
to its neighbours asking them about their state (compromising
probability) and
```

```
                % "nb_set" will receive this request
                % "nb_set" will transmit a packet to  sick i
carrying their state (compromising probability)
                % Sick sensor i will receive these packets and
perform some computation processing to re-order
                % its neighbors according to their
compromising probability
                % and it can assign its qualified neighbors
                if
(qlf_candd_no>0&&Communication_cost_in_Nb_qualified_selct==1)
                    for ses_index=qlf_candd_index;
                    % "ses_index" is the index of one of
the qualified neighbour
                    distat= topo_distance_AS(i,ses_index,
node);
                    ETX=((distat^alhpa)*etxPA+ecct)*
No_of_bits;
                    % the energy consumed in the
transmitter
                    ERX=eccr*No_of_bits;
                    % the energy consumed in the receiver
                    part2=part2+2*(ETX+ERX);
total_power_consumption_in_comm_per_iteration=total_power_
consumption_in_comm_per_iteration+2*(ETX+ERX);
                    node(i,29)=node(i,29)+ETX;
                    node(ses_index,30)=node
(ses_index,30)+ERX;
                    node(ses_index,29)=node
(ses_index,29)+ETX;
                    node(i,30)=node(i,30)+ERX;
                    total_distance=total_distance+2*
distat;
                    total_processs=total_processs+4;
                    distance_energy_matrix(counter1,:)=
[distat  ETX ];
                    counter1=counter1+1;
                end
            end
            %%%disp(['qlf_candd_no
',num2str(total_power_consumption_in_comm_per_iteration)])

   %-------------------------------------------------------------
            %CCCCCCCCCCCCCCCCCCCCCCCCCCCCCCCCCCCCCCCCCCCCCCCCCC
CCCCCCCCCCCCCC
            %CCCCCCCCCCCCCCCCCCCCCCCCCCCCCCCCCCCCCCCCCCCCCCCCCC
CCCCCCCCCCCCCC
```

```
                %RRRRRRRRRRRRRRRRRRRRRRRRRRRRRRRRRRRRRRRRRRRRRR
RRRRRRRRRRRRR
                %RRRRRRRRRRRRRRRRRRRRRRRRRRRRRRRRRRRRRRRRRRRRRR
RRRRRRRRRRRRR
                % Naive scheme in column 6 and 10 in matrix
node for both
                % cases (tau=1) and (tau<1)

%-----------------------------------------------%
                node(i,6)=node(i,3);     % Pr_BSe_comp_naive,
Naive scheme, the BSe of s_i
                %disp(['1- node(i,3)=    ',num2str(node(i,3)),'
node(i,6)=    ',num2str(node(i,6)),   '      ' num2str(i)])
                node(i,10)=1-node(i,3); % Pr_reliab_naive,
Naive scheme, data reliability
                %disp(['before Sk_N ID and Comp  ' num2str(i)
'  '  num2str(node(i,3))    '  Sk_N nb_no  ' num2str(nb_no)  '
Sk_N nb_no qlf_candd_no  '  num2str(qlf_candd_no)])
                %%%disp('SSSSSSSSSSSSSSSSSSSSSSSSSSSSSSSSSSSSSS
SSSS')
                %%%disp(['sick i=  ', num2str(i),'  BSe
',num2str(node(i,6)),'  sum Naive=  ',num2str(node(i,6)+node
(i,10))])

%------------------------------------------------------------%
                %RRRRRRRRRRRRRRRRRRRRRRRRRRRRRRRRRRRRRRRRRRRRRR
RRRRRRRRRRRRR
                %RRRRRRRRRRRRRRRRRRRRRRRRRRRRRRRRRRRRRRRRRRRRRR
RRRRRRRRRRRRR



                %RRRRRRRRRRRRRRRRRRRRRRRRRRRRRRRRRRRRRRRRRRRRRR
RRRRRRRRRRRRR
                %RRRRRRRRRRRRRRRRRRRRRRRRRRRRRRRRRRRRRRRRRRRRRR
RRRRRRRRRRRRR
                % Proposed Scheme Proposed Scheme   Proposed
Scheme

%-------------------------------------------------%
                % data cannot be recovered, if one data part
is lost.
                % If there is enough qualified candidates,
data is distributed stored. Otherwise, data is stored locally.
                if (qlf_candd_no > 0)
                    if (prod(qlf_candd_set)< Pr_up)
                        %%%%disp(['i will be health   ',
num2str(i)])
                        % tau=1, that is m = n = t, then
Pr_BSe_comp =prod_{i=1}^{m}P_{i,j}*P_i
```

```
                        sort_qlf_candd_set=sort(qlf_candd_set);
                        % "sort_qlf_candd_set" is an ordered
version of "qlf_candd_set"
                        %%%disp(['qlf_candd_set=
',num2str(qlf_candd_set')])
                        %%%disp(['sort_qlf_candd_set=
',num2str(sort_qlf_candd_set')])
                        %%%disp(['(prod(qlf_candd_set)=  ',
num2str(prod(qlf_candd_set))])
                        randperm_qlf_candd_set=
qlf_candd_set(randperm(qlf_candd_no));
                        % "randperm_qlf_candd_set" is a
different order version of "qlf_candd_set"
                        randperm_nb_set =
                        nb_set(randperm(nb_no));
                        % "randperm_nb_set" is a different
order version of "nb_set"
                        % in other words the same values exist
in both matrices but
                        % with different order
                        sort_qlf_candd_set_reliab = 1-sort_
qlf_candd_set;      % AMIRAMIRAMIR   ADDED
                        randperm_qlf_candd_set_reliab =
1-randperm_qlf_candd_set;
                        randperm_nb_set_reliab =
1-randperm_nb_set;
                        % note that (qlf_candd_set)  &  (nb_
set) contain the comprised probability
                        % and (randperm_qlf_candd_set_reliab)
& (randperm_nb_set_reliab) contain the reliable probability
which
                        % equal to (1-comprised probability)

                        for  temp_n = 1:length(sort_qlf_candd
_set)
                        % here we select the first t peers
from
                        % "sort_qlf_candd_set" which have
the lowest
                        % compromising probabilties
                        % so that the product of the
compromising probability of these
                        % t peers is lower than the
upperbound threshold
                        % probability.
                        % NOTE: The final value of
"temp_n" refers to the number
                        % of selected peers which can
satisfy the required
                        % security level (Pr_up)
```

```
                              if   prod(sort_qlf_candd_
set(1:temp_n))> Pr_up
                              % if the above condition is
true, this mean that the
                              % number of peers not enough
and the for loop
                              % operates to increase temp_n.
                              % if the condition is false,
the control switch to
                              % "else" statement.
                              %disp(['temp_n 1
',num2str(temp_n),'       the product       ',
num2str(prod(sort_qlf_candd_set(1:temp_n)))])
                              else
                              % "else" mean that  the
"prod(sort_qlf_candd_set(1:temp_n))" is lower than "Pr_up"
                              % and the selected pears is
enough and there is no
                              % need to increase "temp_n"
and the for loop is terminated
                              %disp(['temp_n 2
',num2str(temp_n),'       the product       ',
num2str(prod(sort_qlf_candd_set(1:temp_n)))])
                              break;  % this break will
terminate the for loop
                              end
                              %disp(['temp_n 3
',num2str(temp_n),'       the product       ',
num2str(prod(sort_qlf_candd_set(1:temp_n)))])
                              end   % End of (for  temp_n =
1:length(sort_qlf_candd_set))
                              %disp(['temp_n 4    ',num2str(temp_n),'
the product       ',
num2str(prod(sort_qlf_candd_set(1:temp_n)))])




                              %RRRRRRRRRRRRRRRRRRRRRRRRRRRRRRRRRRRRRRR
RRRRRRRRRRRRRRRRRR
                              %RRRRRRRRRRRRRRRRRRRRRRRRRRRRRRRRRRRRRRRR
RRRRRRRRRRRRRRRRRR
                              % Proposed scheme for tau=1
                              %-------------------------
                              node(i,8)=prod(sort_qlf_candd_
set(1:temp_n))*node(i,3);   % tau=1. comp. prob.
                              node(i,12)=1-prod(sort_qlf_candd_
set(1:temp_n));        % 1- tau=1. reliability prob.
                              %%%node(i,12)=1-node(i,8);
% 2- tau=1. reliability prob.
```

```
                        % BOTH 1 AND 2 CAN BE USED IN
COMPUTING THE DATA
                        % RELIABILITY WITH A VERY SMALL
DIFFERENCE


                        %node(i,12)=prod(randperm_qlf_candd_
set_reliab(1:temp_n));  %YIREN tau=1. reliability prob.
                        %node(i,12)=prod(sort_qlf_candd_set_
reliab(1:temp_n));  %AMIRAMIR tau=1. reliability prob.
                        % chose the first "temp_n" nodes from
the "sort_qlf_candd_set"
                        %which have the lower compromising
probability and multiply
                        %them with each other and with the
compromising probability
                        %of the sick node itself..............
WE MUST USE THE SAME
                        %NODES WHICH USED TO ENHANCE THE
COMPROMOISING PROBABILITY
                        %OF THE SICK NODE .....  TO CACULATE
THE DATA RELIABILITY  OF
                        %THIS SICK NODE      SO THAT THE SUM OF
COLUMNS 8 AND 12 =1
                        % THIS IS THE SAME DONE BY YIREN FOT
TAU>1


                        % Proposed scheme for tau>1
                        %--------------------------
                        % Proposed scheme, m>(nt)/(t+2), n=t
                        m=ceil((temp_n*temp_n)/(temp_n+2));
% this is law exist in [58]
                        node(i,9)=prod(sort_qlf_candd_
set(1:m))*node(i,3);   % tau>1. comp. prob.
                        node(i,13)=1-prod(sort_qlf_candd_
set(1:m));             % 1- tau>1. reliability prob.
                        %%%node(i,13)=1-node(i,9);
% 2- tau>1. reliability prob.
                        % BOTH 1 AND 2 CAN BE USED IN
COMPUTING THE DATA
                        % RELIABILITY WITH A VERY SMALL
DIFFERENCE

                        %%%disp([' sick i=  ', num2str(i),'
BSe  ',num2str(node(i,8)),'  sum proposed tau=1 =  ',num2str
(node(i,8)+node(i,12))])
                        %%%disp([' sick i=  ', num2str(i),'
BSe  ',num2str(node(i,9)),'  sum proposed tau>1 =  ',num2str
(node(i,9)+node(i,13))])
```

```
        %--------------------------------------------------------
                            %RRRRRRRRRRRRRRRRRRRRRRRRRRRRRRRRRRRRRRR
RRRRRRRRRRRRRRRRRR

                            %RRRRRRRRRRRRRRRRRRRRRRRRRRRRRRRRRRRRRRR
RRRRRRRRRRRRRRRRRR



                            %#####################################
##################
                            % Wang's Infocom 09 scheme   Wang's
Infocom 09 scheme

        %---------------------------------------------------------%
                            %%%node(i,7)=prod(nb_set(1:temp_n))*node
(i,3);                  % tau=1   YIREN comp. prob
                            node(i,7)=prod(randperm_nb_
set(1:temp_n))*node(i,3);       % tau=1   AMIRAMIR comp. prob
                            node(i,11)=1-prod(randperm_nb_
set(1:temp_n));             % 1- tau=1 reliability prob.
                            %%%node(i,11)=1-node(i,7);
% 2- tau=1 reliability prob.
                            % BOTH 1 AND 2 CAN BE USED IN
COMPUTING THE DATA
                            % RELIABILITY WITH A VERY SMALL
DIFFERENCE


                            %  Randomly select m nodes
                            node(i,14)=prod(randperm_nb_
set(1:m))*node(i,3);         % tau>1 comp. prob
                            node(i,15)=1-prod(randperm_nb_
set(1:m));                   % 1- tau>1 reliability prob.
                            %%%node(i,15)=1-node(i,14);
% 2- tau>1 reliability prob.
                            % BOTH 1 AND 2 CAN BE USED IN
COMPUTING THE DATA
                            % RELIABILITY WITH A VERY SMALL
DIFFERENCE


                            %%%disp([' sick i=  ', num2str(i),'
BSe  ',num2str(node(i,7)),'  sum Wangs Infocom 09 tau=1 =
',num2str(node(i,7)+node(i,11))])
                            %%%disp([' sick i=  ', num2str(i),'
BSe  ',num2str(node(i,14)),'  sum Wangs Infocom 09 tau>1 =
',num2str(node(i,14)+node(i,15))])
        %---------------------------------------------------------
```

```
%#####################################
##################

%node(i,3)=node(i,8);  % 12/12/2013
% this line make that if the sensor i is healed it will help
% the others during the same round so we can say in the same
% round the sick sensor healed first will help the others
% sick sensors in the same round. I think this not accepted
% because that all sick sensors start the healing process in
% the same round.

heeeleeed=[heeeleeed  i];
%disp(['  hhheeellleeeddd222222  '  num2str(qlf_candd_no)  '   '  num2str(qlf_candd_set')  '  '  num2str(prod(sort_qlf_candd_set(1:temp_n)))  '  '  num2str(prod(sort_qlf_candd_set(1:temp_n))*node(i,3))])




%CCCCCCCCCCCCCCCCCCCCCCCCCCCCCCCCCCCCCCCCCCCC
%CCCCCCCCCCCCCCCCCCCCCCCCCCCCCCCCCCCCCCCCCCCCC
% COMMUNICATION COST
%-----------------------------------
% in this part the sick sensor i will transmit a packet to its qualified neigbours request for help
% "qlf_candd_set" will receive this request
% "qlf_candd_set" will response with a packet carrying their contribution to the sick sensor i
% sick sensor i will receive these contribution and perform
% self-healing process
selected_sensor_indices=0;
% "selected_sensor_indices" is used to store the sensor
% indicies which are used in self healing process to
% satisfy the required security level
for xc=1:temp_n
    %%%disp(['temp_n=  ', num2str(temp_n), '   xc= ', num2str(xc)])
```

```
                                sdd=find(qlf_candd_set==sort_qlf_
candd_set(xc));
                                selected_sensor_indices(xc)=qlf_
candd_index(sdd);
                        end
                        % up to now we can get the sensors
indices which are selected
                        % to perform self-healing and satisfy
the required security
                        % level, these sensors which
contribute in selfhealing by
                        % transmitting data packets for the
sick sensor and hence we
                        % have communication cost due to their
transmission
                        for sens=selected_sensor_indices;
                                distat= topo_distance_AS(i,sens,
node);
                                ETX=((distat^alhpa)*etxPA+ecct)*
No_of_bits;
                                % the energy consumed in the
transmitter
                                ERX=eccr*No_of_bits;
                                % the energy consumed in the
receiver
                                part3=part3+2*(ETX+ERX);

total_power_consumption_in_comm_per_iteration=total_power_
consumption_in_comm_per_iteration+2*(ETX+ERX);
                                node(i,29)=node(i,29)+ETX;
                                node(sens,30)=node(sens,30)+ERX;
                                node(sens,29)=node(sens,29)+ETX;
                                node(i,30)=node(i,30)+ERX;
                                total_distance=total_distance+2*
distat;
                                total_processs=total_processs+4;
                                distance_energy_matrix(counter1,:)
=[distat  ETX ];
                                counter1=counter1+1;
                        end
                        %%%disp(['contribution
',num2str(total_power_consumption_in_comm_per_iteration)])


                        %-----------------------------------%
%CCCCCCCCCCCCCCCCCCCCCCCCCCCCCCCCCCCCCCCCCCCCCCCCCCCCCCCCCCCCCCCC
%CCCCCCCCCCCCCCCCCCCCCCCCCCCCCCCCCCCCCCCCCCCCCCCCCCCCCCCCCCCCCCCC

                        else    % else of  (prod(qlf_candd_set)<
Pr_up))
```

```
                                % this mean that the (qlf_candd_no =
0) which mean that there is no qualified neighbor nodes
                                % and hence there is no self-healing
(cooperation) and any compromised probability will be
                                % the defined compromised probability
defined in columns 3. And any reiablel probability wil be
                                % (1-compromised probability)

                                %%%%disp(['prod(qlf_candd_set)=
',num2str(prod(qlf_candd_set))])
                                %%%%disp(['  qlf_candd_no=
',num2str(qlf_candd_no)])
                                %%%%vvvvv=55555
                                %%%disp([' sick i=  ', num2str(i),'
BSe ',num2str(node(i,8)),'  sum proposed tau=1 =  ',num2str(node
(i,8)+node(i,12))])
                                %%%disp([' sick i=  ', num2str(i),'
BSe ',num2str(node(i,9)),'  sum proposed tau>1 =  ',num2str(node
(i,9)+node(i,13))])
                                %%%disp([' sick i=  ', num2str(i),'
BSe  ',num2str(node(i,7)),'  sum Wangs Infocom 09 tau=1 =
',num2str(node(i,7)+node(i,11))])
                                %%%disp([' sick i=  ', num2str(i),'
BSe  ',num2str(node(i,14)),'  sum Wangs Infocom 09 tau>1 =
',num2str(node(i,14)+node(i,15))])



%CCCCCCCCCCCCCCCCCCCCCCCCCCCCCCCCCCCCCCCCCCCCCCCCCCCCCCCCCCCCCCCCC
%CCCCCCCCCCCCCCCCCCCCCCCCCCCCCCCCCCCCCCCCCCCCCCCCCCCCCCCCCCCCCCCCC
                                % it should be power consumption in
this region for
                                % sensors that cannot heal during this
round but they
                                % consume power when trying to self
heal
                                for ses_index=qlf_candd_index;

                                % "ses_index" is the index of one
of the qualified neighbour
                                distat= topo_distance_AS(i,ses_
index,node);

                                ETX=((distat^alhpa)*etxPA+ecct)*
No_of_bits;
                                % the energy consumed in the
transmitter
                                ERX=eccr*No_of_bits;
                                % the energy consumed in the
receiver
```

```
                                part4=part4+2*(ETX+ERX);
total_power_consumption_in_comm_
per_iteration=total_power_consumption_in_comm_per_
iteration+2*(ETX+ERX);
                           node(i,29)=node(i,29)+ETX;
                           node(ses_index,30)=node
(ses_index,30)+ERX;

                           node(ses_index,29)=node
(ses_index,29)+ETX;

                           node(i,30)=node(i,30)+ERX;
                           total_distance=total_
distance+2*distat;

                           total_processs=total_processs+4;
                           distance_energy_matrix(counter1,:)=
[distat  ETX ];

                           counter1=counter1+1;
                     end
%CCCCCCCCCCCCCCCCCCCCCCCCCCCCCCCCCCCCCCCCCCCCCCCCCCCCCCCCCCCCCCC
%CCCCCCCCCCCCCCCCCCCCCCCCCCCCCCCCCCCCCCCCCCCCCCCCCCCCCCCCCCCCCCC
                     end            %  end of (prod(qlf_candd_
set)< Pr_up))
                else
                     node(i,8)=node(i,3);node(i,7)=node(i,3);
                     node(i,9)=node(i,3);node(i,14)=node(i,3);
                     node(i,12)=1-node(i,3);node(i,11)=
1-node(i,3);

                     node(i,13)=1-node(i,3);node(i,15)=
1-node(i,3);

                     %%% the above four lines has NO any affect
on the

                     %%% results in case of single attack
                     %%% we should ensure the behaviour in case
of many

                     %%% attacks

                     still_sickkkkkkk=still_sickkkkkkk+1;
                     %%%%disp(['    still_sickkkkkkk
',num2str(i), '  else of  ((qlf_candd_no > 0)&&(prod(qlf_
candd_set)< Pr_up))    ',' qlf_candd_no= ',num2str
(qlf_candd_no)])
                     %%%%disp(['  qlf_candd_no=
',num2str(qlf_candd_no)])
                     %%%disp([' after Sk_N ID and Comp  '
num2str(i)  '  '   num2str(node(i,3))   '  Sk_N nb_no  '
num2str(nb_no)  '  Sk_N nb_no qlf_candd_no  '
num2str(qlf_candd_no)])
                     %%%disp(['nb_no_set  ', num2str(nb_set')])
                     %%%disp(['qlf_candd_set  ',
num2str(qlf_candd_set')])
```

```
                  %%%disp(['final- node(i,3)=
',num2str(node(i,3))   '     '     num2str(i)])
                  %%%disp('**************************QWEQ
WQQWQWQWWQWQWQWQWQWQW*******************************')


                  %CCCCCCCCCCCCCCCCCCCCCCCCCCCCCCCCCCCCCCCC
CCCCCCCCCCCCC
                  %CCCCCCCCCCCCCCCCCCCCCCCCCCCCCCCCCCCCCCCC
CCCCCCCCCCCCC
                  %CCCCCCCCCCCCCCCCCCCCCCCCCCCCCCCCCCCCCCCC
CCCCCCCCCCCCCC
                  % I think we should have power consumption
in this
                  % part, hence the sensor reach this point
has the follwing
                  % data:
                  % 1- It consumes power only during
neighbours select
                  % ."nb_no"
                  % 2- it has no qualified neighbours
"qlf_candd_no=0"
                  % so it does not  consume any power
during qualified
                  % neighbours selection, but it tries to
select the
                  % qualified neighbours so we must
consider this power
                  % consumption during the trying of
getting its
                  % qualified neigbours
                  % To solve the above problem, we will use
the part
                  % of power consumption during neighbours
selection as
                  % the following
                  for sens_index=nb_set_index;
                      if sens_index>0
                          distat= topo_distance_AS(i,sens_
index,node);
                          % The above line mean that the
sensor try to send to its
                          % neighbours again trying to find
any qualified neighbours
                      else
                          distat=sensor_TR;
                      end

                      ETX=((distat^alhpa)*etxPA+ecct)*
No_of_bits;
```

```
                          % the energy consumed in the
transmitter
                          ERX=eccr*No_of_bits;
                          % the energy consumed in the receiver,
hence
                          % the "qlf_candd_no=0", therevieving
power not
                          % consider
                          part5=part5+2*ETX;
total_power_consumption_in_comm_per_iteration=total_power_
consumption_in_comm_per_iteration+2*ETX;
                          node(i,29)=node(i,29)+2*ETX;
                          % Only "i" transmits two times as we
assume trying
                          % to get its qualified neighbours
                          %node(sens_index,30)=node
(sens_index,30)+ERX;
                          % No qualified neighbours to receive

                          %node(sens_index,29)=node
(sens_index,29)+ETX;
                          % No qualified neighbours to transmit

                          %node(i,30)=node(i,30)+ERX;
                          % "i" not receive any thing

                          total_distance=total_distance+
2*distat;
                          total_processs=total_processs+2;
                          distance_energy_matrix(counter1,:)=
[distat  ETX ];
                          counter1=counter1+1;

                     end
                     % the above block of power consumption to
substitute
                     % the power consumption of the sick sensor
which has no
                     % qualified neighbours "qlf_candd_no=0".
In reality we
                     % know that this sensor consumes power in
communication
                     % to get its qualified neighbours but they
are not exist
                     % In this case we assume that the sick
sensor transmits
                     % two times trying two get qualified
neighbours, and
                     % there is no receiving because....
("qlf_candd_no=0")
```

```
                     %CCCCCCCCCCCCCCCCCCCCCCCCCCCCCCCCCCCCCCCCCCCC
CCCCCCCCCCCCCC
                     %CCCCCCCCCCCCCCCCCCCCCCCCCCCCCCCCCCCCCCCCCCCC
CCCCCCCCCCCCCC
                     %CCCCCCCCCCCCCCCCCCCCCCCCCCCCCCCCCCCCCCCCCCCC
CCCCCCCCCCCCCC


            end    %  end of ( (qlf_candd_no > 0)
         end            %  end of (for i = 1:sick_node)

         disp([' in round    ' num2str(healing) '  there are
'  num2str(length(heeeleeed)) '  sick sensors are healled'])

         node(heeeleeed,3)=node(heeeleeed,8);
         disp ([' the healed sensors BSe=  ',
num2str(node(heeeleeed,3)')])

         sick_node=find(node(:,3)>=0.3);
         sick_node_after_heal(healing)=length(sick_node);
         disp(['the No. of sick_node after Healing round
No.  ', num2str(healing),  '  is  '  ,num2str(length(sick_
node))  ])
disp('*****************************************************')

         current_sick_node_after_healing=find
(node(:,3)>=0.3);
         % the above line compute the number of
(current_sick_node_after_healing)
         % after the healing started
%%%%%%%Plot_Circle_range_Around_Sensor(node,healing,adv_TR,
sensor_TR,single_attack,many_attacks_serial,many_attacks_
parallel,No_of_parallel_attacks)
 f=6;
      end    %End of (if healing==1)




energy_consumed_all_heal_round(healing)=total_power_consumption_
in_comm_per_iteration;
        % this will store the amount of energy consumed in
each round
        % separetely (individually)
        total_Comm_distance_all_heal_round(healing)=total_
distance;
        total_processs_all_heal_round(healing)=total_processs;

        total_No_of_nibrs_all_heal_round(healing)=total_
neighbours;   %1/4/2016
```

```
        total_No_of_quilfd_all_heal_round(healing)=total_
qualified_neigbours;  % 1/4/2016
total_sick_node_all_round_before_heal(healing)=length(current_
sick_node_before_healing);
total_sick_node_all_round_after_heal(healing)=length(current_
sick_node_after_healing);
        five_power_parts_all_heal_round(healing,:)=[part1
part2  part3  part4  part5];
        sum_power_parts_all_heal_round(healing)=sum([part1
part2  part3  part4  part5]);
        moving_distance_all_heal_round(healing)=sum(sqrt((node
(:,22)).^2+(node(:,23)).^2));

        healing

        total_Nb_before_mobility_all_round(healing)=length
(find(node(:,24)));
        quilfid_Nb_before_mobility_all_round(healing)=length(
find(node(:,25)));
        total_Nb_after_mobility_all_round(healing)=length
(find(node(:,26)));
        quilfid_Nb_after_mobility_all_round(healing)=length
(find(node(:,27)));

        [coverage_matrixx] =COVERAGE_MEASUREMENT_1(node,
sensor_TR,node_no);
        coverage_matrix_all_round(healing,:)=coverage_matrixx;
        % this function used to give the number of neighbours
for each
        % sensor, and store it in the "coverage_matrixx" which
has a length
        % of 1x500, and this will refer to the coverage in the
network,




        bse_all_values(healing)=mean(node(:,8));%5-8-2016
        bse(healing)=sum(node(:,8))./find_over_column_by_
Amir(node(:,8));%5-8-2016
```

```
    end    % End of (for  healing=1:heal_rnds)
    five_power_parts_all_itt(iTime,:)=mean(five_power_parts_
all_heal_round);
    sum_power_parts_all_itt(iTime,:)=sum_power_parts_all_
heal_round;
    still_sickkkkkkk_for_all_itt(iTime)=still_sickkkkkkk;
    moving_distance_for_all_itt(iTime,:)=moving_distance_
all_heal_round;

    coverage_matrix_all_itt(iTime,:)=mean(coverage_matrix_
all_round);

    energy_consumed_for_all_itt(iTime,:)=energy_consumed_all_
heal_round;

    total_Comm_distance_for_all_itt(iTime,:)=total_Comm_
distance_all_heal_round;
    total_processs_for_all_itt(iTime,:)=total_processs_all_
heal_round;
    % the above equation stores the power consumption in
communication for
    % all healing rounds and for all iteration times

    total_No_of_nibrs_all_itt(iTime,:)=total_No_of_nibrs_all_
heal_round;
    total_No_of_quilfd_all_itt(iTime,:)=total_No_of_quilfd_
all_heal_round;

total_sick_node_before_heal_all_itt(iTime,:)=total_
sick_node_all_round_before_heal;

total_sick_node_after_heal_all_itt(iTime,:)=total_sick_
node_all_round_after_heal;

    adversary_location_all_itt(iTime,:)=[node(1,16)
node(1,17)];
    power_consump_in_Comm_all_itt(iTime)=sum(node(:,29))+sum
(node(:,30));

total_Nb_before_mobility_all_itt(iTime,:)=total_Nb_
before_mobility_all_round;

quilfid_Nb_before_mobility_all_itt(iTime,:)=quilfid_Nb_
before_mobility_all_round;

total_Nb_after_mobility_all_itt(iTime,:)=total_Nb_after_
mobility_all_round;

quilfid_Nb_after_mobility_all_itt(iTime,:)=quilfid_Nb_
after_mobility_all_round;
```

```
    [coverage_matrix_over_point,coverage_th]
=COVERAGE_MEASUREMENT(node,sensor_TR,node_no,maxx,maxy);

coverage_matrix_over_point_all_itt=coverage_matrix_over_point_
all_itt+coverage_matrix_over_point;
    % the above function and line are very time consuming,
because at each round the above function computes the coverage
of (500*500)
    % points in the network, it is very time consuming, So we
will transfer this function so that instead to be computed
each round
    % it will be computed each iteration


    Neighbours_for_each_sensoe_index(iTime,:)=node(:,4);
    qlfd_Nibrs_for_each_sensoe_index(iTime,:)=node(:,5);



    Iter_node_set=Iter_node_set+node;              % sum all
the node matrices nTime
    %      node(not_sick_node,:)=[];     % Amir
    Iter_node_dataset=[Iter_node_dataset;node];     % store
all the node matrices
    %size(Iter_node_dataset)
    self_healing=[self_healing;sick_node_befor_heal,0,sick_
node_after_heal];
    % this matrix stores in each raw the following as shown in
the following
    % vector: [sick_node_befor_heal,0,sick_node_after_heal]



    col_6(iTime,:)=node(:,6);      col_10(iTime,:)=node(:,10);

    col_8(iTime,:)=node(:,8);      col_12(iTime,:)=node(:,12);
    col_9(iTime,:)=node(:,9);      col_13(iTime,:)=node(:,13);

    col_7(iTime,:)=node(:,7);      col_11(iTime,:)=node(:,11);
    col_14(iTime,:)=node(:,14);     col_15(iTime,:)=node(:,15);

    bse_all_values_itt(iTime,:)=bse_all_values; %5-8-2016
    bse_itt(iTime,:)=bse;                        %5-8-2016
```

```
    %figure;hold on; for jjjj=1:500; plot(node(jjjj, 1),
node(jjjj, 2), 'b.', 'MarkerSize', 7);
        %text(node(jjjj, 1),node(jjjj,
2),num2str(node(jjjj,3)));     end              plot(node(1, 16),
node(1, 17), 'r.', 'MarkerSize', 20)%;text(node(jj,
1),node(jj, 2),num2str(jj));            vvv=10;
end        % End of (for iTime=1:nTime)



disp('RRRRRRRRRRRRRRRRRRRRR    RESULTS    RRRRRRRRRRRRRRRRRRRRRR
RRRRRRRRRRRRRRRRRRRRRRRRRRRRRR')
disp(['motion steps=    ', num2str(motion_steps),'    single
attacks=',num2str(single_attack),'    many attacks
serial=',num2str(many_attacks_serial),'   many attacks
parallel=',num2str(many_attacks_parallel),'    selected healed
to move=   ',num2str(select_heal_move),'    folded network
enabled  '])
disp(['(iteration=  ',num2str(nTime),')--(heal rounds= ',
num2str(heal_rnds),')--(sensor range= ',num2str(sensor_TR),')-
-(ADV rangr= ',num2str(126.13)])
disp(['Comm_cost_in_Nb_selct=    ',num2str(Communication_cost_
in_Nb_selct),'   Comm_cost_in_Nb_qualified_selct=
',num2str(Communication_cost_in_Nb_qualified_selct)])
disp('=======================================================
=========')
average_power_consumption_in_Comm_per_itt=mean(power_consump_
in_Comm_all_itt);
disp(['average_power_consumption_in_Comm_per_itt=
',num2str(average_power_consumption_in_Comm_per_itt)])
disp('                                                ')
mean_five_power_parts_per_iteration=mean
(five_power_parts_all_itt);
disp(['mean_five_power_parts_per_iteration=     ',
num2str(mean_five_power_parts_per_iteration)])
disp('                                                ')
mean_consumed_power_per_itt=sum(energy_consumed_for_all_itt)./
find_over_column_by_Amir(energy_consumed_for_all_itt);
disp(['mean_consumed_power_per_itt=
',num2str(mean_consumed_power_per_itt)])
mean_consumed_power_per_itt_over_all_values=mean
(energy_consumed_for_all_itt);
disp(['mean_consumed_power_per_itt_over_all_values=
',num2str(mean_consumed_power_per_itt_over_all_values)])
disp('                                                ')
mean_sum_power_parts_per_ittr=sum(sum_power_parts_all_itt)./
find_over_column_by_Amir(sum_power_parts_all_itt);
disp(['mean_sum_power_parts_per_ittr=
',num2str(mean_sum_power_parts_per_ittr)])
```

```
mean_sum_power_parts_per_ittr_over_all_values=mean
(sum_power_parts_all_itt);
disp(['mean_sum_power_parts_per_ittr_over_all_values=
',num2str(mean_sum_power_parts_per_ittr_over_all_values)])
disp('=========================================================
=========')
mean_moving_distance_for_per_itt=sum(moving_distance_for_
all_itt)./find_over_column_by_Amir
(moving_distance_for_all_itt);
disp(['mean_moving_distance_for_per_itt=
',num2str(mean_moving_distance_for_per_itt)])
mean_moving_distance_per_itt_over_all_values=mean
(moving_distance_for_all_itt);
disp(['mean_moving_distance_per_itt_over_all_values=
',num2str(mean_moving_distance_per_itt_over_all_values)])
disp('                                               ')
mean_Comm_distance_per_itt=sum(total_Comm_distance_for_all_
itt)./find_over_column_by_Amir(total_Comm_
distance_for_all_itt);
disp(['mean_Comm_distance_per_itt=
',num2str(mean_Comm_distance_per_itt)])
mean_Comm_distance_per_itt_over_all_values=mean
(total_Comm_distance_for_all_itt);
disp(['mean_Comm_distance_per_itt_over_all_values=
',num2str(mean_Comm_distance_per_itt_over_all_values)])
disp('=========================================================
=========')
mean_No_of_process_per_itt=sum(total_processs_for_all_itt)./
find_over_column_by_Amir(total_processs_for_all_itt);
disp(['mean_No_of_process_per_itt=
',num2str(mean_No_of_process_per_itt)])
mean_No_of_process_per_itt_over_all_values=mean
(total_processs_for_all_itt);
disp(['mean_No_of_process_per_itt_over_all_values=
',num2str(mean_No_of_process_per_itt_over_all_values)])


disp('=========================================================
=========')
mean_total_sick_nodes_before_heal=sum(total_sick_node_before_
heal_all_itt)./find_over_column_by_Amir(total_sick_node_
before_heal_all_itt);
disp(['mean_total_sick_nodes_before_heal=
',num2str(mean_total_sick_nodes_before_heal)])
mean_total_sick_nodes_before_heal_over_all_values=mean
(total_sick_node_before_heal_all_itt);
disp(['mean_total_sick_nodes_before_heal_over_all_values=
',num2str(mean_total_sick_nodes_before_heal_over_all_values)])
disp('                                               ')
```

```
mean_total_sick_nodes_after_heal=sum(total_sick_node_after_
heal_all_itt)./find_over_column_by_Amir(total_sick_node_
after_heal_all_itt);
disp(['mean_total_sick_nodes_after_heal=
',num2str(mean_total_sick_nodes_after_heal)])
mean_total_sick_nodes_after_heal_over_all_values=mean
(total_sick_node_after_heal_all_itt);
disp(['mean_total_sick_nodes_after_heal_over_all_values=
',num2str(mean_total_sick_nodes_after_heal_over_all_values)])

disp('=========================================================
=========')


mean_of_total_No_of_nibrs=sum(total_No_of_nibrs_all_itt)./
find_over_column_by_Amir(total_No_of_nibrs_all_itt);
disp(['mean_of_total_No_of_nibrs=                     ',num2str
(mean_of_total_No_of_nibrs./mean_total_sick_nodes_before_
heal)])
mean_of_total_No_of_nibrs_over_all_values=mean(total_No_of_
nibrs_all_itt);
disp(['mean_of_total_No_of_nibrs_over_all_values=
',num2str(mean_of_total_No_of_nibrs_over_all_values./mean_
total_sick_nodes_before_heal_over_all_values)])
disp('                                 ')
mean_total_No_of_quilfd_nibrs=sum(total_No_of_quilfd_all_
itt)./find_over_column_by_Amir(total_No_of_quilfd_all_itt);
disp(['mean_total_No_of_quilfd_nibrs=
',num2str(mean_total_No_of_quilfd_nibrs./mean_total_sick_
nodes_before_heal)])
mean_of_total_No_of_quilfd_nibrs_over_all_values=mean
(total_No_of_quilfd_all_itt);
disp(['mean_of_total_No_of_quilfd_nibrs_over_all_values= ',
num2str(mean_of_total_No_of_quilfd_nibrs_over_all_values./
mean_total_sick_nodes_before_heal_over_all_values)])
disp('                                 ')
disp(['ratio qualified/total neibs=
',num2str(mean_total_No_of_quilfd_nibrs./
mean_of_total_No_of_nibrs)])
disp(['ratio of quilf /total neigb all values=
',num2str(mean_of_total_No_of_quilfd_nibrs_over_all_values./
mean_of_total_No_of_nibrs_over_all_values)])
disp('=========================================================
=========')
mean_Neighbours_for_each_sensoe_index=sum(Neighbours_for_each_
sensoe_index)./find_over_column_by_Amir(Neighbours_for_
each_sensoe_index);
disp(['mean_Neighbours_for_each_sensoe_index=
',num2str(mean_Neighbours_for_each_sensoe_index)])
```

```
mean_Neighbours_for_each_sensoe_index_over_all_values=mean
(Neighbours_for_each_sensoe_index);
disp(['mean_Neighbours_for_each_sensoe_index_over_all_values=
',num2str(mean_Neighbours_for_each_sensoe_index_over_all_
values)])
mean_qlfd_Nibrs_for_each_sensoe_index=sum(qlfd_Nibrs_for_each_
sensoe_index)./
find_over_column_by_Amir(qlfd_Nibrs_for_each_sensoe_index);
disp(['mean_qlfd_Nibrs_for_each_sensoe_index=
',num2str(mean_qlfd_Nibrs_for_each_sensoe_index)])
mean_qlfd_Nibrs_for_each_sensoe_index_over_all_values=mean
(qlfd_Nibrs_for_each_sensoe_index);
disp(['mean_qlfd_Nibrs_for_each_sensoe_index_over_all_values=
',num2str(mean_qlfd_Nibrs_for_each_sensoe_index_over_all_
values)])
disp('=====================================================
=========')
mean_total_Nb_before_mobility_all_itt=mean(total_Nb_before_
mobility_all_itt);
disp(['mean_total_Nb_before_mobility_all_itt=
',num2str(mean_total_Nb_before_mobility_all_itt)])
mean_total_Nb_before_mobility_all_itt_over_all_values=sum
(total_Nb_before_mobility_all_itt)./find_over_column_by_Amir
(total_Nb_before_mobility_all_itt);
disp(['mean_total_Nb_before_mobility_all_itt_over_all_values=
',num2str(mean_total_Nb_before_mobility_all_
itt_over_all_values)])
mean_quilfid_Nb_before_mobility_all_itt=mean(
quilfid_Nb_before_mobility_all_itt);
disp(['mean_quilfid_Nb_before_mobility_all_itt=
',num2str(mean_quilfid_Nb_before_mobility_all_itt)])
mean_quilfid_Nb_before_mobility_all_itt_over_all_values=sum
(quilfid_Nb_before_mobility_all_itt)./
find_over_column_by_Amir(quilfid_Nb_before_mobility_all_itt);
disp(['mean_quilfid_Nb_before_mobility_all_itt_over_all_
values=  ',num2str(mean_quilfid_Nb_before_mobility_all_itt_
over_all_values)])
disp('                                                    ')
mean_total_Nb_after_mobility_all_itt=mean(total_Nb_after_
mobility_all_itt);
disp(['mean_total_Nb_after_mobility_all_itt=
',num2str(mean_total_Nb_after_mobility_all_itt)])
mean_total_Nb_after_mobility_all_itt_over_all_values=sum
(total_Nb_after_mobility_all_itt)./find_over_column_by_Amir
(total_Nb_after_mobility_all_itt);
disp(['mean_total_Nb_after_mobility_all_itt_over_all_values=
',num2str(mean_total_Nb_after_mobility_all_itt_over_
all_values)])
mean_quilfid_Nb_after_mobility_all_itt=mean(
quilfid_Nb_after_mobility_all_itt);
```

```
disp(['mean_quilfid_Nb_after_mobility_all_itt=
',num2str(mean_quilfid_Nb_after_mobility_all_itt)])
mean_quilfid_Nb_after_mobility_all_itt_over_all_values=sum
(quilfid_Nb_after_mobility_all_itt)./find_over_column_by_Amir
(quilfid_Nb_after_mobility_all_itt);
disp(['mean_quilfid_Nb_after_mobility_all_itt_over_all_values=
',num2str(mean_quilfid_Nb_after_mobility_all_itt_over_
all_values)])




mean_coverage_matrix_per_itt=sum(coverage_matrix_all_itt)./
find_over_column_by_Amir(coverage_matrix_all_itt);
meancoverage_matrix_per_itt_over_all_values=mean
(coverage_matrix_all_itt);
mean_coverage_matrix_over_point_per_itt=coverage_matrix_over_
point_all_itt./nTime;




net_healing=[self_healing(:,1) self_healing(:,(2+healing):
(2*healing+1))];
% "self_healing" matrix stores in each raw the following as
shown in
  % the following vector: [sick_node_befor_heal,0,sick_node_
after_heal], if
  % the healing round="10", then the vector length will be
"21"
% "net_healing" matrix stores in each raw the following as
shown in this
  % vector: [self_healing(:,1) self_healing(:,(2+healing):(2*h
ealing+1))]
average_net_healing=mean(net_healing);


mean_col_6=sum(col_6)./find_over_column_by_Amir(col_6);
mean_col_10=sum(col_10)./find_over_column_by_Amir(col_10);

mean_col_8=sum(col_8)./find_over_column_by_Amir(col_8);
mean_col_12=sum(col_12)./find_over_column_by_Amir(col_12);
mean_col_9=sum(col_9)./find_over_column_by_Amir(col_9);
mean_col_13=sum(col_13)./find_over_column_by_Amir(col_13);

mean_col_7=sum(col_7)./find_over_column_by_Amir(col_7);
mean_col_11=sum(col_11)./find_over_column_by_Amir(col_11);
mean_col_14=sum(col_14)./find_over_column_by_Amir(col_14);
mean_col_15=sum(col_15)./find_over_column_by_Amir(col_15);
```

```
mean_bse_all_values_itt=mean(bse_all_values_itt); %5-8-2016
mean_bse_itt=mean(bse_itt);                        %5-8-2016




%***********************************************************
*********************************
%***********************************************************
**************************
%*************** PLOTTING  PLOTTING  PLOTTING ***************
%------------------------------------------------------------
% mean of Iter_node_set
Iter_node_set=Iter_node_set/nTime;

nb_no_comp_reliab_set=unique(Iter_node_dataset(:,4));
% the above line compute the unique values of total neighbours
sensors
No_of_quif_set=unique(Iter_node_dataset(:,5));   %  AMIEAMIR
% the above line compute the unique values of qualified
neighbours sensors


nb_no_comp_reliab_set=[nb_no_comp_reliab_set,zeros(length
(nb_no_comp_reliab_set),10)];
% the above line to make "nb_no_comp_reliab_set" has a 11
columns
No_of_quif_set=[No_of_quif_set,zeros(length(No_of_quif_set),
10)];   %  AMIEAMIR
% the above line to make "No_of_quif_set" has a 11 columns

[ total_sick_node_before_heal_all_itt
adversary_location_all_itt];
%     plotting_part_code

%save('D:\PHD\1\__6_paper\results\Serial many attack\without_
mobility')
%save('D:\PHD\1\__6_paper\results\Serial many attack\
with_mobility_1_Heal_Move_to_center_of_sicks')
%save('D:\PHD\1\__6_paper\results\Serial many attack\with_
mobility_1_Heal_Move_to_the ADV location')
%save('E:\Amir_PHD\1\__6_paper\results\Serial many attack\
with_mobility_all_Healed_ will move randomly within the
cluster')

%save('D:\PHD\1\__6_paper\results\Parallel many attack\
without_mobility')
%save('D:\PHD\1\__6_paper\results\Parallel many attack\
with_mobility_1_Heal_Move_to_center_of_sicks')
```

```
%save('D:\PHD\1\__6_paper\results\Parallel many attack\with_
mobility_1_Heal_Move_to_the ADV location')
%save('E:\Amir_PHD\1\__6_paper\results\Parallel many attack\
with_mobility_all_Healed_ will move randomly within the
cluster')


%save('E:\Amir_PHD\1\__6_paper\results\Single attack\
without_mobility')
%save('E:\Amir_PHD\1\__6_paper\results\Single attack\
with_mobility_1_Heal_Move_to_center_of_sicks')
save('D:\PHD\1\__6_paper\results\Single attack\with_
mobility_1_Heal_Move_to_the ADV location')
%save('E:\Amir_PHD\1\__6_paper\results\Single attack\with_
mobility_all_Healed_ will move randomly within the cluster')
```

References

1. 3GPP TR 36.814 V1.2.1. "Further advancements for EUTRA: Physical layer aspects". *Technical Specification Group Radio Access n/w.* Jun. 2009.
2. I. P802.16j/D9. "Draft amendment to IEEE standard for local and metropolitan area network part 16: Air interface for fixed and mobile broadband wireless access systems: Multihop relay specification". May 2009.
3. Y. Yang, H. Hu, J. Xu, and G. Mao. "Relay technologies for WiMAX and LTE Advanced mobile systems". *IEEE Comm. Magazine* 47(10), 100–105, Oct. 2009.
4. E. H. Callaway. *Wireless Sensor Networks: Architectures and Protocols.* CRC Press, 2003.
5. H. M. Ammari. *The Art of Wireless Sensor Networks.* Springer, 2014.
6. I. Stojmenovic, Ed. *Handbook of Sensor Networks: Algorithms and Architectures.* Vol. 49. John Wiley & Sons, 2005.
7. T. M. Y. Vo and J. Talim. "Random distribution for data survival in unattended wireless sensor networks". In: *Sensor Technologies and Applications (SENSORCOMM), Fourth International Conference*, pp. 468–471, IEEE, 2010.
8. S. K. V. L. Reddy. *Data Security in Unattended Wireless Sensor Networks.* PhD Dissertation, University of Ottawa, 2013.
9. R. Di Pietro, L. V. Mancini, C. Soriente, A. Spognardi, and G. Tsudik. "Catch me (if you can): Data survival in unattended sensor networks". *Proc. of PERCOM '08*, Washington, DC, USA, IEEE Computer Society, pp. 185–194, 2008.
10. D. Ma and G. Tsudik. "Extended abstract: Forward-secure sequential aggregate authentication". *Proc. IEEE Symp. on Security and Privacy*, Oakland, CA, USA, pp. 86–91, May 2007.
11. R. D. Pietro, L. V. Mancini, C. Soriente, A. Spognardi, and G. Tsudik. "Playing hide-and-seek with a focused mobile adversary in unattended wireless sensor networks". *Ad Hoc Networks* 7(8), 1463–1475, 2009.
12. R. D. Pietro, L. V. Mancini, C. Soriente, A. Spognardi, and G. Tsudik. "Data security in unattended wireless sensor networks". *IEEE Transactions on Computers* 58(11), 1500–1511, 2009.
13. R. D. Pietro, L. V. Mancini, C. Soriente, A. Spognardi, and G. Tsudik. "Maximizing data survival in unattended wireless sensor networks against a focused mobile adversary". *IACR Cryptology e-Print Archive*, 1–23, 2008.
14. Information Processing Technology Office (IPTO) Defense Advanced Research Projects v Agency (DARPA), BAA 07-46 LANdroids Broad Agency Announcement, URL: http://www:darpa:mil=IPTO=solicit=open=BAA—07-46 P IP:pdf, 2007.

15. T. Dimitriou and A. Sabouri. "Pollination: A data authentication scheme for unattended wireless sensor networks". In: *Proc. IEEE Trust, Security and Privacy in Computing and Communications*, pp. 409–416, 2011.

16. R. D. Pietro, L. V. Mancini, C. Soriente, A. Spognardi, and G. Tsudik. "Collaborative authentication in unattended wireless sensor networks". *Proc. 2nd ACM Conference on Wireless Network Security*, Zurich, Switzerland, pp. 237–244, 16–19 Mar. 2009.

17. R. Di Pietro, D. Ma, C. Soriente, and G. Tsudik. "POSH: Proactive co-operative self-healing in unattended wireless sensor networks". *Proc. IEEE Symp. on Reliable Distributed Systems*, Napoli, Italy, pp. 185–194, Oct. 2008.

18. R. D. Pietro, G. Oligeri, C. Soriente, and G. Tsudik. "Securing mobile unattended WSNs against a mobile adversary". In: *Proc. IEEE Symp. on Reliable Distributed Systems*, pp. 11–20, 2010.

19. M. A. Santos, C. B. Margi, M. A. Jr, G. F. Pereira, and B. T. Oliveira. "Implementation of data survival in unattended wireless sensor networks using cryptography". *Proc. IEEE Local Computer Networks (LCN), Sense App*, Denver, Washington USA, IEEE Computer Society, pp. 961–967, 10–14 Oct. 2010.

20. Y. Wang, G. Attebury, and B. Ramamurthy. "A survey of security issues in wireless sensor networks". *IEEE Commun. Surveys Tutorials* 8, 2–23, 2006.

21. G. Padmavathi and D. Shanmugapriya. "Survey of attacks, security mechanisms and challenges in wireless sensor networks". *International Journal of Computer Science and Information Security (IJCSIS)*, 4(1&2), 2009.

22. Z. Ruan, X. Sun, W. Liang, D. Sun, and Z. Xia. "CADS: Co-operative anti-fraud data storage scheme for unattended wireless sensor networks". *Inform. Technol. J.* 9, 1361–1368, 2010.

23. W. Ren, J. Zhao, and Y. Ren. "MSS: A multi-level data placement scheme for data survival in wireless sensor networks". Fifth International Conference on Wireless Communications, Networking and Mobile Computing, IEEE, 2009.

24. V. Erceg, L. Greenstein, S. Tjandra, S. Parkoff, A. Gupta, B. Kulic, A. A. Julius, and R. Bianchi. "An empirically based path loss model for wireless channels in suburban environments". *IEEE Journal on Selected Areas in Communications* 17(7), 1205–1211, 1999.

25. A. Goldsmith. *Wireless Communications*. Cambridge University Press, 2005.

26. F. Owen and C. Pudney. "Radio propagation for digital cordless telephones at 1700 MHz and 900 MHz". *IEEE Electronics Letters* 25(1), 52–53, 1989.

27. T. Rappaport. *Wireless Communications—Principles and Practice*, 2nd ed. Englewood Cliffs, NJ: Prentice-Hall, 2001.

28. S. Seidel, T. Rappaport, S. Jain, M. Lord, and R. Singh. "Path loss, scattering, and multipath delay statistics in four European cities for digital cellular and microcellular radiotelephone". *IEEE Transactions on Vehicular Technology* 40(4), 721–730, 1991.

29. A. F. De Toledo, and A. M. D. Turkmani. "Propagation into and within buildings at 900, 1800, and 2300 MHz". In: *Proc. on IEEE Vehicular Technology Conference (VTC)*, 633–636, 1992.

30. A. F. De Toledo, A. M. D. Turkmani, and J. D. Parsons. "Estimating coverage of radio transmission into and within buildings at 900, 1800, and 2300 MHz". *IEEE Personal Communications Magazine* 5(2), 40–47, 1998.

31. T. S. Rappaport. *Wireless Communications*. Chaps. 3 and 4. Upper Saddle River, NJ: Prentice-Hall, 1996.

32. J. Garg, K. Gupta, and P. K. Ghosh. "Performance analysis of MIMO wireless communications over fading channels—A review". *International Journal of Advanced Research in Electrical, Electronics and Instrumentation Engineering* 2(4), Apr. 2013.

33. J. B. Andersen, T. S. Rappaport, and S. Yoshida. "Propagation measurements and models for wireless communications channels". *IEEE Communications Magazine*, 42–49, Jan. 1995.

34. P. M. Shankar. *Introduction to Wireless Systems*. John Wiley & Sons, 2002.

35. A. Sendonan, E. Erkip, and B. Hahang. "User cooperation diversity techniques. Part 1, System description". *IEEE Transactions on Communications*, 51(11), 1927–1938, Nov. 2003.

36. Y. Sitti. "Error performance analysis of cooperative systems with receiver diversity: Effect of shadowing in source–destination link". *Signal Processing and Communications Applications (SIU)*, 2011 IEEE 19th Conference, pp. 482–485, 20–22 Apr. 2011.

37. A. Singh. "Cooperative spectrum sensing in multiple antenna based cognitive radio network using an improved energy detector". *IEEE Communications Letters* 16(1), 64–67, Jan. 2012.

38. Y. Nasser. "Effect of mobility on the performance of amplify-and-forward cooperation in SC-FDMA systems". *Computing, Networking and Communications (ICNC)*, 2012 International Conference, pp. 798–803, Jan. 30 2012–Feb. 2 2012.

39. L. Zheng and D. N. C. Tse. "Diversity and multiplexing: A fundamental tradeoff in multiple antenna channels". *IEEE Transactions on Information Theory* 49(5), 1073–1096, May 2003.

40. M. Jankiraman. *Space–Time Codes and MIMO Systems*. Artech House, 2004.

41. B. Vucetic and J. Yuan. *Space–Time Coding*. John Wiley, 2003.

42. D. Gesbert, M. Shafi, D. Shiu, P. J. Smith, and A. Naguib. "From theory to practice: An overview of MIMO space–time coded wireless systems". *IEEE Journal on Selected Areas in Communications*. 21(3), 281–301, Apr. 2003.

43. L. Zheng and D. N. C. Tse. "Diversity and multiplexing: A fundamental tradeoff in multiple-antenna channels". *IEEE Transactions on Information Theory* 49(5), 1073–1096, May 2003.

44. E. C. der Meulen. "Three-terminal communication channels". *Advances in Applied Probability* 3(1), 120–154, 1971.

45. T. M. Cover and A. A. E. Gamal. "Capacity theorems for the relay channel". *IEEE Transactions on Information Theory* 25(5), 572–584, Sept. 1979.

46. K. J. Rayliu, A. K. Sadek, Weifengsu, and Andres Kwasinski. *Cooperative Communications and Networking*. Cambridge University Press, ISBN-13 978-0-511 -46548-2, 2009.

47. J. N. Laneman, G. W. Wornell, and D. N. C. Tse. "An efficient protocol for realizing cooperative diversity in wireless networks". *Proc. IEEE ISIT*, Washington, DC, p. 294, June 2001.

48. A. Sendonaris, E. Erkip, and B. Aazhang. "User cooperation diversity part I and part II". *IEEE Transactions on Communications* 51(11), 1927–1948, Nov. 2003.

49. A. Wyner and J. Ziv. "The rate-distortion function for source coding with side information at the decoder". *IEEE Transactions on Information Theory* 22(1), 1–10, Jan. 1976.

50. T. E. Hunter and A. Nosratinia. "Cooperative diversity through coding". In: *Proc. IEEE International Symposium Information Theory (ISIT)*, Laussane, Switzerland, p. 220, Jul. 2002.

51. A. Stefanov and E. Erkip. "Cooperative coding for wireless networks". *IEEE Transactions on Communications* 52(9), 1470–1476, Sept. 2004.

52. T. E. Hunter and A. Nosratinia. "Diversity through coded cooperation". *IEEE Transactions on Wireless Communications* 5(2), 283–289, Feb. 2006.

53. W. Su and X. Liu. "On optimum selection relaying protocols in cooperative wireless networks". *IEEE Transactions on Communications* 58(1), 52–57, Jan. 2010.

54. J. N. Laneman, D. N. C. Tse, and G. W. Wornell. "Cooperative diversity in wireless networks: Efficient protocols and outage behavior". *IEEE Transactions on Information Theory* 50(12), 3062–3080, 2004.

55. S. S. Ikki and M. H. Ahmed. "Performance analysis of decode-and-forward incremental relaying cooperative diversity networks over Rayleigh fading channels". Vehicular Technology Conference, 2009.

56. S. S. Ikki and M. H. Ahmed. "Performance analysis of incremental-relaying cooperative-diversity networks over Rayleigh fading channels". *IET Communications* 5(3), 337–349, Feb. 2011.

57. A. H. Bastami and A. Olifat. "Optimal incremental relaying in cooperative diversity systems". *IET Communications* 7(2), 152–168, Jan. 2013.

58. Q. Zhou and F. Lau. "Two incremental relaying protocols for cooperative networks". *IET Communications* 2(10), 1272–1278, Nov. 2008.

59. H. Long, K. Zheng, W. Wang, and F. Wang. "Approximate performance analysis of the incremental relaying protocol and modification". In: *Proc. IEEE 70th VTC-Fall*, Anchorage, AK, USA, pp. 1–5, Sep. 2009.

60. J. Kuang, C. Hu, H. Long, K. Zheng, and W. Wang. "Selective fractional incremental relaying protocol in cooperative systems". International Conference on Communications, Circuits and Systems (ICCCAS), 2010.

61. M. M. Fareed, M.-S. Alouini, and H.-C. Yang. "Efficient incremental relaying for packet transmission over fading channels". *IEEE Transactions on Wireless Communications* 13(7), 3609–3620, Jul. 2014.

62. E. Beres and R. Adve "Selection cooperation in multi-source cooperative net- works". *IEEE Transactions on Wireless Communications*, vol. 7, pp. 118–127, Jan. 2008.

63. A. Bletsas, A. Khisti, D. Reed, and A. Lippman. "A simple cooperative diversity method based on network path selection". *IEEE Journal on Selected Areas in Communications* 24, 659–672, Mar. 2006.

64. A. Bletsas, A. Khisti, and M. Win. "Cooperative communications with outage-optimal opportunistic relaying". *IEEE Transactions on Wireless Communications* 6(9), 3450–3460, Sept. 2007.

65. A. Adinoyi, Y. Fan, H. Yanikomeroglu, and V. Poor. "On the performance of selective relaying". In: *Proc. IEEE Vehicular Technology Conference (VTC) Fall*, Sept. 2008.

66. A. S. Ibrahim, A. K. Sadek, W. Su, and K. J. Liu. "Cooperative communications with relay selection: When to cooperate and whom to cooperate with". *IEEE Transactions on Wireless Communications* 7(7), 2814–2827, Jul. 2008.

67. E. Beres and R. Adve. "Selection cooperation in multi-source cooperative networks". *IEEE Transactions on Wireless Communications* 7, 118–127, Jan. 2008.

68. C.-K. Toh. *Ad Hoc Wireless Networks: Protocols and Systems.* Prentice Hall Publishers, 2002.

69. C. Siva Ram Murthy and B. S. Manoj. *Ad Hoc Wireless Networks: Architectures and Protocols.* Prentice Hall PTR, May 2004.

70. H. Ilhan, I. Altunbas, and M. Uysal. "Performance analysis and optimization of relay-assisted vehicleto-vehicle (v2v) cooperative communication". In: *Signal Processing, Communication and Applications Conference, 2008. SIU 2008. IEEE 16th*, pp. 1–4, Apr. 20–22, 2008.

71. J. Santa, A. Moragon, and A. F. Gomez-Skarmeta. "Experimental evaluation of a novel vehicular communication paradigm based on cellular networks". In: *Intelligent Vehicles Symposium, 2008 IEEE*, pp. 198–203, Jun. 4–6, 2008.

72. M. L. Sichitiu and M. Kihl. "Inter-vehicle communication systems: A survey". *IEEE Communications Surveys & Tutorials* 10, 88–105, second quarter 2008.

73. Y. Zou, J. Zhu, B. Zheng, and Y.-D. Yao. "An adaptive cooperation diversity scheme with best-relay selection in cognitive radio networks". *IEEE Transactions on Signal Processing* 58(10), Oct. 2010.

74. Y. Zou, B. Zheng, and Y.-D. Yao. "Outage probability analysis of cognitive transmissions: Impact of spectrum sensing overhead". *IEEE Transactions on Wireless Communications* 9(8), 2676–2688, June 2010.

75. E. Peh and Y. C. Liang. "Optimization for co-operative sensing in cognitive radio networks". *IEEE Wireless Communications and Networking Conference*, pp. 27–32, Mar. 2007.

76. A. Nosratinia, T. E. Hunter, and A. Hedayat. "Cooperative communication in wireless networks". *IEEE Communications Magazine* 42(10), 74–80, 2004.

77. M. Dohler, Y. Li. *Cooperative Communications Hardware, Channel & Phy*. John Wiley & Sons, ISBN 978-0-470-99768-0, 2010.

78. J. L. Massey. "An introduction to contemporary cryptology". *Proc. IEEE* 76(5), 533–549, May 1988.

79. G. Kapoor and S. Piramithu. "Vulnerabilities in some recently proposed RFID ownership transfer protocols". *IEEE Communications Letters* 14(3), 260–262, Mar. 2010.

80. A. Barenghi, L. Breveglieri, I. Koren, and D. Naccache. "Fault injection attacks on cryptographic devices: Theory, practice, and countermeasures". *Proc. IEEE* 100(11), 3056–3076, Nov. 2012.

81. X. Zhou, L. Song, and Y. Zhang, Eds. *Physical Layer Security in Wireless Communications*. CRC Press, 2013.

82. M. Bloch and J. Barros. *Physical-Layer Security: From Information Theory to Security Engineering*. Cambridge University Press, 2011.

83. E. Jorswieck, A. Wolf, and S. Gerbracht. *Secrecy on the Physical Layer in Wireless Networks*. In-Tech Publishers, 2010.

84. R. Liu and W. Trappe. *Securing Wireless Communications at the Physical Layer*. Norwell, MA: Springer, 2009.

85. C. E. Shannon. "Communication theory of secrecy systems". *The Bell System Technical Journal* 28(4), 656–715, Oct. 1949.

86. A. D. Wyner. "The wire-tap channel". *The Bell System Technical Journal* 54(8), 1355–1387, Jan. 1975.

87. I. Csiszár and J. Körner. "Broadcast channels with confidential messages". *IEEE Transactions on Information Theory* IT-24(3), 339–348, May 1978.

88. S. K. Leung-Yan-Cheong and M. E. Hellman. "The Gaussian wire-tap channel". *IEEE Transactions on Information Theory* IT-24(4), 451–456, Jul. 1978.

89. E. Tekin and A. Yener. "Achievable rates for the general Gaussian multiple access wire-tap channel with collective secrecy". In: *Proc. 44th Annu. Allerton Conf. Commun. Contr. Comput.* 2006.

90. E. Tekin and A. Yener. "The multiple access wire-tap channel: Wireless secrecy and cooperative jamming". In: *Proc. Information Theory and Applications Workshop*, 2007.

91. E. Tekin and A. Yener. "The general Gaussian multiple access and two-way wire-tap channels: Achievable rates and cooperative jamming". *IEEE Transactions on Information Theory* 54(6), 2735–2751, Jun. 2008.

92. E. Tekin and A. Yener. "The Gaussian multiple access wire-tap channel". *IEEE Transactions on Information Theory* 54(12), 5747–5755, Dec. 2008.

93. L. Lai and H. El Gamal. "The relay-eavesdropper channel: Cooperation form secrecy". *IEEE Transactions on Information Theory* 54(9), 4005–4019, Sept. 2008.

94. R. Bassily and S. Ulukus. "Deaf cooperation and relay selection strategies for secure communication in multiple relay networks". *IEEE Transactions on Signal Processing* 61(6), 1544–1554, Mar. 2013.

95. E. Ekrem and S. Ulukus. "Cooperative secrecy in wireless communications". In: *Securing Wireless Communications at the Physical Layer*, W. Trappe and R. Liu, Eds. New York: Springer-Verlag, pp. 143–172, 2009.

96. X. Tang, R. Liu, P. Spasojevic, and H. V. Poor. "The Gaussian wiretap channel with a helping interferer". In: *Proc. IEEE Int. Symp. Inf. Theory*, Toronto, ON, Canada, pp. 389–393, Jul. 2008.

97. X. He and A. Yener. "Providing secrecy with structured codes: Tools and applications to two-user Gaussian channels". *IEEE Transactions on Information Theory* 60(4), 2121–2138, 2014.

98. X. He and A. Yener. "Secure degrees of freedom for Gaussian channels with interference: Structured codes outperform Gaussian signaling". In: *Proc. IEEE Global Telecommun. Conf.*, 2009.

99. X. He. *Cooperation and Information Theoretic Security in Wireless Networks*. PhD Dissertation, Dept. Electr. Eng., Pennsylvania State Univ., State College, PA, USA, 2010.

100. O. O. Koyluoglu, H. E. Gamal, L. Lai, and H. V. Poor. "Interference alignment for secrecy". *IEEE Transactions on Information Theory* 57(6), 3323–3332, Jun. 2011.

101. A. S. Motahari, S. Oveis-Gharan, M. A. Maddah-Ali, and A. K. Khandani. "Real interference alignment: Exploiting the potential of single antenna systems". *IEEE Transactions on Information Theory* 60(8), 4799–4810, 2014.

102. I. Krikidis, J. Thompson, and S. McLaughlin. "Relay selection for secure cooperative networks with jamming". *IEEE Transactions on Wireless Communications* 8(10), 5003–5011, Oct. 2009.

103. D. H. Ibrahim, E. S. Hassan, and S. A. El-Dolil. "A new relay and jammer selection schemes for secure one-way cooperative networks". *Wireless Personal Commu.* 72(2), DOI: 10.1007/s11277-013-1384-5, 2013.

104. R. Bassily and S. Ulukus. "Deaf cooperation and relay selection strategies for secure communication in multiple relay networks". *IEEE Transactions on Signal Processing* 61(6), 1544–1554, Mar. 2013.

105. D. H. Ibrahim, E. S. Hassan, and S. A. El-Dolil. "Improving physical layer security in two-way cooperative networks with multiple eavesdroppers". *Proc. of IEEE INFOS*, Egypt, 2014.

106. D. H. Ibrahim, E. S. Hassan, and S. A. El-Dolil. "Relay and jammer selection schemes for improving physical layer security in two-way cooperative networks". *Computers & Security Journal* 50, 47–59, 2015.

107. R. Bassily and S. Ulukus. "Secure communication in multiple relay networks through decode-and-forward strategies". *Journal of Communications and Networks* 14(4), 352–363, Aug. 2012.

108. Y. Oohama. "Coding for relay channels with confidential messages". In: *Proc. IEEE Information Theory Workshop (ITW)*, Cairns, Australia, pp. 87–89, 2001.

109. X. He and A. Yener. "Cooperation with an untrusted relay: A secrecy perspective". *IEEE Transactions on Information Theory* 56(8), 3807–3827, Aug. 2010.

110. X. He and A. Yener. "Two-hop secure communication using an untrusted relay". *EURASIP J. Wireless Commun. Network.*, 2009.

111. X. He and A. Yener. "Strong secrecy and reliable Byzantine detection in the presence of an untrusted relay". *IEEE Transactions on Information Theory* 59(1), 177–192, Jan. 2013.

112. Y. Liang and H. V. Poor. "Multiple access channels with confidential messages". *IEEE Transactions on Information Theory*. 54(3), 976–1002, Mar. 2008.

113. R. Liu, I. Maric, R. D. Yates, and P. Spasojevic. "The discrete memoryless multiple access channel with confidential messages". In: *Proc. IEEE Int. Symp. Information Theory*, 2006.

114. E. Ekrem and S. Ulukus. "Effects of cooperation on the secrecy of multiple access channels with generalized feedback". In: *Proc. Conf. Information Sciences Systems*, 2008.

115. E. Ekrem and S. Ulukus. "Secrecy in cooperative relay broadcast channels". *IEEE Transactions on Information Theory* 57(1), 137–155, Jan. 2011.

116. E. D. Silva, A. L. D. Santos, L. C. P. Albini, and M. Lima. "Identity-based key management in mobile ad hoc networks: Schemes and applications". *IEEE Wireless Communication* 15, 46–52, Oct. 2008.

117. R. Liu, I. Maric, P. Spasojevic, and R. D. Yates. "Discrete memoryless interference and broadcast channels with confidential messages: Secrecy rate regions". *IEEE Transactions on Information Theory* 54, 2493–2507, Jun. 2008.

118. M. Bloch, J. Barros, M. R. D. Rodrigues, and S. W. McLaughlin. "Wireless information-theoretic security". *IEEE Transactions on Information Theory* 54, 2515–2534, June 2008.

119. J. Barros and M. R. D. Rodrigues. "Secrecy capacity of wireless channels". *Proc. IEEE Int. Symp. Inf. Theory*, Seattle, USA, pp. 356–360, July 2006.

120. Y. Liang, H. V. Poor, and L. Ying. "Wireless broadcast networks: Reliability, security, and stability". *Proc. IEEE Inf. Theory Appl. Work*, San Diego, CA, USA, pp. 249–255, Feb. 2008.

121. L. Dong, Z. Han, A. P. Petropulu, and H. V. Poor. "Secure wireless communications via cooperation". *Proc. Allerton Conf. Commun. Cont. Comp.*, Urbana-Champaign, IL, USA, Sept. 2008.

122. S. Yang and J.-C. Belfiore. "Towards the optimal amplify-and-forward cooperative diversity scheme". *IEEE Transactions on Information Theory* 53, 3114–3126, Sept. 2007.

123. L. Dong, Z. Han, A. Petropulu, and H. V. Poor. "Improving wireless physical layer security via cooperating relays". *IEEE Transactions on Signal Processing* 58, 1875–1888, Mar. 2010.

124. L. Lai and H. El Gamal. "The relay–eavesdropper channel: Cooperation for secrecy". *IEEE Transactions on Information Theory* 54, 4005–4019, Sept. 2008.

125. I. Krikidis. "Opportunistic relay selection for cooperative networks with secrecy constraints". *IET Communications* 4, 1787–1791, 2010.

126. E. Beres and R. Adve. "Selection cooperation in multi-source cooperative networks". *IEEE Transactions on Wireless Communications* 7, 118–127, Jan. 2008.

127. O. Simeone and P. Popovski. "Secure communications via cooperating base stations". *IEEE Communications Letter* 12, 188–190, Mar. 2008.

128. P. Popovski and O. Simeone. "Wireless secrecy in cellular systems with infrastructure-aided cooperation". *IEEE Transactions on Information Forensics and Security* 4, 242–256, Jun. 2009.

129. T. Wang and G. B. Giannakis. "Mutual information jammer-relay games". *IEEE Transactions on Information Forensics and Security* 3, 290–303, Jun. 2008.

130. E. S. Hassan. "Energy-efficient hybrid opportunistic cooperative protocol for single-carrier frequency division multiple access-based networks". *IET Communications* 6(16), 2602–2612, 2012.

131. A. Y. Al-nahari, I. Krikidis, A. S. Ibrahim, M. I. Dessouky, and F. E. Abd El-Samie. "Relaying techniques for enhancing the physical layer secrecy in cooperative networks with multiple eavesdroppers". *Transactions on Emerging Telecommunications Technologies*, DOI: 10.1002/ett.2581, Nov. 2012.

132. Y. Liang, H. V. Poor, and S. Shamai. "Secure communication over fading channels". *IEEE Transactions on Information Theory* 54(6), 2470–2492, Jun. 2008.

133. L. Dong, Z. Han, A. P. Petropulu, and H. V. Poor. "Amplify-and forward based cooperation for secure wireless communications". In: *Proc. IEEE Int. Conf. Acoustics, Speech, and Signal Processing*, Taipei, Taiwan, Apr. 2009.

134. B. Rankov and A. Wittneben. "Achievable rate regions for the two-way relay channel". In: *Proc. IEEE Int. Symp. Information Theory*, Seattle, WA, Jul. 2006.

135. B. Rankov and A. Wittneben. "Spectral efficient protocols for half duplex fading relay channels". *IEEE Journal on Selected Areas in Communications* 25(2), 379–389, Feb. 2007.

136. J. Chen, R. Zhang, L. Song, Z. Han, and B. Jiao. "Joint relay and jammer selection for secure decode-and-forward two-way relay networks". In: *Proc. of Communications (ICC), IEEE International Conference*, 2011.

137. N. Zhou, X. Chen, C. Li, and Q. Lai. "Relay selection for physical layer security in decode-and-forward two-way relay networks". *Journal of Information & Computational Science*, DOI: 10.12733/jics20102372, 5821–5828, Dec. 2013.

138. J. Chen, R. Zhang, L. Song, Z. Han, and B. Jiao. "Joint relay and jammer selection for secure two-way relay networks". *IEEE Transactions on Information Forensics and Security* 7(1), 310–320, Feb. 2012.

139. I.F. Akyildiz, W. Su, Y. Sankarasubramaniam, and E. Cayirci. "Wireless sensor networks: A survey". *Computer Networks* 38(4), 393–422, 2002.

140. K. Al Agha, M.-H. Bertin, T. Dang, A. Guitton, P. Minet, T. Val, and J.-B. Viollet. "Which wireless technology for industrial wireless sensor networks? The development of OCARI technology". *IEEE Transactions on Industrial Electronics* 56(10), 4266–4278, 2009.

141. M. S. Mahmoud and Y. Xia. *Networked Filtering and Fusion in Wireless Sensor Networks*. CRC Press, 2014.

142. J. Yick, B. Mukherjee, and D. Ghosal. "Wireless sensor network survey". *Computer Networks* 52, 2292–2330, 2008.

143. S. Toumpis and T. Tassiulas. "Optimal deployment of large wireless sensor networks". *IEEE Transactions on Information Theory* 52, 2935–2953, 2006.

144. J. Yick, G. Pasternack, B. Mukherjee, and D. Ghosal. "Placement of network services in sensor networks". *Int. J. Wireless and Mobile Computing (IJWMC), Self-Organization Routing and Information, Integration in Wire-less Sensor Networks (Special Issue)* 1, 101–112, 2006.

145. D. Pompili, T. Melodia, and I. F. Akyildiz. "Deployment analysis in underwater acoustic wireless sensor networks". WUWNet, Los Angeles, CA, 2006.

146. I. F. Akyildiz and E. P. Stuntebeck. "Wireless underground sensor networks: Research challenges". *Ad-Hoc Networks* 4, 669–686, 2006.

147. M. Li and Y. Liu. "Underground structure monitoring with wireless sensor networks". *Proc. the IPSN*, Cambridge, MA, 2007.

148. I. F. Akyildiz, D. Pompili, and T. Melodia. "Challenges for efficient communication in underwater acoustic sensor networks". *ACM Sigbed Review* 1(2), 3–8, 2004.

149. J. Heidemann, Y. Li, A. Syed, J. Wills, and W. Ye. "Underwater sensor networking: Research challenges and potential applications". *Proc. the Technical Report*, USC/ Information Sciences Institute, 2005.

150. I. F. Akyildiz, T. Melodia, and K. R. Chowdhury. "A survey on wireless multi-media sensor networks". *Computer Networks* 51, 921–960, 2007.

151. D. Estrin, R. Govindan, J. Heidemann, and S. Kumar. "Next century challenges: Scalable coordination in sensor networks". ACM MobiCom'99, Washington, USA, pp. 263–270, 1999.

152. B. Atwood, B. Warneke, and K. S. J. Pister. "Preliminary circuits for smart dust". *Proc. Southwest Symp. Mixed-Signal Design*, pp. 87–92, 2000.

153. R. Sharma and N. T. S. Kumar. "Review paper on wireless sensor networks". In: *Proc. of the Intl. Conf. on Recent Trends in Computing and Communication Engineering, (RTCCE)*, pp. 978–981, 2013.

154. N. Bulusu, D. Estrin, L. Girod, and J. Heidemann. "Scalable coordination for wireless sensor networks: Self-configuring localization systems". International Symposium on Communication Theory and Applications (ISCTA 2001), Ambleside, UK, Jul. 2001.

155. J. M. Kahn, R. H. Katz, and K. S. J. Pister. "Next century challenges: Mobile networking for smart dust". *Proc. of the ACM MobiCom'99*, Washington, USA, pp. 271–278, 1999.

156. Y. H. Nam et al. "Development of remote diagnosis system integrating digital telemetry for medicine". *International Conference IEEE-EMBS*, Hong Kong, pp. 1170–1173, 1998.

157. N. Noury, T. Herve, V. Rialle, G. Virone, E. Mercier, G. Morey, A. Moro, and T. Porcheron. "Monitoring behavior in home using a smart fall sensor". *IEEE-EMBS Special Topic Conference on Microtechnologies in Medicine and Biology*, pp. 607–610, Oct. 2000.

158. B. Sibbald. "Use computerized systems to cut adverse drug events: Report". *CMAJ: Canadian Medical Association Journal* 164(13), 1878–1878, 2001.

159. E. M. Petriu, N. D. Georganas, D. C. Petriu, D. Makrakis, and V. Z. Groza. "Sensor based information appliances". *IEEE Instrumentation and Measurement Magazine*, 31–35, Dec. 2000.

160. C. Herring and S. Kaplan. "Component-based software systems for smart environments". *IEEE Personal Communications*, 60–61, Oct. 2000.

161. J. M. Rabaey, M. J. Ammer, J. L. da Silva Jr., D. Patel, and S. Roundy. "Pico radio supports ad hoc ultra-low power wireless networking". *IEEE Computer Magazine*, 42–48, 2000.

162. G. J. Pottie, W. J. Kaiser. "Wireless integrated network sensors". *Communications of the ACM* 43(5), 551–558, 2000.

163. E. Shih, S. Cho, N. Ickes, R. Min, A. Sinha, A. Wang, and A. Chandrakasan. "Physical layer driven protocol and algorithm design for energy-efficient wireless sensor networks". *Proc. of ACM MobiCom'01*, Rome, Italy, pp. 272–286, Jul. 2001.

164. G. Hoblos, M. Staroswiecki, and A. Aitouche. "Optimal design of fault tolerant sensor networks". *IEEE International Conference on Control Applications*, Anchorage, AK, pp. 467–472, Sept. 2000.

165. J. Rabaey, J. Ammer, J. L. da Silva Jr., and D. Patel. "Pico-radio: Ad-hoc wireless networking of ubiquitous low-energy sensor monitor nodes". *Proc. of the IEEE Computer Society Annual Workshop on VLSI (WVLSI'00)*, Orlando, Florida, pp. 9–12, Apr. 2000.

166. C. Intanagonwiwat, R. Govindan, and D. Estrin. "Directed diffusion: A scalable and robust communication paradigm for sensor networks". *Proc. of the ACM Mobi-Com'00*, Boston, MA, pp. 56–67, 2000.

167. L. Li and J. Y. Halpern. "Minimum-energy mobile wireless networks revisited". *IEEE International Conference on Communications ICC'01*, Helsinki, Finland, Jun. 2001.

168. A. Savvides, C. Han, and M. Srivastava. "Dynamic fine-grained localization in ad-hoc networks of sensors". *Proc. of ACM MobiCom'01*, Rome, Italy, pp. 166–179, Jul. 2001.

169. W. Du, R. Wang, and P. Ning. "An efficient scheme for authenticating public keys in sensor networks". *Proc. of the 6th ACM International Symposium on Mobile Ad Hoc Networking and Computing*, ACM, 2005.

170. A. S. Wander, N. Gura, H. Eberle, V. Gupta, and S. C. Shantz. "Energy analysis of public-key cryptography for wireless sensor networks". In: *Third IEEE International Conference on Pervasive Computing and Communications*, pp. 324–328, Mar. 2005.

171. K. Ren, W. Lou, K. Zeng, and P. J. Moran. "On broadcast authentication in wireless sensor networks". *IEEE Transactions on Wireless Communications* 6(11), 4136–4144, 2007.

172. K. C. Barr and K. Asanović. "Energy-aware lossless data compression". *ACM Transactions on Computer Systems (TOCS)* 24.3, 250–291, 2006.

173. Texas Instruments Inc., "Msp430 Family of Ultra-low power 16-bit RISC Processors", http://www.ti.com. Accessed Nov. 2, 2012.

174. D. Ma, C. Soriente, and G. Tsudik. "New adversary and new threats: Security in unattended sensor networks". *IEEE Network*, Mar. 2009.

175. D. Ma and G. Tsudik. "Security and privacy in emerging wireless networks". *IEEE Wireless Communications, Special Issue on Security and Privacy in Emerging Wireless Networks*, 2010.

176. R. Ostrovsky and M. Yung. "How to withstand mobile virus attacks". In: *Proc. of the Tenth Annual ACM Symposium on Principles of Distributed Computing*, ACM, pp. 51–59, July 1991.

177. Y. Frankel, P. Gemmell, P. D. MacKenzie, and M. Yung. "Proactive RSA". In: *Annual International Cryptology Conference*, Springer Berlin Heidelberg, pp. 440–454, August 1997.

178. T. Rabin. "A simplified approach to threshold and proactive RSA". In: *Annual International Cryptology Conference*, Springer Berlin Heidelberg, pp. 89–104, Aug. 1998.

179. R. Gennaro, S. Jarecki, H. Krawczyk, and T. Rabin. "Robust and efficient sharing of RSA functions". In: *Proc. Annual Int. Cryptology Conference (CRYPTO)*, pp. 157–172, 1996.

180. R. Gennaro, S. Jarecki, H. Krawczyk, and T. Rabin. "Robust threshold DSS signatures". In: *International Conference on the Theory and Applications of Cryptographic Techniques*, Springer Berlin Heidelberg, pp. 354–371, May 1996.

181. D. Ma and G. Tsudik. "DISH: Distributed self-healing in unattended sensor networks", *Proc. of SSS '08*, Detroit, MI, USA, pp. 47–62, Nov. 2008.

182. S. Basagni, K. Herrin, D. Bruschi, and E. Rosti. "Secure pebblenets". *Proc. of ACM International Symposium on Mobile Ad Hoc Networking & Computing*, Long Beach, CA, pp. 156–163, Oct. 2001.

183. C. Karlof and D. Wagner. "Secure routing in wireless sensor networks: Attacks and countermeasures". *Ad Hoc Networks* 1(2), 293–315, 2003.

184. X. Chen, K. Makki, K. Yen, and N. Pissinou. "Sensor network security: A survey". *IEEE Communications Surveys & Tutorials* 11(2), 52–73, 2009.

185. P. Turaga and Y. A. Ivanov. "Diamond sentry: Integrating sensors and cameras for real-time monitoring of indoor spaces". *IEEE Sensors Journal* 11(3), 593–602, Mar. 2011.

186. W. Jang, W. M. Healy, and M. J. Skibniewski. "Wireless sensor networks as a part of a web-based building environmental monitoring system". *Automation in Construction* 17(6), 729–736, Aug. 2008.

187. Z. Sun, P. Wang, M. C. Vuran, M. A. Al-Rodhaan, A. M. Al-Dhelaan, and I. F. Akyildiz. "Border sense: Border patrol through advanced wireless sensor networks". *Ad Hoc Networks* 9, 468–477, 2011.

188. M. Clure, D. R. Corbett, and D. W. Gage. "The DARPA LANdroids program". Defense Advanced Research Projects Agency (DARPA). In: Unmanned Systems Technology XI, Orlando, FL. *SPIE Proceedings* 7332, Apr. 2009.

189. W. Lou, W. Liu, and Y. Fang. "SPREAD: Enhancing data confidentiality in mobile ad hoc networks". *Proc. IEEE INFOCOM '04*, Hong Kong, pp. 2404–2413, Mar. 2004.

190. R. D. Pietro, D. Ma, C. Soriente, and G. Tsudik. "Self-healing in unattended wireless sensor networks". *ACM Transactions on Sensor Network*, Article 39, 19 pages, Mar. 2010.

191. A. Kamra, V. Misra, J. Feldman, and D. Rubenstein. "Growth codes: Maximizing sensor network data persistence". *Proc. of SIGCOMM '06*, New York: ACM, pp. 255–266, 2006.

192. V. Gianuzzi. "Data replication effectiveness in mobile ad-hoc networks". *Proc. of PE-WASUN '04*, New York: ACM, pp. 17–22, 2004.

193. C. Chen and Y. Tsai. "Location privacy in unattended wireless sensor networks upon the requirement of data survivability". *IEEE Journal on Selected Areas in Communications* 29(7), 1480–1490, 2011.

194. P. Kamat, Y. Zhang, W. Trappe, and C. Ozturk. "Enhancing source location privacy in sensor network routing". *Proc. IEEE ICDCS 2005*, Columbus, Ohio, USA, pp. 599–608, Jun. 2005.

195. K. Mehta, D. Liu, and M. Wright. "Location privacy in sensor networks against a global eavesdropper". *Proc. IEEE ICNP 2007*, Beijing, China, pp. 314–323, Oct. 2007.

196. B. H. Liu, W. C. Ke, C. H. Tsai, and M. J. Tsai. "Constructing a message-pruning tree with minimum cost for tracking moving objects in wireless sensor networks is NB-complete and an enhanced data aggregation structure". *IEEE Transactions on Computers* 57(6), 849–863, Jun. 2008.

197. Y. Ren, V. Oleshchuk, and F. Y. Li. "Optimized secure and reliable distributed data storage scheme and performance evaluation in unattended WSNs". *Computer Communications Journal Elsevier (COMCOM)*, 1–11, Aug. 2012, URL: http://dx.doi.org/10.1016/j.comcom. 2012.08.001.

198. J. Newsome, E. Shi, D. Song, and A. Perrig. "The sybil attack in sensor networks: Analysis & defenses". *Proc. of the 3rd Int. Symposium on Information Processing in Sensor Networks*, Berkeley, California, USA, pp. 259–268, 26–27 Apr. 2004.

199. A. Perrig, J. Stankovic, and D. Wagner. "Security in wireless sensor networks". *Magazine Communications of the ACM—Wireless Sensor Networks* 47(6), 53–57, 2004.

200. I. Reed and G. Solomon. "Polynomial codes over certain finite fields". *Journal of the Society for Industrial and Applied Mathematics* 8(2), 300–304, 1960.

201. M. Conti, R. D. Pietro, L. V. Mancini, and A. Mei. "Mobility and cooperation to thwart node capture attacks in MANETs". *EURASIP Journal on Wireless Communications and Networking*, 2009.

202. A. Becher, E. Becher, Z. Benenson, and M. Dornseif. "Tampering with motes: Real-world physical attacks on wireless sensor networks". In: *Proceeding of the 3rd International Conference on Security in Pervasive Computing (SPC)*, pp. 104–118, 2006.

203. A. S. Elsafrawey, E. S. Hassan, and M. I. Dessouky. "Cooperative hybrid self-healing scheme for secure and data reliability in unattended wireless sensor networks". *IET Information Security*, doi: 10.1049/iet-ifs.2014.0267, 2015.

204. A. R. Silva, M. Moghaddam, M. Liu., "Case study on the reliability of unattended outdoor wireless sensor systems". *9th Annual IEEE International Systems Conference (SysCon)*, pp. 785–791, 13–16 Apr. 2015.

205. Y. Ren, V. I. Zadorozhny, V. A. Oleshchuk, and F. Y. Li. "A novel approach to trust management in unattended wireless sensor networks". *IEEE Transactions on Mobile Computing* 13(7), 1409–1423, 2014.

206. A. S. Elsafrawey, E. S. Hassan, and M. I. Eldosoki. "A new cooperative hybrid self-healing scheme for secure and data reliability in UWSNs". *Proc. of the 30th National Radio Science Conference (30th NRSC)*, NTI, Cairo, Egypt, 15–16 April 2013.

207. M. Batalin, M. Rahimi, Y. Yu, D. Liu, A. Kansal, G. Sukhatme, W. Kaiser, M. Hansen, G. J Pottie, M. Srivastava, and D. Estrin. "Call and response: Experiments in sampling the environment". *Proc. of 2nd Annual Conference on Sensors and Systems*, Baltimore, MD, USA, Nov. 2004.

208. M. H. Rahimi, H. Shah, G. S. Sukhatme, J. Heidemann, and D. Estrin. "Studying the feasibility of energy harvesting in mobile sensor networks". *Proc. of the IEEE International Conference on Robotics and Automation*, Taipei, Taiwan, Sept. 2003.

209. T. Camp, J. Boleng, and V. Davies. "A survey of mobility models for ad hoc network research". *Wireless Communications & Mobile Computing (WCMC): Special Issue on Mobile Ad Hoc Networking: Research, Trends and Applications* 2(5), 483–502, 2002.

210. S. Batabyal and P. Bhaumik. "Mobility models, traces and impact of mobility on opportunistic routing algorithms: A survey". *IEEE Communications Surveys & Tutorials* 17(3), 1679–1707, 2015.

211. S. Madi and H. Al-Qamzi. "A survey on realistic mobility models for vehicular ad hoc networks (VANETs)". *10th IEEE International Conference on Networking, Sensing and Control (ICNSC)*, pp. 333–339, 2013.

212. A. S. Elsafrawey, E. S. Hassan, and M. I. Dessouky. "Improving UWSNs security and data reliability using a cluster controlled mobility scheme". *Proc. of IEEE INFOS*, Egypt, 2014.

213. A. S. Elsafrawey, E. S. Hassan, and M. I. Dessouky. "Analytical analysis of a cluster controlled mobility scheme for data security and reliability in UWSNs". *Proc. of NRSC*, 2015.

214. A. Howard, M. Mataric, and G. S. Sukhatme. "An incremental self-deployment algorithm for mobile sensor networks". *Autonomous Robots-Special Issue on Intelligent Embedded Systems* 13(2), 113–126, 2002.

215. S. Poduri and G. Sukhatme. "Constrained coverage for mobile sensor networks". *Proc. of the IEEE International Conference on Robotics and Automation*, April 2004.

216. R. D. Pietro, G. Oligeri, C. Soriente, and G. Tsudik. "United we stand: Intrusion-resilience in mobile unattended WSNs". *IEEE Transactions Mobile Computing (TMC)* 12(7), 1456–1468, 2013.

217. R. D. Pietro, G. Oligeri, C. Soriente, and G. Tsudik. "Intrusion-resilience in mobile unattended WSNs". *Proc. of 29th Annual Joint Conference of the IEEE Computer and Communications Societies (INFOCOM'10)*, pp. 2303–2311, 2010.

218. T. Iida, K. Emura, A. Miyaji, and K. Omote. "An intrusion and random-number-leakage resilient scheme in mobile unattended WSNs". *Advanced Information Networking and Applications Workshops (WAINA), 26th International Conference on. IEEE*, 2012.

219. B. Liu, P. Brass, O. Dousse, P. Nain, and D. F. Towsley. "Mobility improves coverage of sensor networks". In: *6th ACM International Symposium on Mobile Ad Hoc Networking and Computing (MobiHoc'05)*, pp. 300–308, 2005.

220. G. Wang, G. Cao, T. F. L. Porta, and W. Zhang. "Sensor relocation in mobile sensor networks". In: *24th Annual Joint Conference of the IEEE Computer and Communications Societies (INFOCOM'05)*, pp. 2302–2312, 2005.

221. R. Dutta, Y. D. Wu, and S. Mukhopadhyay. "Constant storage self-healing key distribution with revocation in wireless sensor network". In: *IEEE International Conference on Communications (ICC'07)*, pp. 1323–1328, 2007.

222. F. Sivrikaya and B. Yener. "Time synchronization in sensor networks: A survey". *IEEE Network* 18(4), 45–50, 2004.

223. E. Kiesling, C. Strauss, A. Ekelhart, B. Grill, and C. Stummer. "Simulation-based optimization of information security controls: An adversary-centric approach". *Simulation Conference (WSC)*, pp. 2054–2065, 2013.

224. W. Z. Khan, M. Y. Aalsalem, and M. N. M. Saad. "Detection of masked replication attack in wireless sensor networks". *8th Iberian Conference on Information Systems and Technologies (CISTI)*, pp. 1–6, 2013.

225. C. Tang and P. K. McKinley. "Energy optimization under informed mobility". *IEEE Transaction on Parallel and Distributed Systems* 17(9), 947–962, 2006.

226. Z. Pala, K. Bicakci, and M. Turk. "Effects of node mobility on energy balancing in wireless networks". *Computers & Electrical Engineering* 41, 314–324, 2015.

227. Y. Yan and Y. Mostofi. "Utilizing mobility to minimize the total communication and motion energy consumption of a robotic operation". *IFAC Proceedings* 45(26), 180–185, 2012.

228. D. K. Goldenberg et al. "Towards mobility as a network control primitive". *Proc. of the 5th ACM International Symposium on Mobile Ad Hoc Networking and Computing*, ACM, pp. 163–174, 2004.

229. M. Grossglauser and D. N. C. Tse. "Mobility increases the capacity of ad hoc wireless networks". *IEEE/ACM Transactions on Networking* 10(4), 477–486, 2002.

230. S. A. Jafar. "Too much mobility limits the capacity of wireless ad hoc networks". *IEEE Transactions on Information Theory* 51(11), 3954–3965, 2005.

231. M. Schwager, J. McLurkin, J. J. E. Slotine, and D. Rus. "From theory to practice: Distributed coverage control experiments with groups of robots". *Springer Tracts in Advanced Robotics* 54, 127–136, Mar. 2009.

232. Q. Wang, X. Wang, and X. Lin. "Mobility increases the connectivity of k-hop clustered wireless networks". *Proc. of the 15th Annual International Conference on Mobile Computing and Networking, (MOBICOM)*, pp. 121–131, 2009.

233. S. Čapkun, J.-P. Hubaux, and L. Buttyán. "Mobility helps security in ad hoc networks". *Proc. of the Fourth ACM International Symposium on Mobile Ad Hoc Networking & Computing*, ACM, pp. 46–56, 2003.

234. A. T. Hoang and M. Motani. "Exploiting wireless broadcast in spatially correlated sensor networks". *IEEE International Conference on Communications (ICC 2005)*, vol. 4, 2005.

235. C. C. Ooi and C. Schindelhauer. "Minimal energy path planning for wireless robots". *Mobile Networks and Applications* 14(3), 309–321, 2009.

236. J. Bachrach and C. Taylor. "Localization in sensor networks". *Handbook of Sensor Networks: Algorithms and Architectures*, 2005.

237. K. Römer, P. Blum, and L. Meier. "Time synchronization and calibration in wireless sensor networks". *Handbook of Sensor Networks*. pp. 199–237, Chichester: John Wiley and Sons, 2005.

238. Y. Mei, Y. H. Lu, Y. C. Hu, and C. G. Lee. "A case study of mobile robot's energy consumption and conservation techniques". *Proc. of the 12th International Conference on Advanced Robotics and Automation (ICAR'05)*, pp. 492–497, 2005.

239. P. A. Tipler. *Physics for Scientists and Engineers*. Third edition. Worth Publishers, 1991.

Index

Page numbers followed by f and t indicate figures and tables, respectively.